AA001101

MATERIALS RESEARCH SOCIETY
SYMPOSIUM PROCEEDINGS VOLUME 1079

Materials and Processes for Advanced Interconnects for Microelectronics

March 24-28, 2008
San Francisco, California, USA

Printed from e-media with permission by:

Curran Associates, Inc.
57 Morehouse Lane
Red Hook, NY 12571
www.proceedings.com

ISBN: 978-1-60560-864-8

Some format issues inherent in the e-media version may also appear in this print version.

CAMBRIDGE UNIVERSITY PRESS
Cambridge, New York, Melbourne, Madrid, Cape Town,
Singapore, São Paulo, Delhi, Tokyo, Mexico City

Cambridge University Press
32 Avenue of the Americas, New York, NY 10013-2473, USA

www.cambridge.org

Materials Research Society
506 Keystone Drive, Warrendale, PA 15086
http://www.mrs.org

©Materials Research Society 2008

This publication is in copyright. Subject to statutory exception
and to the provisions of relevant collective licensing agreements,
no reproduction of any part may take place without the written
permission of Cambridge University Press.

First published 2008

CODEN: MRSPDH

ISBN: 978-1-60560-864-8

Cambridge University Press has no responsibility for the persistence or
accuracy of URLs for external or third-part Internet Web sites referred to
in this publication and does not guarantee that any content on such Web sites
is, or will remain, accurate or appropriate.

Additional copies of this publication are available from:

Curran Associates, Inc.
57 Morehouse Lane
Red Hook, NY 12571 USA
Phone: 845-758-0400
Fax: 845-758-2634
Email: curran@proceedings.com
Web: www.proceedings.com

TABLE OF CONTENTS

Molecular-level Manipulation Technology for Low-k Dielectrics Controlling the Physical and Chemical Structures Toward 32nm-node BEOLs 1
Y. Hayashi

UV Curing Effects of Low-k Materials under Reactive Conditions 11
D. Moore, C. Waldfried, P. Sakthivel, O. Escorcia

Moisture Adsorption in Plasma-Damaged Porous Low-k Dielectrics 18
E. Vinogradova, C.E. Smith, D.W. Mueller, A.J. McKerrow, R. Reidy

Effects of Plasma Surface Treatment on the Self-forming Barrier Process in Porous SiOCH 24
S. Chung, J. Koike, Z. Tokei

Formation of Mn Oxide with Thermal CVD and its Diffusion Barrier Property Between Cu and SiO_2 30
K. Neishi, S. Aki, J. Iijima, J. Koike

Extendibility Study of a PVD Cu Seed Process with Ar+ Rf-Plasma Enhanced Coverage for 45nm Interconnects 35
A.H. Simon, T. Bolom, T.J. Tang, B. Baker, C. Peters, B. Rhoads, P.L. Flaitz, S. Sankaran, S. Grunow

Self-formation of Ti-rich Layers at Cu(Ti)/low-k Interfaces 41
K. Kohama, K. Ito, S. Tsukimoto, K. Mori, K. Maekawa, M. Murakami

Selective Oxidation and Resistivity Reduction of Cu-Mn Alloy Films for Self-forming Barrier Process 47
J. Iijima, Y. Fujii, K. Neishi, J. Koike

Al Diffusion in Polycrystalline Cu 53
F. Gstrein, H. Kennel, A. Budrevich, B. Miner, J. Plombon, E. Andideh

Sacrificial Passivation of Nanoscale Metal Powders for Transient Liquid Phase Bonding 59
N.S. Bosco, B. Manhat, J. Janczak-Rusch

Nanoindentation of Lead Free Solders for Harsh Environments 70
V.F. Marques, P. Grant, C. Johnston

Effects of Pulse Duration and Polarity on the Electromigration Behavior of Copper Interconnects under Pulsed Current Stress 79
M.K. Lim, C.L. Gan, Y.C. Ee, C.M. Ng, B.C. Zhang, J.B. Tan

Pd Segregation at (001) B2-NiSi/Si Epitaxial Interface Studied by Density Functional Theory 85
D. Kim, H. Seo, Y. Kim

Electromigration of Cu Interconnect Lines Prepared by a Plasma-based Etch Process 89
G. Liu, Y. Kuo

Numerical Analysis of Packaging-Induced Failures in Cu/Low-k Interconnects 95
A.P. Karmarkar, X. Xu, X. Lin, G. Rollins, V. Moroz, X. Lin

Off-Angular Deposition Compensation for PVD Selective Re-sputtering Process 101
H. Yang, F. Zhang, K. Nelson, J.M. Tseng, J. Forster, A. Sunddarrajan, A. Bhatnagar, N. Kumar, P. Gopalraja

TVS Measurements of Metal Ions in Low-k Dielectrics: Effect of H_2O Uptake 106
I. Cioti, Z. Tokei, G. Mangraviti, G. Beyer

Patterned Wafers Backside Thinning for 3-D Integration and Multilayer Stack Achievement by Direct Wafer Bonding 112
B. Charlet, A. Chiteboun, M. Zussy, L. Bally, P. Leduc, M. Assous

Dependence of Thermal Stability and Ni(Pt)Si/Si on Crystal Orientation 119
K. Okubo, K. Kawamura, S. Akiyama, Y. Kotaka, T. Itani, H. Watatani, K. Yanai, M. Nakaishi, M. Kase

Modelling and Characterization of Ultrasonic Consolidation Process of Aluminium Alloys 125
A.M. Siddiq, E. Ghassemieh

Growth and Integration of High-Density CNT for BEOL Interconnects 133
A.R. Negreira, D.J. Cott, A.S. Verhulst, S. Esconjauregui, N. Chiodarelli, J.E. Weis, C.M. Whelan, G. Groeseneken, M. Heyns, S. De Gendt, P.M. Vereecken

Mechanical Integrity Study of Air Gap Structures Assisted by FE Simulations 144
S. Moreau, F. Gaillard, J. Barbe, R. Gras, G. Passemard, J. Torres

Dielectric Recovery of Plasma Damaged Organosilicate Low-k Films 150
H. Shi, J. Bao, H. Huang, J. Liu, R.S. Smith, Y. Sun, P.S. Ho, M.L. McSwiney, M. Moinpour, G.M. Kloster

Changes of UV Optical Properties of Plasma Damaged Low-k Dielectrics for Sidewall Damage Scatterometry 156
P. Marsik, A. Urbanowicz, K. Vinokur, Y. Cohen, M.R. Baklanov

Using a Barrier Layer to Inhibit Ti/Oxide Reaction to Reduce RC Delay and Improve Electromigration in Al-Cu/Ti/W Interconnect for High Power Analog and Mixed Signal Applications 162
W.J. Murphy, T.C. Lee, J. Chapple-Sokol, D.A. Delibac, Z.X. He, S.E. Luce, S.A. Mongeon, D.C. Thomas, D.S. Vanslette, T.D. Sullivan

Semiconductor Film Bonding Technology and Application in Two-axis Hall Sensor Fabrication 169
K. Koh, T. Matushita, K. Hohkawa

Organization of Magnetic/Noble Metal Heterostructures by an Applied External Magnetic Field 175
N. Pazos-Perez, D. Baranov, M. Hilgendorff, J. Perez-Juste, L.M. Liz-Marzan, M. Giersig

In-situ Early Stage Electromigration Study in Al Line Using Synchrotron Polychromatic X-ray Microdiffraction 182
K. Chen, N. Tamura, K.N. Tu

Plasma Modification of Si-O-Si Bond Structure in Porous SiOCH Films 188
F.N. Dultsev, A.M. Urbanowicz, M.R. Baklanov

Cobalt Silicidation on Sub 100nm Hole Patterned Vertical Diode Formed by Silicon Epitaxial Growth and Its Electrical Properties 194
M.Y. Lee, K.B. Lee, H.S. Lee, S.J. Chae, I.K. Han, H.S. Kang, S.W. Park

From Process Assumptions to Development to Manufacturing 200
T. Standaert, A. Gabor, A. Simon, A. Lisi, C. Peters, C. Child, D. Kioussis, E. Engbrecht, F. Chen, F. Baumann, G. Lembach, H. Wendt, J. Choi, J. Linville, K. Chanda, K. Kumar, K. Davis, L. Economikos, L. Nicholson, M. Chae, N. Lustig, O. Bravo, P. McLaughlin, R.P. Srivastava, R. Filippi, S. Sankaran, T. Bloom, V. Menon, V. McGahay, W. Li, W. Tseng, W. Landers, Y. Choi, G. Biery, T. Gow

Copper CMP with Composite Polymer Core - Silica Shell Abrasives: A Defectivity Study 212
S. Armini, C.M. Whelan, M. Moinpour, K. Maex

Author Index

Mater. Res. Soc. Symp. Proc. Vol. 1079 © 2008 Materials Research Society 1079-N01-01

Molecular-level Manipulation Technology for Low-k Dielectrics Controlling the Physical and Chemical Structures toward 32nm-node BEOLs

Yoshihiro Hayashi

LSI Fundamental Research Laboratory, NEC Electronics, 1120, Shimokuzawa, Sagamihara, 228-1198, Japan

ABSTRACT

Low-k materials in ULSI devices are required not only to lower the parasitic capacitances, but also to have mechanical stability with damage-less interfaces. By plasma-polymerization (PP) process using ring-type siloxane precursor, a new self-organized porous SiOCH film is developed with preserving the original hexagonal silica-backbone, so called as a molecular-pore-stack (MPS) SiOCH film. The hydrocarbon-rich MPS film has high endurance to the process damages. A density-modulated MPS film with low-k cap is obtained with reinforced interfaces by plasma co-polymerization (PcP) process in the one-step deposition scheme using the ring-type and a linear-type siloxane, applicable for 32nm-nodes and beyond at a low manufacturing cost.

INTRODUCTION

To reduce the parasitic capacitance in scaled-down ULSI interconnects [1], a low-k dielectric film of SiOCH with k~3 had been introduced in 90nm-nodes. For further scaling, porous low-k films have been implemented into 65nm-nodes ULSIs [9, 10] as shown in Fig. 1. The porous low-k films contain tiny pores with k~1 in the SiOCH matrix, reducing the effective film density or eventually the dielectric constant (k). But, the pores have potential issues to degrade the film stability [2] and the dielectric reliability, if we don't care the pore structure as illustrated in Fig. 2. Namely, the porous low-k film is easy to adsorb moisture (H_2O) of k~80. Especially, in case of the pores clustered to the film surface just like tiny channels, moisture is immersed easily into the film. The moisture uptake is a serious problem to increase the effective k-value [3]. Secondary, the porous low-k film is very sensitive to chemical/physical attacks to release the hydrocarbon (-CHx) from the low-k matrix, increasing the k-value. To overcome these issues, physical and chemical structure of the porous films should be controlled precisely.

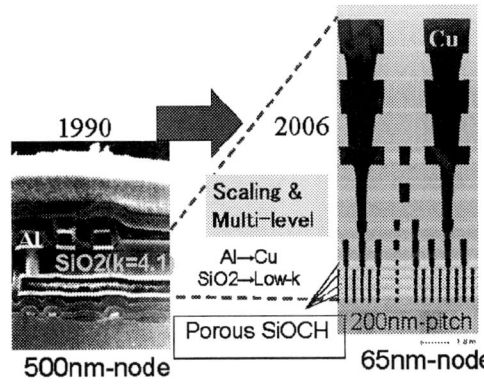

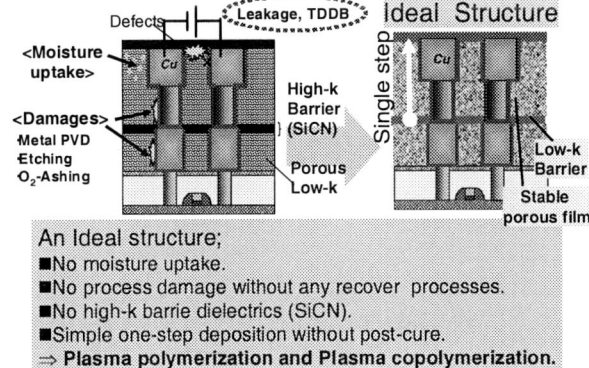

Fig. 1 Structure evolution of LSI interconnects from 500nm to 65nm-nodes during 16 years.

Fig. 2 Issues on the pours film integration and an ideal structure desired.

1

Most of the dielectric films have been deposited by plasma-CVD process, so it is better to control the porous structure by the plasma-CVD or the modified ones. Meanwhile, we developed a modified plasma-CVD process, so called as "plasma polymerization (PP) process" at 1999 as shown in Fig. 3. The idea is quite simple that siloxane monomers with side-chains of unsaturated hydrocarbon (USHC) are vaporized, and introduced into He plasma to deposit a siloxane film through the USHC polymerization. This idea was confirmed firstly by using divinyl siloxane benzocyclobuten (DVS-BCB) to deposit the PP BCB film [4-6]. Next, we expected that the PP process could be possible for deposition of a self-organized porous film when ring-type siloxane molecules with USHC side-chains was used as a precursor [7]. We also considered that the PP process could extend to a plasma copolymerization (PcP) process by introducing two types of monomers with USHC to reinforce the film properties [8, 9].

In this paper, first a PP process is explained to deposit a self-organized porous SiOCH film, so-called as a "Molecular Pore Stack"(MPS) SiOCH film. Then, a newly-developed modulated PcP process is elucidated for the density-modulation in the MPS film. Finally, based on the modulated PcP process with sophisticated molecular structure design [10-15], cost-effective full low-k Cu interconnects is described.

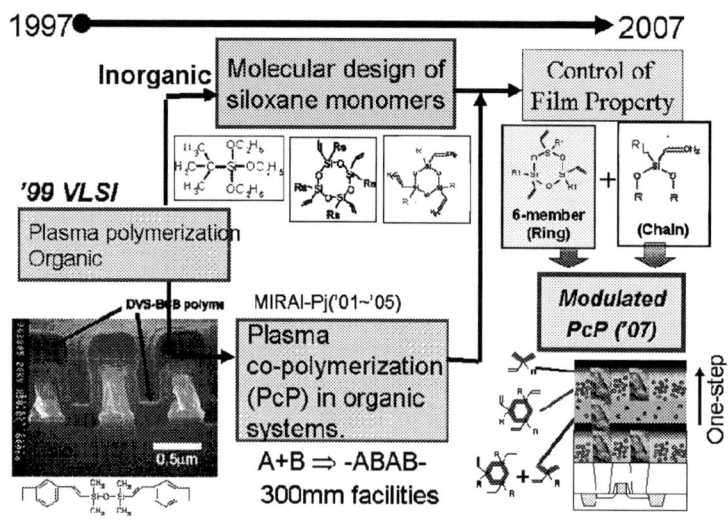

Fig. 3 Physical and chemical structural control of low-k SiOCH films by plasma polymerization (PP) and plasma copolymerization (PcP) processes.

EXPERIMENTAL

Several types of ring-type siloxane monomers were prepared as shown in Table I [17]. The siloxane was vaporized with He carrier gas, and was introduced into RF plasma chamber vacuumed down to 0.1~0.5 torr as shown in Fig. 4. The film was deposited on a silicon wafer heated at 350°C by 100W~400W. For the PcP or the modulated PcP, a ring-type and a liner-type siloxanes were introduced at the same time with the mixing ratio fixed or modulated, respectively. The chemical structure and the composition film was analyzed by FTIR, Raman spectroscopy, and RBS. The pore size and silica-backbone-structure were examined by small-angle X-ray scattering (SAXS) and solid-state NMR (Si^{29}).

Table I Ring-type siloxane monomers used in this experiment [17].

Type	Ring	Ring	Ring	Ring
Symbol	4V4Rs	3V3Rs	3V3R$_L$	3E3R$_L$
Monomer structure				
Member of SiO ring	8	6	6	6
Side-chain 1 (R1)	USHC (Vinyl)	USHC (Vinyl)	USHC (Vinyl)	SHC (Ethyl)
Side-chain 2 (R2)	SHC Alkyl (Rs): -CH3	SHC Alkyl (Rs): -CH3	SHC Alkyl (R$_L$)	SHC Alkyl (R$_L$)

Molecular weight : R$_L$>R$_S$

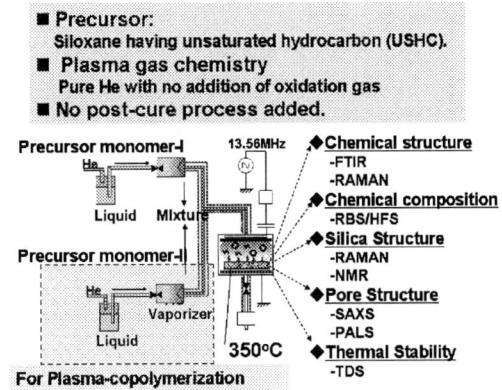

Fig. 4 Low-k deposition system for the PP and the PcP processes, which is based on a conventional plasma-CVD system.

RESULTS and DISCUSSION
MPS SiOCH film by plasma-polymerization (PP) process

Chemical and physical structure of SiOCH films depends on molecular structure of the precursor monomers [7]. Namely, in case of the siloxane monomers without USHC side-chains, the precursor monomers were decomposed in He plasma before the film deposition, whose composition was much different from that of the precursor. No porous SiOCH film was obtained from linear-type siloxane monomers. On contrast, when ring-type siloxane monomer with USHC was used, porous SiOCH film was deposited. Important finding is that the porous structure was reflected to the original ring structure. Fig. 5 shows the pore diameter distribution and the Raman spectroscopy of the SiOCH films from the precursors of 6-member or 8-member rings such as 3V3R$_L$ and 4V4Rs, which have the silica rings of 0.36nm and 0.45nm diameters, respectively. The pore diameter distribution of the SiOCH film from the 6-member ring (3V3R$_L$) was smaller than that form the 8-member ring (4V4Rs). By Raman spectroscopy, it is also found that the SiOCH film from 3V3R$_L$ had hexagonal silica network while the SiOCH film from 4V4Rs composed of poly- or octahedral siloxane network. Solid-state NMR spectroscopy of Si^{29} indicates that the SiOCH film from 3V3R$_L$ consisted of cyclic siloxane networks more than a conventional plasma-CVD SiOCH film. These experimental results indicate that the ring-type siloxane molecules with USHC side-chains are polymerized in He plasma chamber to deposit a self-organized porous SiOCH films on the Si-wafer.

The ring-type siloxane molecules have side-chains of saturated hydrocarbon (SHC) as well as USHC. Both of the side-chains are very important for control the plasma reaction or eventually the deposition rate and/or the chemical composition [17]. Fig. 6 shows the deposition characteristics of the SiOCH films from the 6-member hexagonal ring-type siloxanes as listed in Table I. From comparison of 3V3Rs with 3E3Rs, the USHC side-chain is confirmed again to enhance the deposition rate keeping the k-value low in He plasma. Comparing 3V3R$_L$ with 3V3Rs, the larger SHC improves the deposition rate and enriches the carbon-comment in the film. By PALS analysis [16], the SiOCH film from 3V3R$_L$ had closed pores in Fig. 7.

Namely, by PP process using the ring-type siloxane with USHC side-chains, self-organized porous SiOCH film is obtained without post-cure process such as UV or EB radiation [18].

Hereafter, the porous SiOCH film from 3V3R$_L$ is called as MPS SiOCH film. The MPS-SiOCH film of k~2.5 is featured as a hydrocarbon-rich SiOCH film with 0.36nm-diameter pores, while the conventional PECVD SiOCH films are oxygen-rich or carbon-poor as shown in Table II.

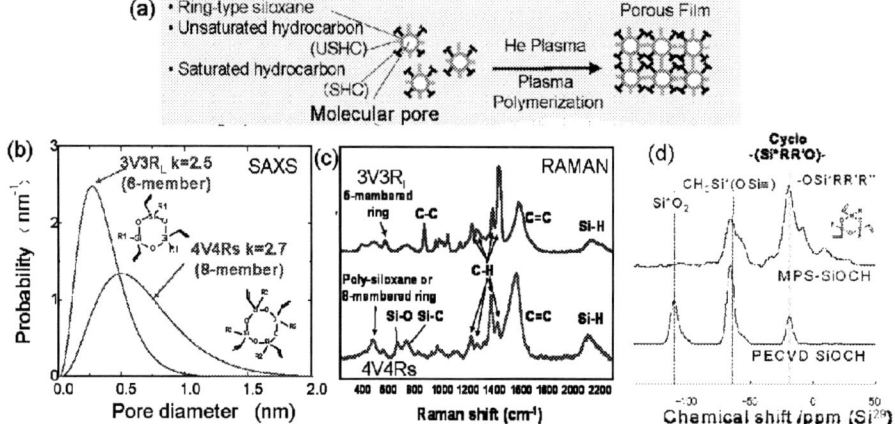

Fig. 5 Characteristics of the SiOCH films by the PP process using 6-member (3V3R$_L$) or 8-menber (4V4Rs) ring-type siloxanes; (a) schematic illustration of the plasma reaction process, (b) pore-diameter distribution by SAXS, (c) Raman and (d) Solid-state NMR (Si^{29}) spectroscopes.

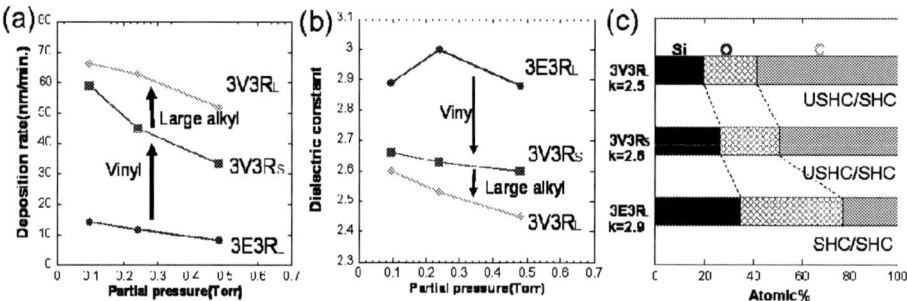

Fig. 6 Characteristics of the SiOCH films by the PP process using 6-member ring-type siloxanes having different side-chain chemicals; (a) the deposition rate and (b) the k-vales as a function of the deposition pressure, and (c) the chemical composition in the films (RBS) [17].

Table II Properties of SiOCH films.

	MPS-SiOCH	Porous-SiOCH	Rigid-SiOCH
Precursor	Ring-type siloxane (6-member)	Chain-type Siloxane + Oxidizer	Chain-type Siloxane + Oxidizer
k-value	2.5	2.7	3.1
Pore size (nm)	0.36 Closed Pore	0.54 Closed Pore	None porous
Density (g/cm³)	1.13	1.28	1.48
Modulus	2.8 (GPa)	5.6 (GPa)	12 (GPa)
Si:O:C	1:0.81:2.96	1:1.58:0.86	1:1.6:0.8
	C-rich	O-rich	

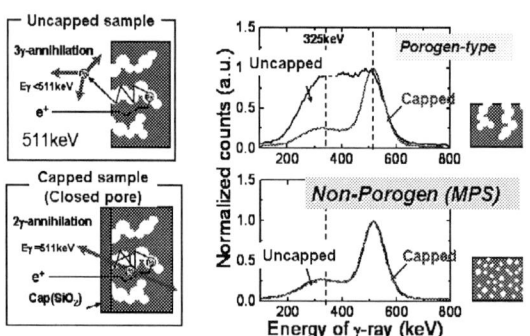

Fig. 7 PALS of bare and capped films of porgen-type and non-porogen-type porous SiOCH films. The MPS film had no difference between the bare and the capped films, indicative of closed pores in the film [16].

4

Stability of MPS SiOCH film

The MPS SiOCH film, which has carbon-rich hexagonal silica backbone with sub-nanometer pores, is superior stability against moisture uptake and plasma-induced damages. Fig. 8 shows the k-value increment of the SiOCH films as a function of the pore diameter after water dipping and PCT test (125°C, 100%-humidity). In case of the dipping in water, no k-value increment was observed, indicative of no water immersion into the pores within 1nm-dieameters. After sever PCT test, the k-value increased due to the moisture uptake into the pores over 0.6nm-diameters. No k-value increment was observed for the MPS film even though the PCT test.

Fig. 9 shows the k-value increment after PVD Ta barrier metal deposition as a function of the DC bias to the Ta target [19]. Note that, in this RF PVD system, lower DC bias to the target decrease the substrate bias. In case of the conventional carbon-poor SiOCH films in Table II, the k-values increased over -40V biases, while k-value of the MPS film was unchanged even at 0V bias or the maximum substrate bias. By FTIR analysis, the conventional carbon-poor SiOCH films lost a lot of hydrocarbons by the Ta ion bombardment, resulting in alternation of $Si-CH_3$ bonds to Si-H bond. In case of the MPS film with large hydrocarbon side-chains, the hydrocarbon content decreased by the Ta bombardment, but the $Si-CH_3$ bonds did not decrease due to protection of the silica backbone by the long hydrocarbon chains from the direct bombardment of Ta ions. Fig. 10 shows XSEM of the Cu damascene interconnects as a function of O_2 plasma ashing time duration. Here, the damage layer of hydrocarbon depletion was detected by dipping in diluted HF solution [20]. It is found that the hydrocarbon-rich MPS film had the strongest endurance to the O_2 plasma.

The MPS SiOCH film of k~2.5 has been implemented as inter-metal dielectric (IMD) film for 45nm-node ULSI with the 140nm-pitched lines [12] as shown in Fig. 11. The effective k-value of the interconnect module (k_{eff}) including high-k SiCN cap was 2.95, decreasing the signal delay of a interconnect-loaded ring-oscillator by 15% to a 65nm-node structure with k_{eff}=3.4. The stable MPS SiOCH film is a strong candidate for 45nm-node and beyond.

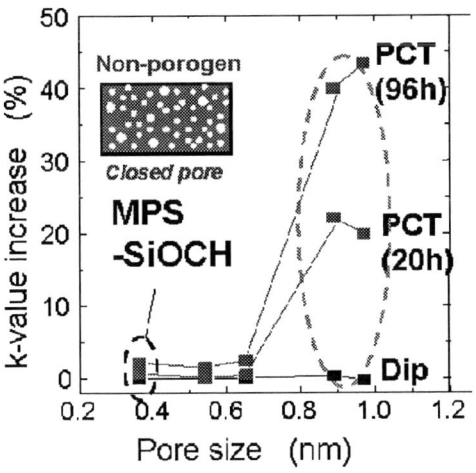

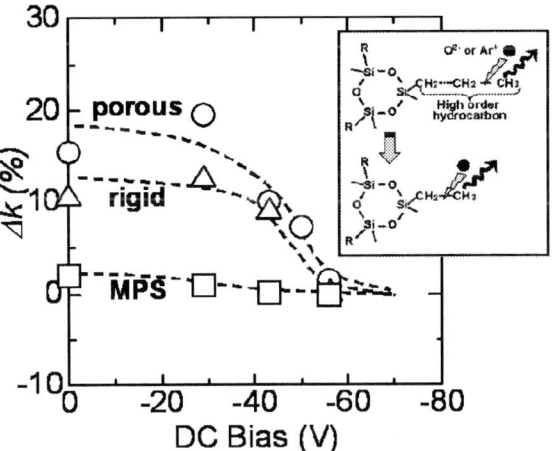

Fig. 8 Effect of the pore size on the k-value increment of SiOCH films after water-dipping, pressure-cooker test of 125°C under 100% humidity. No increment of the k-value was observed in the MPS film even after PCT.

Fig. 9 Damages on the SiOCH films by Ta barrier PVD as a function of target DC bias. The 0V of target bias derives the hardest ion bombardment to the SiOCH films [19]. The MPS had the strongest endurance to the Ta ion bombardment by PVD.

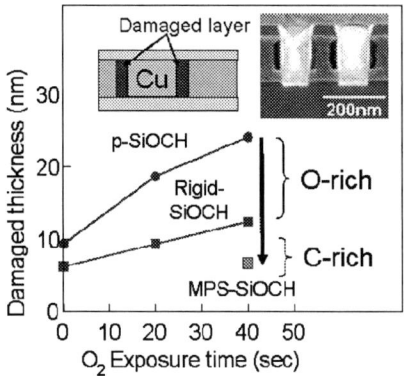

Fig. 10 Damaged thickness of the SiOCH films by extended O2 plasma exposure after the SiOCH etching. The damaged thickness was determined as the etched thickness by dipping diluted HF solution [20]

Fig. 11 XTEM images of Cu dual damascene interconnect in the MPS-SiOCH for 45nm-node ULSIs. The line pitch and the via-diameter are 140nm and 70nm, respectively [12].

Density-modulated MPS SiOCH film for 32nm-nodes ULSIs

Toward 32/22nm-nodes, the interconnect pitch is scaled down while the power supply voltage is kept a constant due to the lower limitation of V_{th} for the scaled-down MOSFETs, thus increasing the electric field between the interconnects. Under the high electric field, interface among the interconnect modules should be reinforced without any introduction of defects as shown in Fig. 12. We have innovated a modulated plasma-copolymerization (PcP) process to reinforce the interface.

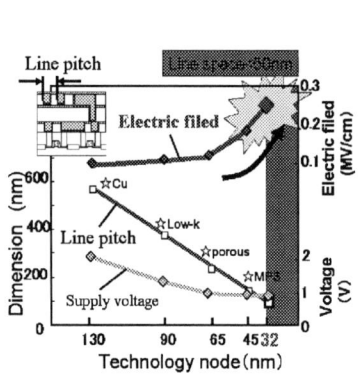

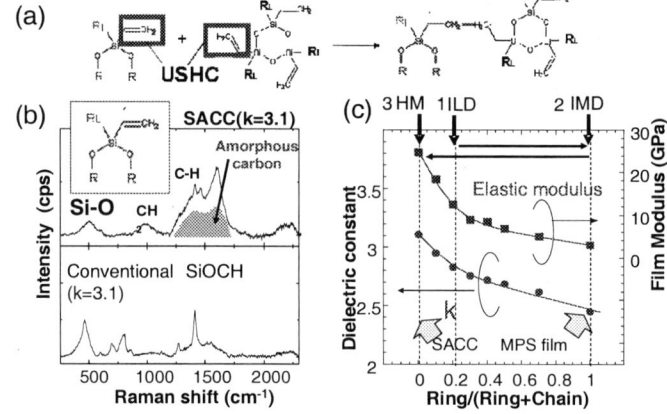

Fig. 12 Scaling issue toward 32nm/22nm-node integration such as the electrical field strengthen between the narrow-pitched lines under non-scaled power supply voltage.

Fig. 13 A modulated PcP process for a density-modulated MPS SiOCH film using a mixed gas of the ring-type (3V3R$_L$) and a linear-type siloxane with side-chains of USHC and SHC; (a) the PcP reaction scheme, (b) Raman spectroscopy of a silica-amorphous-carbon composite (SACC) film from 100% injection of the linear-type siloxane as well as a conventional SiOCH film, and (c) the film properties by PcP process, and. For the density-modulated MPS film, the mixing ratio is changed between 0% and 100% [14].

The modulated PcP process uses two type of siloxane monomers such as the hexagonal ring-type ($3V3R_L$) and a linear-type, both of which have the side-chains of USHC (vinyl) and SHC (R_L) [14] as shown in Fig. 13(a). The ring-type and the linear-type siloxanes are copolymerized through the USHC side-chains in He plasma. In case of the 100% injection of the linear-type siloxane, a silica-amorphous carbon composite (SACC) film was deposited with k=3.1, detected by Raman spectroscopy in Fig. 13(b). The SACC film had extremely high modulus of 26 GPa. By mixing the monomer gas of the linear-type to the ring-type, the k-value and the modulus of the MPS film increased as shown in Fig. 13(c). Conversely, this fact indicates us that the MPS film properties could be controlled by changing the mixing ratio such as 20% ⇒ 100% ⇒ 0% (100% of linear-type) of the ring-type in the one deposition scheme. By the modulated PcP process, a Cu dual damascene (DD) module with 100nm-pitched lines was fabricated with a density-modulated MPS film [14], in which the MPS IMD film was sandwiched between the copolymerized MPS via-ILD and the SACC of surface protection hard-mask (HM) as shown in Fig. 14. In other words, the upper and bottom interfaces of the MPS film were reinforced by the SACC and the copolymerized MPS with SACC. The interconnect module had k_{eff}=2.95 including high-k SiCN capping layer on the Cu lines.

Fig. 15 shows Pool-Frenkel (PF) plots of the leakage currents between the 50nm-spced lines in the moderated MPS film as well as the MPS films with the SiO_2-HM or the SiOCH-HM [14]. These HM layers were deposited by multi-step process with air break. The modulated MPS structure reduced the leakage current very much. The active energies were similar between the three structures, but the pre-exponential factor, which was related to the defect density, was 1/2000 smaller than the multi-step MPS films. The high endurance of the modulated MPS film might be related to suppression of the defects at the interface by the modulated PcP process.

Namely, it is confirmed that the modulated MSP structure is desirable for the scaled-down interconnects to keep dielectric endurance by the modulated PcP process using ring-type and linear-type siloxane monomers.

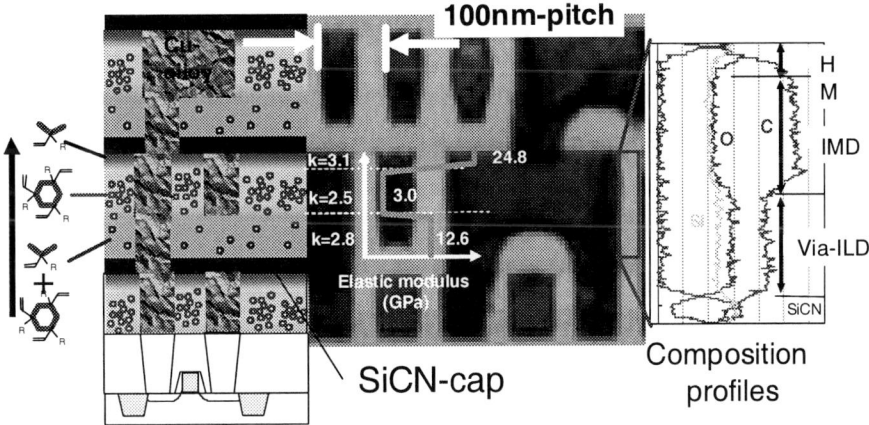

Fig. 14 A 100nm-pitched Cu DD interconnects in the density-modified MPS SiOCH films by modified PcP process. Here, the gas mixing ratio of the ring and linear types is changed during the deposition such as 20%, 100% and 0% of the ring-type. As the result, the MPS IMD film is sandwiched between the copolymerized MPS via-ILD and the thin SACC of surface protection hard-mask (HM) layer. By RBS analysis, the compositional profile of the modulated MPS film in the film thickness direction is modulated by corresponding to the mixing ratio modulation [14].

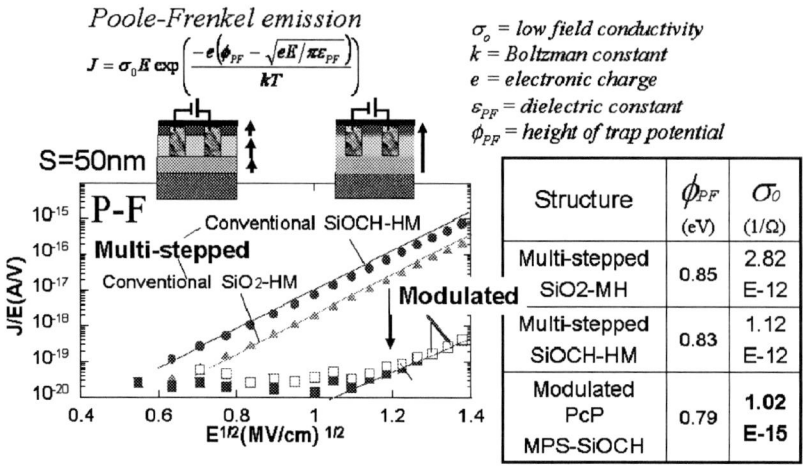

Fig. 16 Pool-Frenkel plots of the 50nm-spaced Cu lines in the modulated MPS films by the modulated PcP process as well as the multi-stepped MPS films with SiO₂-HM and SiOCH-HM [14]

Ultimate Full Low-k Interconnect toward 22nm-node and beyond

We found that the SACC had excellent barrier property against the Cu diffusion even though the k-value was 3.1, which could be utilized as a low-k capping layer, instead of high-k SiCN layer, on Cu lines [15] as shown in Fig. 17. So, we incorporated the SACC-cap in the modulated MSP deposition scheme, in which the gas mixing ration was changed as 100% ⇒ 80% ⇒ 0% ⇒ 100% injection of the linear-type siloxane to the ring-type one for the capping layer, the via-ILD, the IMD and the surface protection HM layer, respectively. Fig. 18 shows XTEM micrograph and RBS compositional profiles of the Cu interconnects in the modulated MPS film with the SACC capping layer seamlessly. Here, before the SACC deposition, the Cu line surface was treated by anti-oxidation coating in the same reaction chamber. The effective k-value of the interconnect module was reduced from 2.95 with the SiCN-cap to 2.75 with the SACC-cap. No degradation was observed in the dielectric endurance between the lines. Moreover, the EM reliability was improved due to simplicity in the deposition process.

Fig. 18 summarizes the structural evaluation of the Cu interconnect modules from the MPS-SiOCH film to the intimate full low-k MPS SiOCH films. All of these MPS-based SiOCH films are based on the PP or the modulated PcP processes using the hexagonal ring-type and the linear-type siloxanes. The in-situ density-modulation in the MPS SiOCH film accomplishes robust low-k module to reinforce the interfaces without defect introduction under simple and cost-effective process.

CONCLUSIONS

By the modulated PcP process, physical and chemical structures in porous SiOCH films is manipulated by molecular level. The self-organized pore structure was obtained by the PP process using the hexagonal ring-type siloxane with USHC side-chains. The MPS film with hydro carbon-rich hexagonal silica matrix has high endurance to the process damage. By in-situ modulated PcP process using the ring-type and the linear-type siloxanes, a full low-k module of k_{eff}=2.75 with low-k SACC cap was realized with good mechanical strength, adhesion and reliability, at a low manufacturing cost. The modulated PcP process is a key technology for scaled-down ULSI fabrication toward 32/22nm-nodes integrations.

ACKNOWLEDGMENTS

This research has been done in LSI Fundamental Research Laboratory, NEC ELECTRONICS and NEC. The author thanks especially to Drs. F. Ito, H. Yamamoto, J. Kawahara, H. Ohtake, N. Inoue, M. Tagami, M. Tada, M. Ueki, K. Hijioka, M. Abe, H. Watanabe and Y. Mochizuki, all in NEC group. The DVS-BCB monomer was provided by Dow Chemical Corporation, and the liner/ring-types siloxane precursors were supplied by TOSOH CORPORATION and TOSOH FINECHEM CORPORATION. A part of researches on the plasma-polymerization technology had been supported by NEDO through MIRAI project. The PALS analysis has been done by Drs. T. Ohdaira and R. Suzuki1, National Institute of Advanced Industrial Science and Technology. A part of solid-state NMR spectroscopy was supported by Drs. Riko Miyoshi and Keiko Matsuda, TORAY RESEARCH CENTER. Inc. The author thanks to all of these research supports and discussions.

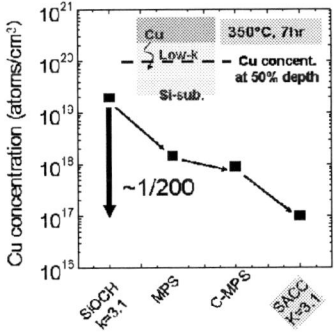

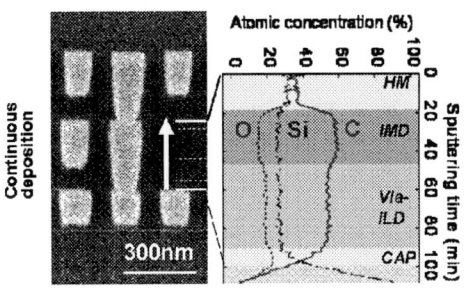

Fig. 17 Cu concentration in SiOCH films after 7 hr-annealing at 350°C with PVD Cu on the film top surface. The SACC of k=3.1 has barrier property of Cu diffusion referred to the conventional SiOCH film of k=3.1 [15]

Fig. 18 XSEM and the compositional profiles in an ultimate full low-k Cu interconnects by the modulated PcP process. Here, SACC film with k=3.1,instead of high-k SiCN is deposited on the Cu lines, followed by the modulated MPS film deposition [15].

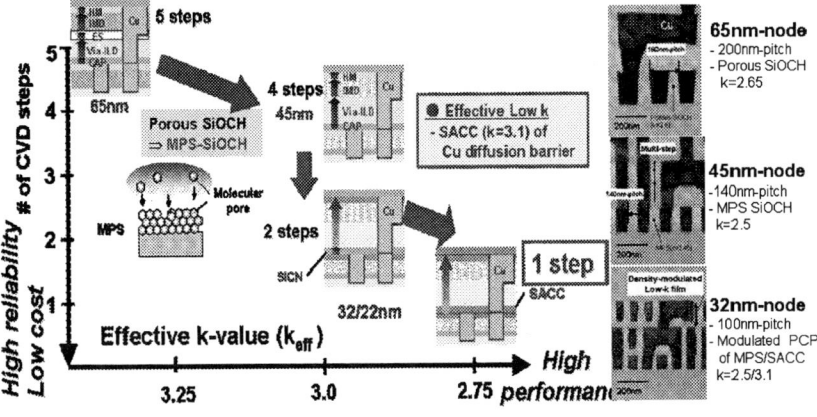

Fig. 19 The structural evolution of the low-k/Cu interconnect modules from the MPS SiOCH, the modulated MPS and the intimate full low-k MPS SiOCH films toward simple and cost-effective process.

9

REFERENCES

1. Y. Hayashi, IEEE Intl. Interconnect Tech. Conf., (San Francisco, USA), 2002, p145 (2002).
2. K. Hijioka, M. Tagami, F. Itoh, H. Ohtake, T. Takeuchi, S. Sato, and Y. Hayashi, Jap. J. Appl. Phys., **43**, 4B, p1807 (2004).
3. F. Itoh, K. Hijioka, T. Takeuchi, and Y. Hayashi, Proc. of Advanced Metallization Conf.(San Diego, UAS), p381 (2004)..
4. J. Kawahara , A. Nakano, S. Saito, K. Kinoshita, T. Onodera, and Y. Hayashi, 1999 Symp. VLSI Technol. (Kyoto, JPN) Dig. p.45 (1999).
5. J. Kawahara , A. Nakano, K. Kinoshita, Y. Harada, M. Tagami, M. Tada and Y. Hayashi, Plasma Source Sci., Technol. 12, pS80 (2003).
6. H. Ohtake, S. Saito, M. Tada, T. Onodera and Y. Hayashi, IEEE Trans. Semiconduct. Manufact., **18**, 4, p672 (2005).
7. Y. Hayashi, Y. Harada, F. Itoh, T. Takeuchi, M. Tada, M. Tagami, H. Ohtake, K. Hijioka, S. Saito, T. Onodera, D. Hara and K. Tokudome, IEEE Intl. Interconnect Tech. Conf. (San Francisco, USA), p225 (2004).
8. J. Kawahara, A. Nakano, N. Kunimi, K. Kinoshita, Y. Hayashi, A. Ishikawa, Y. Seino, T. Ogata, Y. Sonoda, T. Yoshino, T. Goto, S. Takada, H. Miyoshi, H. Matsuo, and T. Kikkawa, Jpn. J. Appl. Phys., **46**, 7A, p4064 (2007).
9. M. Ueki, M. Narihiro, H. Ohtake, M. Tagami; M. Tada; F. Ito; Y. Harada; M. Abe; N. Inoue; K. Arai; T. Takeuchi; S. Saito; T. Onodera; N. Furutake; M. Hiroi; M. Sekine; Y. Hayashi, 2004 Symp. VLSI Techol. (Honolulu, USA) p60 (2004).
10. M. Tada, T. Tamura, F. Ito, H. Ohtake, M. Narihiro, M. Tagami, M. Ueki, K. Hijioka, M. Abe, N. Inoue, T. Takeuchi, S. Saito, T. Onodera, N. Furutake, K. Arai, M. Sekine, M. Suzuki, Y. Hayashi, IEEE Trans. Electron Devices, **53**, 5, p1169, (2006).
11. M. Tada, H. Ohtake, F. Itoh, M. Narihiro, T. Taiji, Y. Kasama, T. Takeuchi, K. Arai, N. Furutake, N. Oda, M. Sekine, Y. Hayashi, IEEE Trans. Electron Devices, **54**, 4, p797 (2007).
12. M. Tagami; H. Ohtake; M. Tada; M. Ueki; F. Ito; T. Taiji; Y. Kasama; T. Iwamoto; H. Wakabayashi; T. Fukai; K. Arai; S. Saito; H. Yamamoto; M. Abe; M. Narihiro; N. Furutake; T. Onodera; T. Takeuchi; Y. Tsuchiya, Y.; N. Oda; M. Sekine; M. Hane; Y. Hayashi, 2006 Symp. VLSI Techol. (Honolulu, USA) p134 (2006).
13. N. Inoue, M. Tagami; F. Itoh; H. Yamamoto; T. Takeuchi; S. Saito; N. Furutake; M. Ueki; M. Tada; T. Suzuki, and Y. Hayashi, IEEE Intl. Interconnect Tech. Conf. (San Francisco, USA), p181 (2007).
14. M. Tada, M.; H. Yamamoto; F. Ito; M. Narihiro; M. Ueki; N. Inoue; M. Abe; S. Saito; T. Takeuchi; N. Furutake, T. Onodera; J. Kawahara; K. Arai; Y. Kasama; T. Taiji; M. Tohara; M. Sekine; Y. Hayashi, IEEE, Intl. Electron Devices Meeting (IEDM), p351 (2006).
15. M. Ueki, H. Yamamoto, F. Ito, J. Kawahara, M. Tada, T. Takeuchi, S. Saito, N. Furutake, T. Onodera and Y. Hayashi, IEEE, Intl. Electron Devices Meeting (IEDM), p973 (2007).
16. F. Ito, T. Takeuchi, H. Yamamoto, T. Ohdaira, R. Suzuki and Yoshihiro Hayashi, to be published in Advanced Metallization Conf.(AMC 2007, MRS).
17. M. Tada, H. Yamamoto, F. Ito, T. Takeuchi, N. Furutake, and Y. Hayashi, J. Electrochem. Soc., **154**, 7, p 354 (2007).
18. S. Sankaran, et al., IEEE, Intl. Electron Devices Meeting (IEDM), p355 (2006).
19. N. Inoue, N. Furutake, F. Ito, H. Yamamoto, T. Takeuchi and Y. Hayashi, Ext. Abst. 2006 Intl. Conf. Solid State Device and Materials (SSDM), p1026 (2007).
20. Y. Hayashi, H. Ohtake, J. Kawahara, M. Tada, S. Saito, N. Inoue, F. Ito, M. Tagami, M. Ueki, N. Furutake, T. Takeuchi, H. Yamamoto and M. Abe, to be published in IEEE Trans., IEEE Trans. Semiconduct. Manufact (2008).

Mater. Res. Soc. Symp. Proc. Vol. 1079 © 2008 Materials Research Society   1079-N02-04

UV Curing Effects of Low-k Materials under Reactive Conditions

Darren Moore, Carlo Waldfried, Palani Sakthivel, and Orlando Escorcia
Axcelis Technologies, 108 Cherry Hill Drive, Beverly, MA, 01915

ABSTRACT

This work describes the effect of addition of O_2 and NH_3 to the N_2 ambient used during the UV cure process of low-k materials. The effects of O_2 give an acceleration of the cure process, resulting in increased film modulus and shrinkage. The use of NH_3 resulted in a retardation of UV cure, explained by the higher absorption cross section of NH_3 at the wavelengths used.

INTRODUCTION

UV curing of low-k films has received particular attention as a post deposition treatment due to its ability to effectively improve film modulus [1] while removing porogen and retaining porosity, all with low thermal budget and little increase in k value [2]. In general increasing cure time will result in increased modulus, eventually at the disadvantage of an increase in film k. This work will describe the use of oxygen addition to the cure process to achieve an accelerated cure process, enabling a targeted film modulus to be achieved with shorter cure time, with little or no impact on k value.

EXPERIMENT

Three carbon doped SiOCH film types were used for this study. The first was a porous SOG, prepared by a typical spin-on and soft bake process with target k=2.5. The second was a PECVD non-porous film with target k=3.0. The final film was a porous PECVD film formed by the co deposition of a lattive and cyclic organic porogen material. The removal of the porogen achieved the target k=2.5. All films used were of nominal thickness ~ 10kA. The cure process used the Axcelis RapidCure 320FC UV cure tool. The process uses microwave driven electrodless bulbs emitting UV radiation in the 100-400 nm range with the wafer temperature maintained at 400°C with thermal chuck. Standard process conditions used a constant purge of N_2 gas at atmospheric pressure. Oxygen additions to the ambient were set at 50 and 100 ppm in N_2, while ammonia concentrations tested were 50, 100 and 750 ppm. All tests were processed for 3 mins, and compared to the cure process with N_2 only ambient. K values were determined using Al shadow mask techniques. Mechanical properties were determined using an MTS nanoindentor XP. Solid state Si^{29} MAS NMR was carried out on a Bruker Avance NMR spectrometer (11.7 Tesla, 500.2MHz, 4mm $Zr O_2$ rotor @ 10kHz).

DISCUSSION
Effect of UV cure with N_2 ambient

Although cure of dielectric films is well documented, the effects of increasing cure time for the films used in this work will first be summarized. Figure (1) illustrates film shrinkage for cure times of 1, 3, 6 and 13 minutes for all three dielectrics studied. The relative modulus and hardness increases for the SOG 2.5 and CVD 3.0 films, (Figure (2)) show a ~70% increase for the CVD 3.0 and a 400-500% increase for the porous SOG 2.5 film when cured to 13 minutes. Effects of UV cure on the FTIR properties of films are well known and typically show an increase in the SiO network group along with loss of Si-CH$_3$ moieties and reduction of the cage SiO peak. In this work, the post–pre subtracted spectra are used to allow the subtle changes in the SiO network (~1050 cm^{-1}) and SiO cage (~1050 cm^{-1}) region to be measured, while the Si-CH$_3$ (1273 cm^{-1})/SiO peak ratio loss follows Si-CH$_3$ cleavage. Detailed Si^{29} NMR spectroscopy for the SOG 2.5 and CVD 3.0 films is detailed elsewhere [3], showing a conversion of M and D groups to Q and T with increasing cure time.

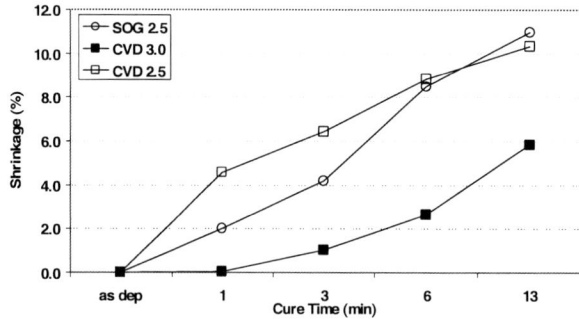

Figure (1): Shrinkage vs cure time of low-k films studied

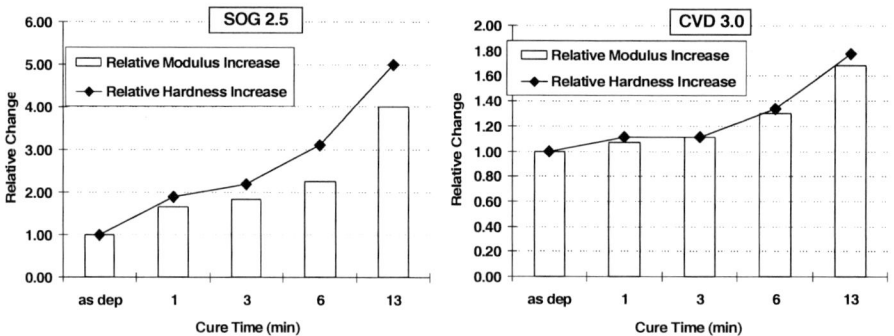

Figure (2): Relative M and H increase vs cure time of SOG 2.5 and CVD 3.0

Effect on UV cure through addition of O$_2$ to the N$_2$ ambient

Addition of 50 and 100ppm O$_2$ to the N$_2$ ambient POR (3min) results in increased shrinkage of up to ~75% for the porous SOG 2.5. Shrinkage increases were less pronounced for the p-CVD 2.5 and CVD 3.0 films (Table (1)). Modulus increases were also significant for the SOG 2.5 film, showing values equivalent to those obtained at twice the time in a pure N$_2$ environment. Increases for the CVD 3.0 were more moderate at ~5% for both modulus and hardness. (CVD 2.5 modulus and hardness results were not measured in this work) Control cures in pure N$_2$ and 100ppm O$_2$/ N$_2$, both with no UV, show identical and negligible shrinkages.

	SOG 2.5			CVD 3.0			CVD 2.5		
	Shrink (%)	M (Gpa)	H (Gpa)	Shrink (%)	M (Gpa)	H (Gpa)	Shrink (%)	M (Gpa)	H (Gpa)
3min (N2)	4.19	2.20	0.22	1.04	8.10	1.11	6.43	-	-
3min 50ppm O2	6.42	2.80	0.30	1.09	8.50	1.16	7.66	-	-
3min 100ppm O2	7.48	3.00	0.30	1.30	8.50	1.16	7.59	-	-
3min (N2) (NO UV)	0.22	-	-						
3min 100ppm O2 (NO UV)	0.24	-	-						

Table (1): Properties of UV cured low-k films in O_2/N_2 atmosphere

Due to their porosity and susceptibility to dielectric damage, both k=2.5 films were examined, and showed no detrimental increase in k value.

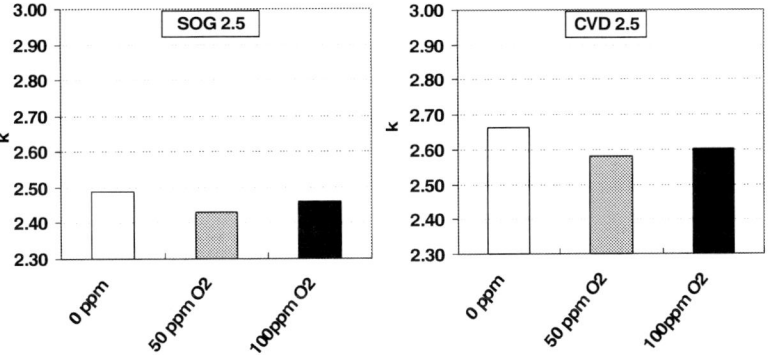

Figure (3): k values of UV cured low-k films in O_2/N_2 atmosphere (3 min cure)

FTIR analysis for these two porous dielectrics show a decrease in Si-CH$_3$/Si-O ratio with cure (Figure (4)), being notable and progressive for the SOG 2.5, while the CVD 2.5 film does not increase with higher concentration of O_2. Figure (5) shows only slight increases in porogen loss (CH region) as compared to the POR 3min N_2 cure.

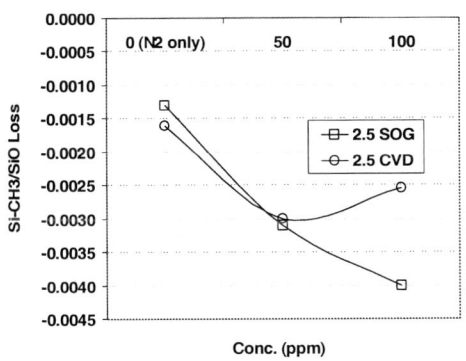

Figure (4): Si-CH$_3$/SiO peak area loss with cure in O_2/ N_2 atmosphere

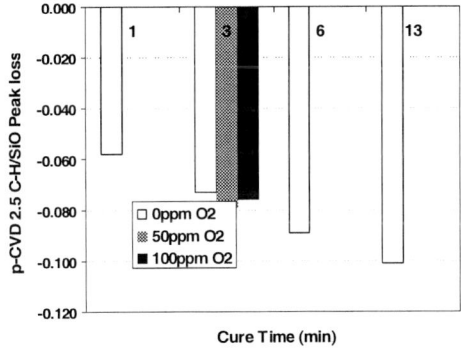

Figure (5): C-H/SiO peak area loss in CVD 2.5 film with O_2/ N_2 cure process

These trends of Si-CH$_3$ loss and film modulus increase occur with complimentary changes as measured by the increases in SiO network (Figure (6)) and decreases in SiO cage peaks. A lower SiO network increase is observed in the CVD 2.5 as compared to the other films studied.

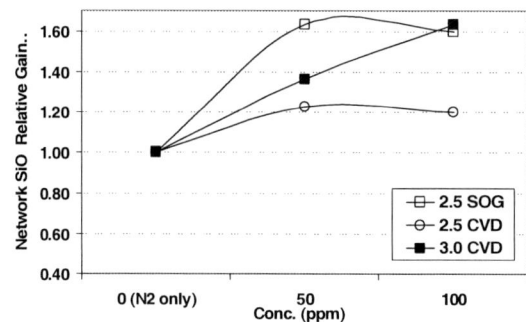

Figure (6): Si-O network in low-k films cured in O_2/N_2 atmosphere

Finally, Si^{29}NMR spectra were obtained for the SOG 2.5 film cured under oxidative conditions, and compared to the 3min POR process. Figure (7) shows the increase in the highly crosslinked Q band that occurs at the expense of a decrease of D and M. There is also a notable increase in the proportion of T_{OH} species.

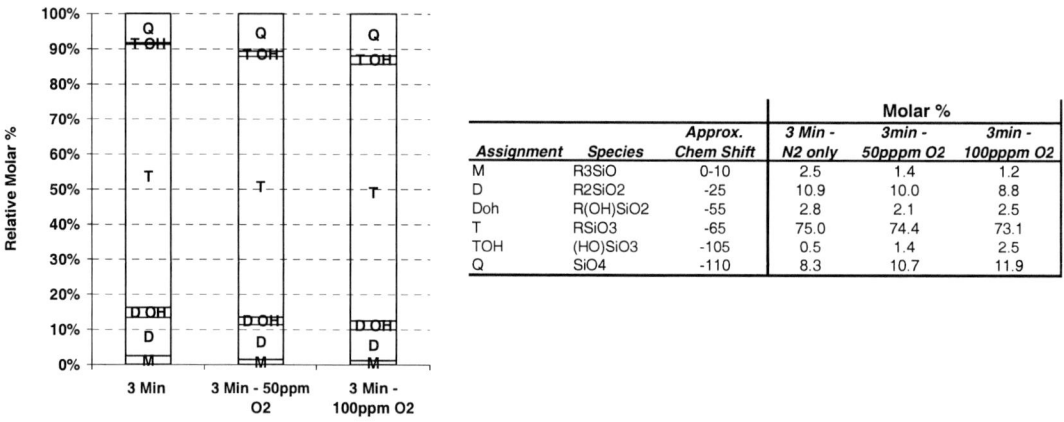

				Molar %	
Assignment	Species	Approx. Chem Shift	3 Min - N2 only	3min - 50pppm O2	3min - 100pppm O2
M	R3SiO	0-10	2.5	1.4	1.2
D	R2SiO2	-25	10.9	10.0	8.8
Doh	R(OH)SiO2	-55	2.8	2.1	2.5
T	RSiO3	-65	75.0	74.4	73.1
TOH	(HO)SiO3	-105	0.5	1.4	2.5
Q	SiO4	-110	8.3	10.7	11.9

Figure (7): Si^{29} NMR of SOG 2.5 cured in O_2/N_2 atmosphere

In order to further examine any detrimental effects of O_2 addition to the UV cure process, SOG 2.5 films of nominal thickness ~ 5kÅ (50% that of those otherwise used in this work) were cured with increasing concentrations of oxygen from 0 to 100ppm. The modulus shows >33% improvement with the lowest concentrations of oxygen used (Table (2)). Further increases in oxygen concentration show no benefit to film modulus and start to show notable increases in film k at 100ppm.

O2 ppm	k	Rel. Modulus	Shrink (%)
0	2.40	1.00	6.69
25	2.41	1.39	9.64
50	2.38	1.33	10.21
75	2.43	1.35	10.44
100	2.51	1.29	10.50

Table (2): Shrinkage, k and modulus of SOG 2.5 cured in O_2/N_2 atmosphere

It is thus proposed that the controlled addition of O_2 can be used to increase lattice SiO crosslinking and yield improvements in film modulus with no effect on k. The maximum concentration of oxygen usable without film damage will be a factor of film type, thickness, presence of porogen, temperature and UV source employed. The mechanism is proposed to be due to the dissociation of oxygen, or ozone, under the UV wavelengths present as in equations (1) and (3) below. The oxygen radical produced facilitates $Si-CH_3$ cleavage, thereby encouraging greater Si-O networking. Further support for this mechanism may be derived from the increase of T_{OH} ($HOSiO_3$) species, formed by $O^{\bullet}$ insertion into the $H-SiO_3$ or cleavage of $R-SiO_3$, potentially being the reason for the increase in k above.

This mechanism is proposed to act on the SiOCH lattice, occurring most readily with the porous SOG 2.5 film. The presence of the cyclic organic porogen in the CVD 2.5 is likely to hinder the catalytic effect of oxygen radical, resulting in the relatively lower shrinkage and SiO network increases. The insensitivity of this film to increasing oxygen concentrations may further support this.

$$O_2 + UV\ (180\text{-}240nm) \rightarrow O^{\bullet} + O^{\bullet} \qquad (1)$$

$$O^{\bullet} + O_2 + M \longrightarrow O_3 + M^* \qquad (2)$$

$$O_3 + UV\ (200\text{-}320nm) \rightarrow O^{\bullet} + O_2 \qquad (3)$$

$$O_3 + O^{\bullet} \rightarrow 2O_2 \qquad (4)$$

Effect of UV cure with additions of NH_3 to the N_2 ambient

Controlled additions of ammonia at concentrations of 50, 100 and 750ppm in N_2 were used in the UV cure of SOG 2.5 and CVD 3.0 films. The CVD 3.0 film, shrinkage results showed little change with respect to the pure N_2 (Table (3)). There was a small increase noted with increasing concentrations of NH_3 for the SOG 2.5 dielectric. This was not reflected with an increase in modulus, both films showing ~8% lower modulus gain compared to the control wafer.

	SOG 2.5			CVD 3.0		
	Shrink (%)	M (Gpa)	H (Gpa)	Shrink (%)	M (Gpa)	H (Gpa)
3min (N2)	4.19	2.6	0.2	1.04	8.3	1.0
3min 50ppm NH3	4.18	2.3	0.2	0.99	7.7	1.0
3min 100ppm NH3	4.43	2.4	0.2	0.77	7.8	1.0
3min 750ppm NH3	4.53	2.4	0.2	0.87	7.7	1.0

Table (3): Properties of UV cured low-k films in NH_3/N_2 atmosphere

FTIR analysis showed a reduction in the extent of SiO network formation and lower SiO cage conversion. Changes in the $Si-CH_3$ / SiO ratio as measured pre and post cure show a marked decrease compared to the POR N_2 cure, representing less $-CH_3$ loss. These effects are again more pronounced for the porous SOG 2.5 dielectric (Figure(8)).

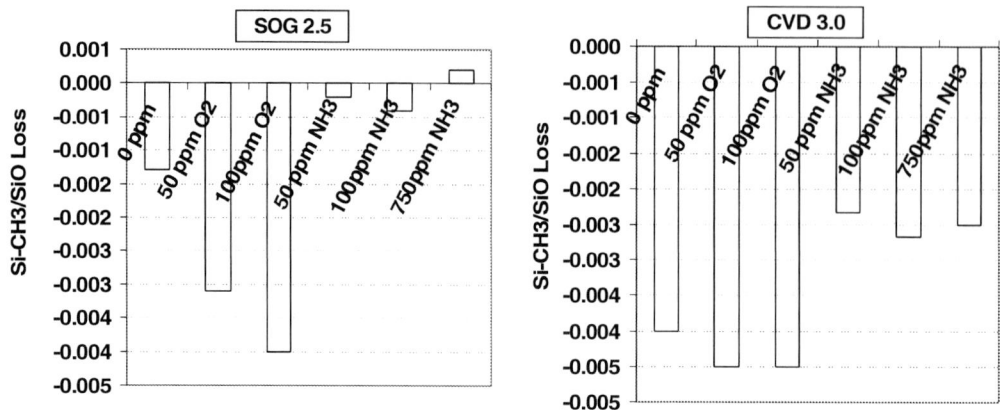

Figure (8): Si-CH$_3$/SiO peak ratio loss of films used in additive cure processes

The additions of NH$_3$ to the N$_2$ ambient is therefore proposed to result in a slight retardation of cure. The contrasting effects of O$_2$ and NH$_3$ are proposed to be due to their different absorption cross sections over the critical wavelengths used (170-220nm), with NH$_3$ known to be orders of magnitude larger than O$_2$[4-5]. These literature values of absorption cross section can be used, along with the path length within the UV cure tool, and the O$_2$ or NH$_3$ concentrations used to yield transmission curves over the wavelengths of interest. Figure (9) illustrates the significant reduction in transmissions thus obtained at λ<220nm in the ammonia atmosphere.

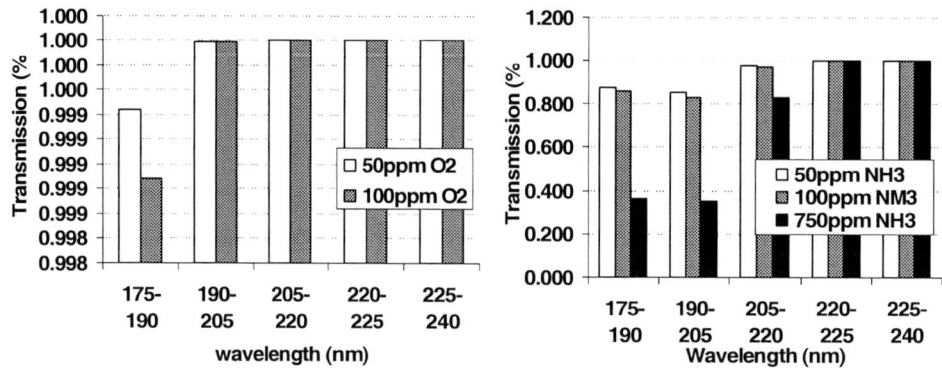

Figure (9): Transmission through O$_2$ and NH$_3$ at varying concentrations

CONCLUSIONS

Controlled additions of O$_2$ to the N$_2$ ambient used in the UV cure of low-k materials has been shown to result in increased in film modulus, hardness and shrinkage. Use of 25-75ppm O$_2$ in N$_2$ results in greater SiO crosslinking and Q moiety formation as shown in Si^{29} NMR. A mechanism of oxygen radical formation and reaction with the SiOCH lattice is proposed. Conversely, addition of NH$_3$ results in a mild retardation of cure, due to its absorption of wavelengths critical for effective UV cure.

REFERENCES

[1]   C. Waldfried, Q. Han, O. Escorcia et al, *Conference Proc. IITC*, p226, (**2002**)

[2]   M.K. Haas , C. Waldfried, O. Escorcia et al, Conference Proc. AMC, p419, (**2006**)

[3]   D. Moore, C. Waldfried, Conference Proc. AMC, (**2007**)

[4]   F.Z. Chen, Planet. *Space Sci.* 47, 261-266 (**1999**)

[5]   K. Yoshino, Planet. *Space Sci.* 40, 185-192 (**1992**)

Mater. Res. Soc. Symp. Proc. Vol. 1079 © 2008 Materials Research Society 1079-N02-09

Moisture Adsorption in Plasma-Damaged Porous Low-k Dielectrics

Ekaterina Vinogradova[1], Casey E Smith[1,2], DW Mueller[1], Andrew J McKerrow[3,4], and Rick Reidy[1]

[1]Materials Science and Engineering, University of North Texas, PO Box 305310, Denton, TX, 76203-5310
[2]Sematech, Austin, TX, 78741
[3]Texas Instruments, Dallas, TX, 75243
[4]Novellus, Tualatin, OR, 97062

ABSTRACT

Plasma etch/ash processes can induce changes in low-k film surface/bulk chemistries and topographies resulting in increased water adsorption, surface roughness, and metal intrusion. After ashing, the altered surface character of the low-k can impact wetting, adhesion, and, consequently, the resistance of subsequently deposited barrier layers. In this work, we describe the use of deuterium oxide as means of measuring moisture penetration into low-k films. Film chemistries have been monitored using grazing angle attenuated total reflectance (GATR) and transmission Fourier transform infrared spectroscopy (FTIR). To study moisture adsorption in porous spin-on and CVD low-k films, unashed and ashed films have been exposed to D_2O liquid and vapor treatments under "dry" nitrogen. The extent of D_2O uptake, removal and exchange reactions has been studied using transmission and GATR FTIR methods because the D_2O and O-D adsorption peaks are distinct from water and O-H as well as other low-k adsorptions. This method can be used to study Si-OH species because deuterium can exchange with hydrogen within silanols under ambient conditions while methyl groups are much less likely to exchange. Three different low-k films, a porous spin-on MSQ (k=2.2), a porous CVD (k=2.3), and an organosilicate glass (OSG, k=2.85) have been used in this work. In FTIR spectra, unashed low-k films show minimal D_2O adsorption. In MSQ hydrogen-ashed films, the data suggest the presence of deuterium oxide and O-D peaks. Further, D_2O adsorption appears to be considerably higher for ashed films as would be expected due to the hydrophobicity of these films. In the CVD films, there does not appear to be as marked a difference. This method can permit the introduction of a chemical "marker" into low-k wet and ambient processes allowing one to distinguish among adsorptions from different aqueous sources.

INTRODUCTION

Integration of porous low-k films is necessary for the continued scaling of integrated circuits;[1] however, plasma etch/ash processes and subsequent wet cleans can deleteriously affect these films through material damage/loss, removal of hydrophobic moieties, and densification. The nature and degree of damage is specific to processing, and can significantly increase dielectric constant, reduce dielectric breakdown voltage, permit metallic specie intrusion, increase surface roughness[2], and enhance moisture adsorption leading to degradation of device performance and reliability. Loss of hydrophobicity can lead to additional material loss and the intrusion of aqueous species during wet cleans.[3] This works attempts to measure moisture adsorption into these films.

18

EXPERIMENTAL

Three sample low-k films were studied: a porous low-k spin-on, a porous low-k CVD film, and for comparison, a low porosity organosilicate glass. The properties of these films are described below.

Table 1 Properties of sample low-k films

Property	Spin-on low-*k*	CVD low-*k*	OSG
Dielectric constant	2.25	2.30	2.85
Refractive index	1.27	1.32	1.40
Porosity	~40%	~25%	~8%
Pore diameter (nm)	3-4	~2	-

Film properties were characterized by Fourier transform infrared spectroscopy (FTIR), and OSG dielectric properties were studied using C-V measurements. FTIR measurements were conducted using transmission and grazing angle attenuated total reflectance (GATR) methods. Moisture adsorption into these films can be studied by introducing liquid and vapor D_2O and monitoring changes in the FTIR spectra.[4] Deuterium oxide has similar properties to water (i.e., boiling temperature, surface tension, and dielectric constant) although it has a larger viscosity and slower diffusivity than H_2O.[5] OSG, MSQ, and CVD low-films were exposed to D_2O liquid and vapor treatments under "dry" nitrogen. The samples were immersed in a 5 ml of liquid deuterium oxide for 24 hours at room temperature. In the vapor phase D2O experiments, 5 ml of liquid D_2O was evaporated at 80°C and the samples have been exposed for 1 hour in a closed vessel. The extent of D_2O uptake, removal and exchange reactions were studied using transmission and GATR FTIR methods because the D_2O and O-D adsorption peaks are distinct from water and O-H as well as other low-k adsorptions (Table 2). This method can be used to study Si-OH species because deuterium can exchange with hydrogen within silanols under ambient conditions while methyl groups are much less likely to exchange.

Table 2 IR band assignments for H_2O, HOD, and D_2O[6]

H_2O (cm^{-1})		D_2O and HOD (cm^{-1})	
3750–3100	OH stretch	2300–2800	OD stretch
2120	association band	1850	HOD association band
		1550	D_2O association band
1635	OH bend	1460	HOD bend
		1200	D_2O bend

RESULTS AND DISCUSSION

Moisture adsorption into porous low-k's presents several issues because of its detrimental effects on these films such as increased dielectric constant, decreased breakdown voltage, and increased susceptibility to cracking[7]. Strategies for prevention of moisture adsorption have included the addition of hydrophobic groups in SiO_2 structures and the use of hydrophobic polymeric films. Detection of moisture within films is difficult at the levels that begin to impact the dielectric properties. Ogawa et al. and other researchers have shown that defectivity in porous low-k dielectrics impacts dielectric breakdown[8] and, Matz has shown that removal of moisture can improve breakdown behavior.[9] However, the detrimental threshold of moisture remains uncertain. Shaw et al. has studied the diffusion of moisture into polymeric and carbon-doped oxide films.[10] Organosilicate glass (OSG) can show no measurable water adsorption

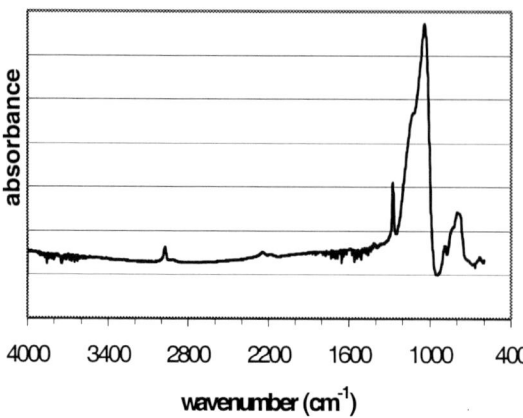

Figure 1 FTIR of OSG film left in ambient showing no evidence water absorption.

FTIR peaks (Fig. 1); yet, can exhibit significant changes in dielectric constant. Previous efforts have examined the impact of ambient atmosphere on an uncapped OSG film in metal-insulator-metal capacitance (MIMCAP) structures exposed to ambient air in a Fab for 6 months, and then baked at 400C for 90 sec in low pressure nitrogen. The dielectric constant of the film was determined by capacitance-voltage measurements on several sizes of metal contacts after the lengthy ambient exposure, immediately after the bake, and then 2 and 7 days after the bake. In Figure 2, smaller Al dot areas are more affected by moisture intrusion than larger dot sizes. After the 400C bake, the dielectric constant decreases considerably, but increases again as it is re-exposed to the atmosphere.

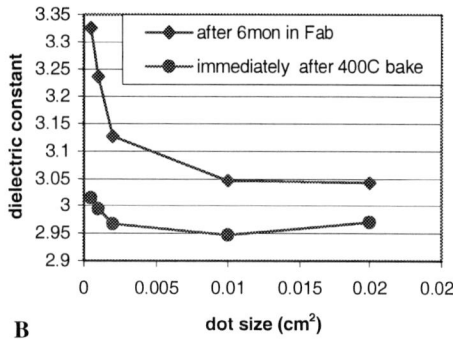

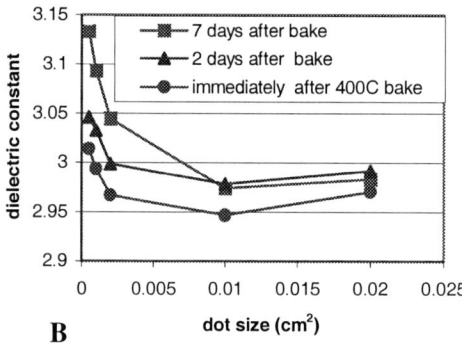

Figure 2 Dielectric constants as a functions of exposure, high temperature bake, contact size in OSG MIMCAPs. In Fig 1A, the dielectric constant of pre-exposed OSG decreases considerably after a 400C bake. In Fig 1B, k increases as the MIMCAP is re-exposed to ambient.

Deuterium oxide does not significantly adsorb into OSG, but there is some evidence (albeit scant) of D_2O and deuterated species within the film. Peaks at 1460 and 1560cm^{-1} are also evident using GATR FTIR methods (not shown). Like OSG, the porous CVD low-k films showed limited D_2O adsorption; however, $H2/N_2$ ashed films evidenced significant moisture uptake. In MSQ films, D_2O adsorption is more pronounced. Figure 4A and B describes the D_2O adsorption in unashed, O_2 and H_2/N_2 ashed porous CVD low-k films using transmission FTIR.

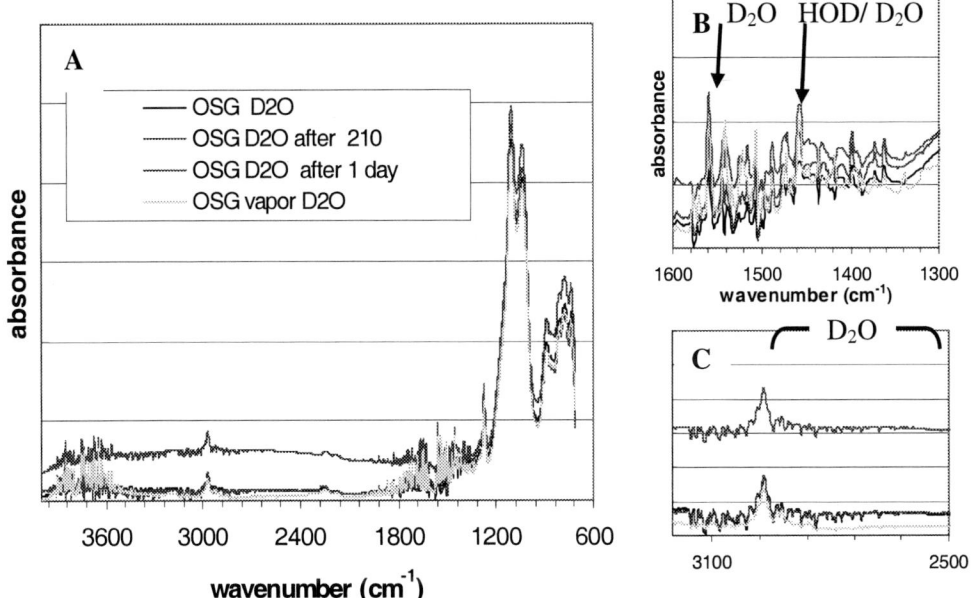

Figure 3 Transmission FTIR of OSG after liquid D_2O exposure, exposure followed by 210C bake for 1hour, 1 day after exposure, and D_2O vapor exposure. One would expect D_2O between 2400-2800cm^{-1} in Fig 3B. Fig 3C suggests D_2O (1560 cm^{-1}) and HOD (1460cm^{-1}), an exchange product. [6]

Deuterium oxide is much more readily adsorbed into reducing ashed CVD films than those exposed to oxidizing ash. It is unclear if any detectable adsorption occurs in the O_2 –ashed sample. After a 210C bakeout, D_2O remains evident in the FTIR spectra indicating difficulty removing the species. Both hydrogen and oxygen ashed spin-on samples significantly adsorbed D_2O although, like the CVD films, the reducing ash sample adsorbed more D_2O. The spin-on films' greater susceptibility to adsorption may be due in part to their higher porosity and larger pores. Regardless of porosity, the persistence of the deuterium oxide indicates that the species can be well adsorbed into both spin-on and CVD films. The presence of the 1460 cm^{-1} peak (H-O-D) suggests an exchange reaction between Si-OH and D_2O resulting in Si-OD. Current efforts are studying the use of this method to quantify hydrogen-deuterium exchange in low-k films, thus, study the availability of SiOH species on the ashed film surface.

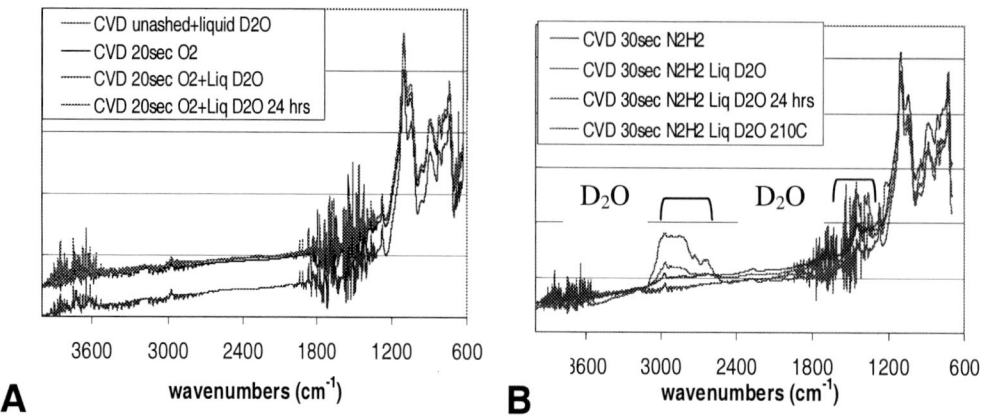

Figure 4 Transmission FTIR of D$_2$O exposed porous CVD films. A-unashed and O$_2$ ashed films show very limited absorption of D$_2$O while in B- H$_2$/N$_2$ films absorb D$_2$O readily.

CONCLUSIONS

D$_2$O appears to be an effective marker to monitor water adsorption in porous films. While more viscous that water, it has a similar volatility and surface tension. The spin-on low-k film used in this study readily adsorbed deuterium oxide although the hydrogen-ashed sample seemed to have adsorbed more D$_2$O than the oxygen-ashed film. Hydrogen-ashed CVD films showed marked adsorption and difficulty removing the moisture by bakeout. In this film, there is some evidence of H$\rightarrow$D exchange as well. The oxygen-ashed and the unashed CVD films did not show any significant adsorption although addition work with GATR FTIR will be conducted to substantiate this conclusion. The oxygen-ashed sample shows some evidence of densification (not shown here) that may contribute to its limited adsorption. While not as porous as the spin-on and CVD films, OSG films can adsorb moisture from the ambient atmosphere that can lead to degradation of their dielectric properties.

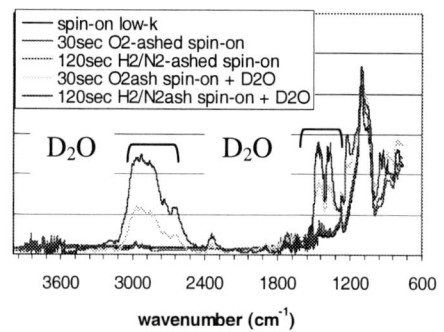

Figure 5 Transmission FTIR of porous spin-on MSQ films after D$_2$O liquid exposure.

ACKNOWLEDGMENTS

The authors would like to acknowledge the support of Sematech, Texas Instruments, and the National Science Foundation (DMR#0316916).

REFERENCES

[1] *International Technology Roadmap for Semiconductors (ITRS)*, 2007 Interconnect Update.
[2] C.E. Smith, D.W. Mueller, P.D. Matz, R. F. Reidy, MRS Proceedings Volume Materials, Technology, and Reliability of Low-k Dielectrics and Copper Interconnects, ed. T.Y. Tsui, Y.-C Joo, A.A. Volinsky, L. Michaelson, M. Lane, 2006, F04-04; C.E. Smith, M. Du, P.D. Matz, D.W. Mueller, R.F. Reidy, Advanced Metallization Conference 2006, p. 425-431.
[3] B.P. Gorman, R.A. Orozco-Teran, Z. Zhang, P.D. Matz, D.W. Mueller, and R.F. Reidy, Journal of Vacuum Science and Technology B, 22, 3 (2004) 1210-1212.
[4] E. Vinogradova, C.E. Smith, D.W. Mueller, R.F. Reidy, *"Application of deuterium exchange to analyze of moisture uptake characteristics of porous low dielectric constant SiCOH films"*, submitted to Electrochemical and Solid State Letters, 2008.
[5] Handbook of Nuclear Chemistry, ed. A. Vértes, S. Nagy, Z. Klencsár, Springer, 2004, p.102
[6] A.A. Christy and P.K. Egeberg, Analyst, 130 (2005) 738–744
[7] J.M. Jacques, T.Y. Tsui, A.J. McKerrow, R. Kraft, MRS Symposium Proceedings, 875, O10.6, (2005)
[8] E.T. Ogawa, J. Kim; G.S. Haase, H.C. Mogul, J.W. McPherson, 2003 IEEE International Reliability Physics Symposium Proceedings 41st Annual, 2003, 166-72
[9] P.D. Matz and R.F. Reidy, Solid State Phenomena 103-104 (2005) 315-322
[10] T.M. Shaw, D. Jimerson, D. Haders, C.E. Murray, A. Grill, D.C. Edelstein, D. Chidambarrao, Advanced Metallization Conference 2003, 2004, 77-84

Mater. Res. Soc. Symp. Proc. Vol. 1079 © 2008 Materials Research Society

Effects of Plasma Surface Treatment on the Self-forming Barrier Process in Porous SiOCH

Seung-Min Chung[1], Junichi Koike[1], and Zsolt Tökei[2]

[1]Dept. of Materials Science, Tohoku University, 6-6-11 Aoba, Aramaki, Aoba-ku, Sendai, 980-8579, Japan

[2]IMEC, Kapeldreef 75, B-3001 Leuven, Belgium

ABSTRACT

Self-forming barrier process was carried out on a porous low-k material with the Cu-Mn alloys. The effects of various surface treatments were investigated in the sample having a pore size of 0.9 nm and a porosity of 25%. Before and after annealing, samples were analyzed in cross section with transmission electron microscopy (TEM) and energy dispersive x-ray spectroscopy (EDS). Concentration profile was also analyzed with time-of-flight secondary ion mass spectroscopy (ToF-SIMS). The results indicated the penetration of Cu into the low-k interior during deposition, followed by the segregation of Cu at the low-k/Si interface during subsequent annealing. Although a diffusion barrier layer was formed and no further Cu penetration was not observed during annealing, initial Cu penetration in the deposition process was detrimental and should be prevented by restoring the plasma damage on the low-k surface.

INTRODUCTION

As the device size shrinks to 32 nm node and beyond, the formation of a thin and conformal barrier layer has become a bottleneck in producing reliable interconnect structures. Various deposition methods have been investigated in combination with the search for new barrier/adhesion materials [1]. Assurance of reliability requires prevention of Cu penetration into a low-k layer and adhesion between the low-k and the diffusion barrier layer [2, 3]. Recently, a self-forming process has been successfully demonstrated in a dual-damascene structure by using a Cu-Mn alloy as a seed layer [4-6]. Although this process appears to be desirable for a future technology node, the previous works used plasma TEOS oxide as a dielectric layer and little information is available for advanced low-k dielectric materials. In the present work, we employed the self-forming barrier process with a Cu-Mn alloy film on a low-k material. Since low-k surface is modified upon plasma exposure, we investigated the effects of various surface treatments on diffusion barrier formation behavior.

EXPERIMENT

Porous low-k films of SiOCH (k=2.5) were synthesized with plasma enhanced chemical vapor deposition (PECVD) on Si wafers. Pore size and porosity were 0.9 nm and 25%, respectively. The low-k films were subjected to various plasma surface treatments in order to understand their influences on the self-forming barrier process with the Cu-Mn alloy. The types of plasma treatment and the corresponding gases are listed in Table 1. Etching was performed with a $CH_4CH_2F_2$ based plasma to investigate the influence of plasma damage during via and trench etching. After etching, either an oxygen or a N_2/H_2 plasma treatment was carried out. It is expected that the application of the oxygen plasma will lead to surface oxidation and form a thin oxide layer, while the N_2/H_2 plasma induces moisture uptake by forming hydrophilic Si-OH bonds. In addition a He+NH$_3$ plasma was also tested, which leads to pore sealing as reported earlier [7]. After the plasma treatment, Cu-8at.%Mn alloy was sputter-deposited to a thickness of 60 nm, followed by copper electroplating to a thickness of 800 nm. The obtained samples were annealed at 300°C for 30 min under a vacuum condition of 1.2 x 10^{-4} Pa to self-form a barrier layer. The microstructure of the samples was observed in cross section with transmission electron microscopy (TEM) before and after annealing. Concentration distribution was investigated with x-ray energy dispersive spectrometry (EDS). TEM samples were prepared by mechanical grinding and ion milling. Depth profiles of constitutive elements were also obtained with time-of-flight secondary ion mass spectroscopy (ToF-SIMS). Since the ion sputtering procedure causes artificial penetration of the constitutive elements in the direction of the ion beam, backside SIMS was performed by ion sputtering the sample from the Si side.

Table 1. Plasma gases used for various surface treatment

Surface treatment	Plasma gas
Etching	$CF_4CH_2F_2$
Pore sealing	$He + NH_3$
Oxidation	Etching + O_2
Moisture uptake	Etching + N_2/H_2

RESULTS AND DISCUSSION

Figure 1 shows cross-sectional TEM images of the samples before annealing. Note that some samples were ion-milled so much that only a small portion of the Cu/Cu-Mn layers remains for the observation. The Cu and the Cu-Mn layers appear in a dark contrast, while the low-k layer appears in a bright contrast. An additional dark-contrasted area can be seen in the low-k layer adjacent to the interface. Though not shown here, the concentration of this area was examined by EDS taken with an electron probe of 1.7 nm in a nominal diameter. A weak Cu peak was observed from this area in all the samples. This indicates that Cu penetrates into the low-k layer during Cu-Mn alloy deposition. The morphology of Cu penetration seems to depend on the type of plasma treatment. In Fig. 1 (a) for the etching treatment, semicircular penetration can be seen in localized areas. In Fig. 1 (b) for the pore sealing treatment, Cu penetration is deeper and wider than in Fig. 1 (a) with occasional sharp penetration of a leaf-like feature. In Fig. 1 (c) for the oxidation treatment, Cu penetration is shallow but uniform in thickness. In Fig. 1 (d) for the moisture uptake treatment, a thick penetrated layer can be seen. The various penetration morphologies reflect the severity and depth of the plasma damage as well as chemical interaction between the Cu atoms and the plasma modified low-k materials.

Figure 2 shows cross-sectional TEM images of the samples after annealing at 300°C for 30 min. In Fig. 2(a) and (c), the Cu penetrated area appears to be lost in contrast and the low-k layer shows a uniform bright contrast. On the other hand, in Fig. 2 (b), the Cu penetrated area appears to spread into a deeper region. In Fig. 2 (d), the Cu penetrated area becomes locally concentrated with a leaf-like feature. The presence of Cu in these regions was confirmed with EDS. It should be noted that Figs. 2 (a), (c) and (d) show a thin interface layer having a non-uniform thickness of 7.5 nm in (a) and 2.5 nm in (c) and (d). EDS spectra taken from the interface layer showed an evident peak of Mn, indicating the formation of a barrier layer containing Mn, as in the case of TEOS-oxide [6].

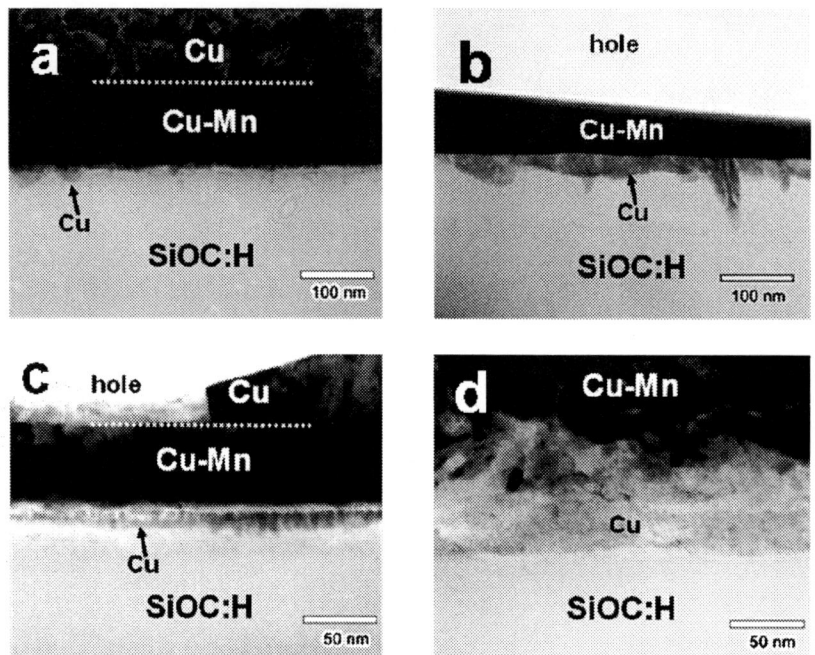

Figure 1. Cross-sectional TEM images of the plasma treated samples before annealing.
(a) plasma etching, (b) pore-sealing, (c) oxidation and (d) moisture uptake

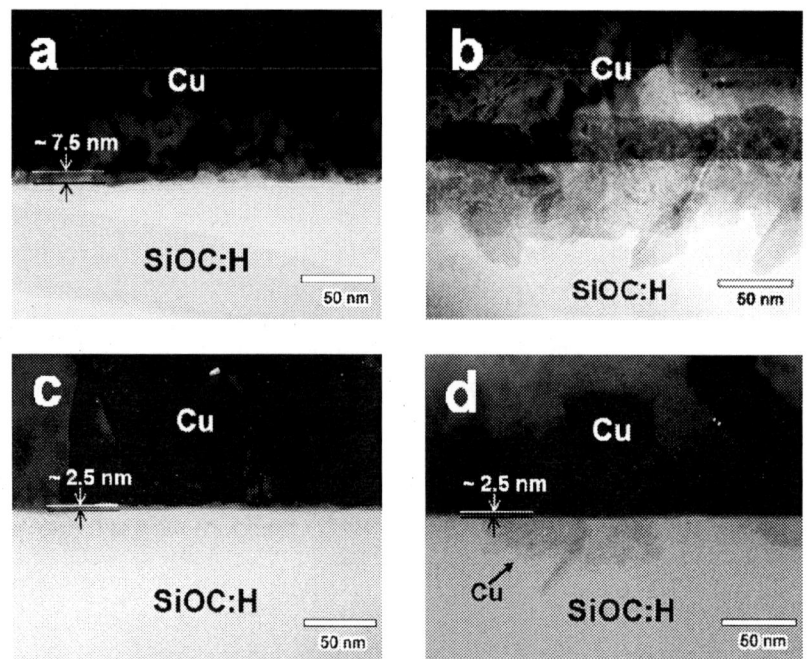

Figure 2. Cross-sectional TEM images after annealing at 300°C for 30 min.
(a) plasma etching, (b) pore sealing, (c) oxidation and (d) moisture uptake.

Figure 3 shows concentration profile of Cu, Mn, Si, O, C and H along the thickness direction of the Cu/Cu-Mn/SiOCH/Si structure before annealing. Although not shown here, Cu concentration within a pristine low-k without surface treatment was approximately 4×10^{18} atoms/cm^3. As shown in Fig. 3 (a), (c) and (d), the plasma surface treated samples also contain a similar amount of Cu in the low-k interior. These results indicate that Cu penetration occurs during PVD deposition of the Cu-Mn seed layer not only into the plasma treated low-k but also into the pristine low-k. An exception can be found in Fig. 3(b) where the Cu concentration of the pore-sealed low-k is approximately 3×10^{17} atoms/cm^3. It has been reported that a modified surface layer does not induce Cu penetration into the low-k interior owing to its higher density and lower porosity than the non-altered porous low-k [8, 9]. However, as shown in the TEM image in Fig. 1 (b), Cu penetration is observed even in the pore-sealed sample. These results indicate that the Cu penetration in the pore-sealed low-k is limited only near the surface region and further penetration to the low-k interior is prevented during the deposition of the Cu-Mn alloy.

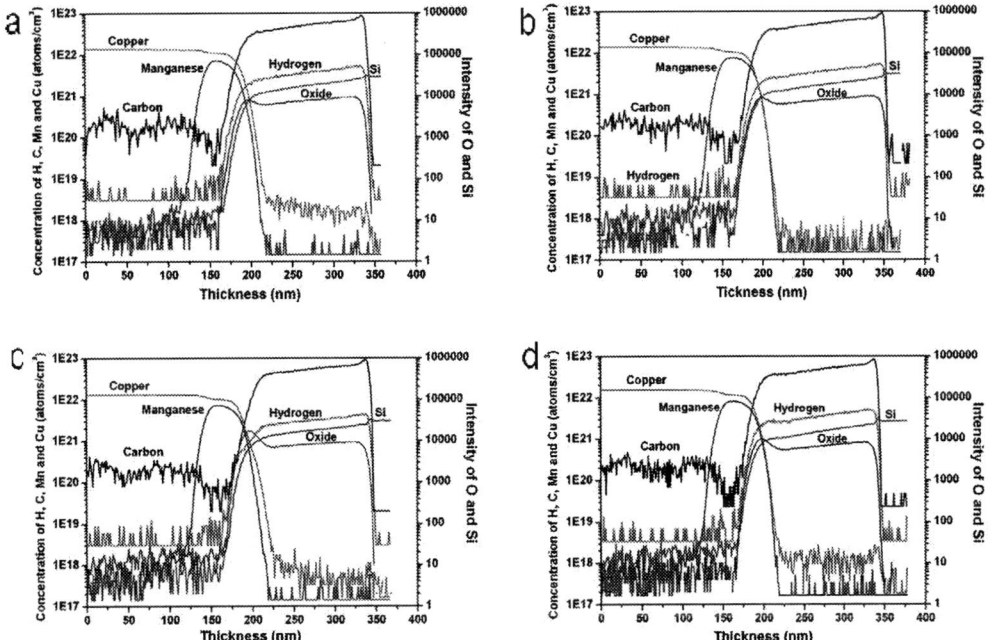

Figure 3. SIMS concentration profile of Cu, Mn, Si, O, C and H before annealing; a sample subjected to surface treatment of (a) plasma etching, (b) pore sealing, (c) oxidation and (d) moisture uptake.

Figure 4 shows concentration profile after annealing. Notice that the Mn distribution shows two separate peaks in the etched, the oxidized, and the moisture uptaken samples, except in the pore-sealed sample. Since the Mn distribution before annealing in Fig. 3 shows a broad single peak, the peak separation is an indication of interface reaction to form a barrier layer. On the other hand, no peak separation in the pore-sealed sample is an indication of no interface reaction, probably because the surface chemistry is altered by the pore-sealing process. All graphs indicate the absence of Mn in the film interior. It should also be noted that Cu peak located beyond the Mn peak toward the low-k side (Figs. 3 (a), (c)) or right on top of the Mn peak (Fig. 3 (b), (d)) moved away from the low-k on the left side of the figure. After annealing no further copper diffusion into the low-k material was detected in all of the plasma treated samples because of the formation of the diffusion barrier. However, comparing with Fig. 3, a change of Cu distribution occurs in the low-k interior. In Fig. 4(a) for the etching treatment, Cu still remains in the low-k. In Fig. 4(b) for the pore-sealing treatment, Cu atoms migrate across the low-k layer and segregate at the Si interface. The Cu content in the low-k material remains negligibly low. In Fig. 4(c) for the oxidation treatment, Cu concentration decreases a little in the low-k film but increases at the Si interface, indicating the segregation of Cu atoms. In Fig. 4 (d) for the moisture uptake treatment, Cu is always present in the low-k and partly segregates at the Si interface.

These results indicate that the pristine low-k material cannot be used due to initial Cu penetration, and any plasma treatment forms a damage layer through which Cu can easily penetrate during sputtering process. The penetrated Cu in low-k material appears to be unstable during annealing and Cu migrates away from the low-k to both interfaces. Mn can form a barrier layer, but the initial penetration of Cu into the low-k material deteriorates the insulating property. Damage restoration is essential to ensure the initial device performance as well as reliability of the self-forming barrier process.

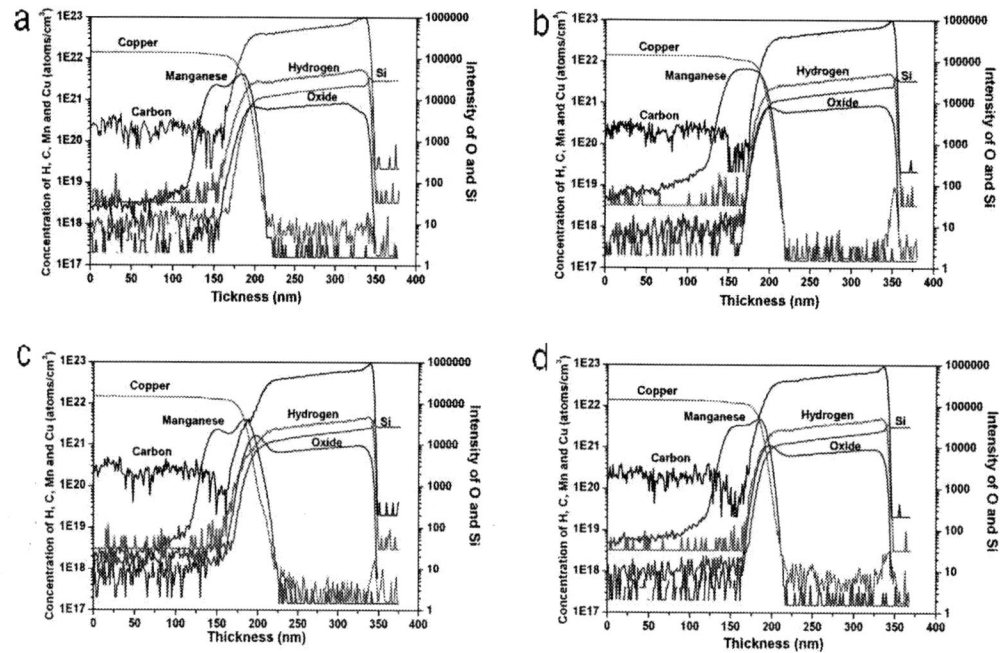

Figure 4. SIMS concentration profile of Cu, Mn, Si, O, C and H after annealing at 300 °C for 30 min; a sample subjected to surface treatment of (a) plasma etching, (b) pore sealing, (c) oxidation and (d) moisture uptake.

CONCLUSIONS

It was found that plasma surface treatment induces copper penetration into the studied SiOCH low-k material when Cu-Mn was deposited on top. The penetration depth depended on the type of the plasma. When the bulk of the low-k material was not modified, copper was unstable in the low-k and moved to interfaces upon annealing. When moisture was present inside the low-k material it stabilized the copper inside the porous dielectric layer. The diffusion barrier was formed with approximate 2-7 nm in thickness, and no significant copper diffusion into the dielectric film was detected during subsequent annealing steps. Damage restoration procedure is needed to successfully implement the self-forming barrier process on the porous low-k material.

REFERENCES

1. W. S. Lau, H. J. Tan, Z. Chen, C. Y. Li, Vacuum 81, 1040 (2007).
2. D. Shamiryan, T. Abell, F. Iacopi and K. Maex, Mater. Today, Jan. 35 (2004).
3. T.Usui and H. Shibata, Nikkei Microdevices 254, 79 (2006).
4. J. Koike, M. Wada, Appl. Phys. Lett. 87, 041911 (2005).
5. M. Haneda, J. Iijima and J. Koike, Appl. Phys. Lett. 90, 252107 (2007).
6. J. Koike et al, J. Appl. Phys. 102, 043527 (2007).
7. A.M. Urbanowicz, M.R. Baklanov, J. Heilen, Y. Travaly, A. Cockburn, Electrochemical and Solid-State Lett. **10**(10), G76-G79 (2007).
8. Y. Travaly, B. Eyckens, L. Carbonel et al, Microelec. Eng. 64, 367 (2002).
9. W. Puyrenier, V. Rouessac, L. Broussous, D. Rebiscoul, A. Ayral, Microporous & Mesoporous Mater. 106, 40 (2007).

Mater. Res. Soc. Symp. Proc. Vol. 1079 © 2008 Materials Research Society 1079-N03-11

Formation of Mn Oxide with Thermal CVD and its Diffusion Barrier Property Between Cu and SiO$_2$

Koji Neishi, Shiro Aki, Jun Iijima, and Junichi Koike
Materials Science, Tohoku University, 6-6-11 Aoba, Aramaki, Aoba-ku, Sendai, 980-8579, Japan

Abstract

A manganese oxide layer was formed by thermal chemical vapor deposition(CVD) on a tetraethylorthosilicate (TEOS) oxide substrate. The thickness of the Mn oxide layer could be varied 2.6 to 10 nm depending on deposition temperature. Heat-treated samples of PVD Cu / CVD Mn oxide /SiO$_2$ indicated no interdiffusion. The CVD Mn oxide was found to be a good diffusion barrier layer.

Introduction

With the increase in packing density of ultra-large scale integrated circuits, geometrical dimensions continue to decrease. As a width of Cu line becomes narrow, a thickness of a barrier layer should be reduced to prevent the increase of the effective resistivity of the interconnect lines [1]. Ta and TaN have been widely used as conventional barrier materials to prevent interdiffusion between Cu line and dielectric insulating layer. Physical vapor deposition (PVD) have been employed as a their deposition method. The barrier layer formation with a PVD technique has become difficult as the technology node is reduced to 45 nm

Without using Ta and TaN, Koike and his colleagues reported the self-formation of a diffusion barrier by direct deposition of a Cu-Mn seed layer on a TEOS oxide substrate. The Mn oxide layer was self-formed at the interface by annealing at elevated temperatures and acted as an excellent barrier layer [2-4]. However, as in the case of the conventional Ta barrier, PVD deposition of the Cu-Mn seed layer will eventually encounter the difficulty in deposition in narrow trenches and vias.

Chemical vapor deposition (CVD) has been known for its good step coverage, and is a candidate technique for the formation of a barrier layer, a Cu seed layer and entire Cu line. In order to take advantage of the excellent reliability of Mn oxide, Mn deposition with CVD as a barrier layer is expected to be very attractive process. Up to date, only a few works have been reported for CVD-Mn layer. Wen-bin et al. reported that the deposition of a metallic Mn layer occurs above the temperature of $410°C$ [5]. Maruyama and Osaki reported that a Mn oxide layer forms above $250°C$ [6]. However, it is desirable to form the Mn barrier layer with CVD at lower temperatures.

In this work, as a first approach of Manganese-based barrier layer formation by CVD, we deposited Mn oxide films using a metal organic precursor by thermal CVD at low temperatures and investigated the formation behavior and diffusion barrier property of the Mn oxide films.

Experimental procedures

Substrates were p-type Si wafers having a plasma TEOS oxides of 100nm in thickness. Bisethylcylopentadienyl manganese (EtCp)$_2$Mn was used as a metal organic manganese precursor. The Mn precursor was vaporized and introduced into the hot- wall reaction chamber with Ar or H$_2$ carrier gas in the thermal-CVD process. The detailed process parameter is shown in Table I. Following the Mn CVD, the sample was transferred to a PVD chamber with keeping vacuum to deposit a Cu overlayer using a DC sputtering system. Two types of the samples were prepared in this study. One is a sample without Cu overlayer to analyze a chemical bonding state of the CVD-Mn layer. Other is a sample with Cu overlayer to investigate a diffusion barrier property of the CVD-Mn layer by annealing at 673K for 100 hours in a vacuum of better than 1.0×10^{-5} Pa.

Transmission electron microscopy (TEM) attached with X-ray energy dispersive spectroscopy (EDS) was used to investigate the microstructure, thickness of the CVD-Mn layer and chemical composition of the sample. X-ray Photoelectron spectroscopy (XPS) was used to analyze to the chemical bonding states of the CVD-Mn layer.

Table I. Experimental Condition in thermal-CVD process

Vaporized temperature of Mn precursor	343K
Temperature of reaction chamber	353K
Carrier gas	Ar or H_2
Flow rate of carrier gas	25 sccm
Substrate temperature	373-673 K
Reaction time	30 minutes

Results and Discussion

Figure 1 shows a cross-sectional TEM image and EDX result of the as-deposited Cu/CVD-Mn/SiO$_2$ with CVD condition at 373 K for 30 minutes with Ar gas flow of 25 sccm. A thin continuous layer is observed at the interface between Cu and SiO$_2$ layer. The CVD-Mn layer does not show a characteristic diffraction contrast of a crystalline structure. Therefore, the CVD-Mn layer is considered to be an amorphous structure. The CVD-Mn layer has a thickness of 3.9 nm. The EDX result of Fig.1 indicates that Cu, Mn, Si, O and C are detected from the CVD-Mn layer. Among these elements, Cu and Si peaks are originated from the neighboring layers of Cu and SiO$_2$. This is because a nominal electron beam size used for TEM-EDX analysis was 2 nm and is nearly the same size as the thickness of the CVD-Mn layer. Beam spread within the sample had lead to the overlap of the Cu and Si peaks. Thus, the CVD-Mn layer is consisted mainly of Mn, O and C.

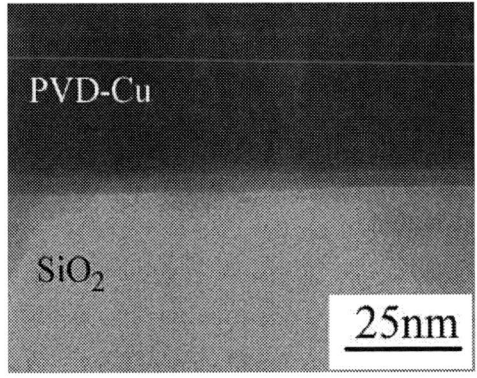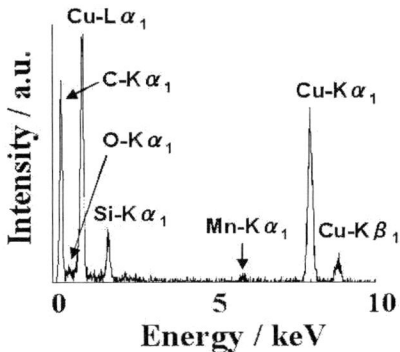

Figure 1. TEM image and EDX result of of as-deposited Cu/CVD-Mn/SiO$_2$ with condition at 373 K with Ar carrier gas.

Figure 2 shows a cross-sectional TEM image and EDX result of the as-deposited Cu/CVD-Mn/SiO$_2$ with CVD condition at 373 K for 30 minutes with H$_2$ gas flow of 25sccm. A thin continuous layer is observed at the interface between Cu and SiO$_2$ layer. It is an amorphous structure with a thickness of 2.6 nm. The thickness of the deposited layer with H$_2$ carrier gas is slightly thinner than that with Ar gas. The CVD-Mn layer is consisted mainly of Mn, O and C.

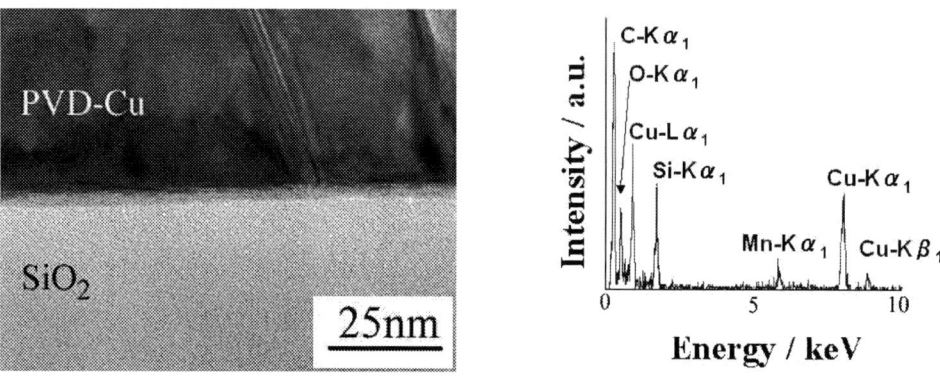

Figure 2. TEM image and EDX result of as-deposited Cu/CVD-Mn/SiO$_2$ with condition at 373K with H$_2$ carrier gas.

Figure 3 shows the XPS spectra of Mn 2p in CVD-Mn/SiO$_2$ deposited at 373 K for 30 minutes with Ar or H$_2$ gas of 25sccm. The XPS results were obtained after brief etching of the CVD-Mn layer surface to remove surface contamination. The peak position indicates that the CVD-Mn layer is not metallic states but in oxide states. This result is in agreement with the results of TEM and EDX. Hereafter, the CVD-Mn layer is called a Mn oxide layer or MnOx.

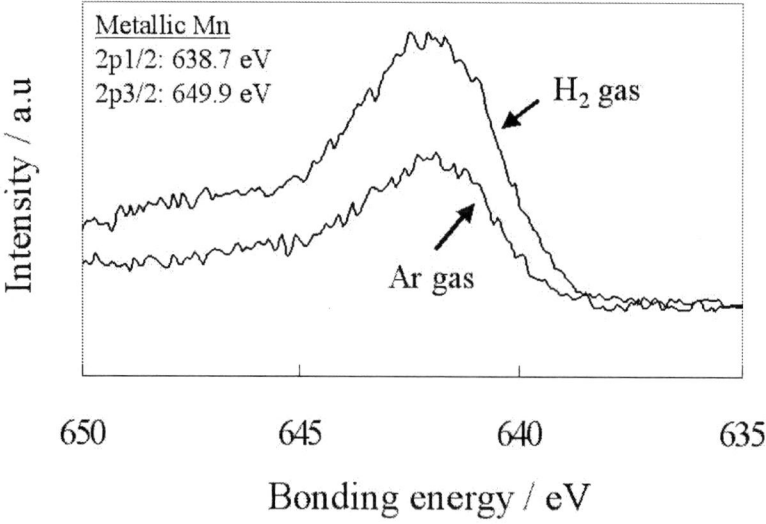

Figure 3. XPS spectra of Mn 2p in MnOx/SiO$_2$ deposited at 373K.

32

Figure 4 shows the thickness of a Mn oxide layer as a function of substrate temperature. Open circles and solid squares correspond to the films deposited with Ar and H_2 carrier gas, respectively. The thickness increases gradually with increasing the substrate temperature. It is also found that the thickness is slightly thinner with H_2 carrier gas than with Ar carrier gas. It should be noted that the thickness of the Mn oxide layer is in the range of 2.6 nm to 10 nm, including the thickness values required in the ITRS road map.

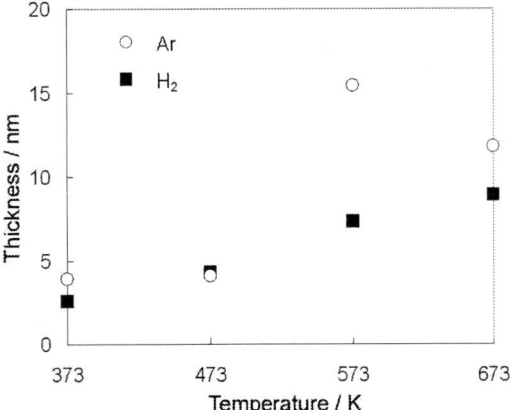

Figure 4. Formation behavior of the MnOx layer.

Figure 5 shows a cross-sectional TEM image and EDX results of PVD-Cu/CVD-MnOx/TEOS-SiO$_2$ after annealing at 673 K for 100 hours. In this sample, the CVD-MnOx layer was deposited at 373 K for 30 min. with H_2 gas. After annealing, the MnOx layer remains to be an amorphous structure. The EDX results confirm no Cu diffusion into TEOS-SiO$_2$ layer after annealing. It is noted that the Si peak in Cu layer is due to sample contamination and is also observed in the upper epoxy layer with the same intensity. The results clearly indicated that the CVD-MnOx acts as a good diffusion barrier. The similar results were obtained for another CVD-MnOx formed at 373 K with Ar gas.

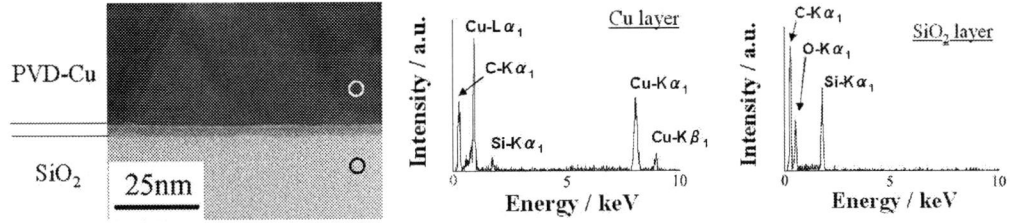

Figure 5. TEM image and EDX results of PVD-Cu/MnOx/SiO$_2$ after annealing at 673 K for 100 hours.

Conclusions

A thin Mn oxide layer was formed at the temperature range of 373 K to 673 K. The MnOx layer was an amorphous having a uniform thickness of 2.6 to 10 nm depending on deposition temperature. The MnOx layer was found to be a good diffusion barrier layer after annealing at 673K for 100 hours. The amorphous structure was thermally stable. The CVD-MnOx process is promising technique to form a conformal and reliable barrier layer for advanced technology node.

Acknowledgements

We thank to K. Matsumoto and H. Sato and H. Ito and S. Hosaka of Tokyo Electron Ltd. for many technical advices and their discussions.

References

1. International Technology Roadmap for Semiconductors, 2006
2. J. Koike and M. Wada, Applied Physics Letter, **87**,041911(2005)
3. M. Haneda, J. Iijima and J. Koike, Applied Physics Letter, **90**, 252107(2007)
4. J. Koike, M. Haneda, J. Iijima, Y. Otsuka, H. Sako and K. Neishi, Journal of Applied Physics, **102**, 043527(2007)
5. S. Wen-bin, K. Durose, A. W. Brinkman and B. K. Tannor, Material Chemistry and Physics, **47**, p.75-77(1997)
6. T. Maruyama and Y. Osaki, Journal of the Electrochemical Society, **142**, p.3137-3141(1995)

Mater. Res. Soc. Symp. Proc. Vol. 1079 © 2008 Materials Research Society

Extendibility Study of a PVD Cu Seed Process with Ar+ Rf-Plasma Enhanced Coverage for 45nm Interconnects

Andrew H. Simon[1], Tibor Bolom[2], Teck Jung Tang[3], Brett Baker[4], Carsten Peters[2], Bryan Rhoads[1], Philip L. Flaitz[1], Sujatha Sankaran[1], and Stephan Grunow[1]

[1]Semiconductor Research and Development Center, IBM Microelectronics, 2070 Route 52, Hopewell Junction, NY, 12533

[2]Advanced Micro Devices, Inc., Hopewell Junction, NY, 12533

[3]Chartered Semiconductor Mfg. Ltd., Hopewell Junction, NY, 12533

[4]IBM T.J. Watson Research Center, Yorktown Heights, NY, 10598

ABSTRACT

We present the results of a systematic benchmarking study, using 45nm-groundrule structures, of a commercially-available ionized PVD Cu technology which employs an in-situ Ar+ radio-frequency (Rf) plasma capability for enhanced coverage, and compare its performance and extendibility against the same seedlayer process operated in conventional low-pressure mode. Studies of single-damascene lines and dual-damascene via structures indicate that the PVD Cu seedlayer with Rf-Plasma enhancement enables a reduction of the PVD Cu seed thickness on the order of 35%, based on studies of Cu voiding, via-yield degradation, and transmission-electron microscopy (TEM). These results illustrate the critical importance of the Rf-plasma resputter capability in extending the PVD Cu process to advanced groundrules at 45nm and beyond.

INTRODUCTION

Since the introduction of Cu as a back end of line (BEOL) wiring material [1] the ability to extend Cu interconnects to smaller groundrules has been critically dependent upon the ability of Cu fill technologies to scale accordingly while maintaining performance and reliability benchmarks[2]. A broad trend in the field has been the development of advanced methods for ionized physical vapor deposition (PVD)[3,4]. In this study, we assess the capability of an enhancement to ionized PVD that has recently become commercially available, the ability to strike an Ar+ in-situ in the Cu PVD chamber in order to enhance sidewall coverage with the Ar+ sputter component[5,6]. In studies at the 45nm groundrule, we report the results of inspection, electrical test, and failure analysis to gauge the extendibility this new capability affords.

EXPERIMENT

All structures in this study were etched in a SiCOH dielectric, at dimensions consistent with semiconductor industry groundrules for back-end-of-line (BEOL) interconnects at the 45nm technology node. The interconnect structures consisted of single-damascene wires with a height:width aspect ratio of approximately 2:1 and a dense line-line spacing of twice the linewidth. In addition, dual-damascene isolated via and via-chain structures were studied using similar linewidths and 3.5:1 aspect-ratio vias.

The structures were degassed and deposited with TaN/Ta barrier layers using a "barrier first" or "punch-through" [7-13] processing scheme. In this scheme an initial barrier deposition is followed sequentially by an in-situ Ar+ sputter-etch and additional barrier deposition steps [14]. The TaN was deposited in the same chamber as the Ta, using N_2 as a reactive sputter gas along with Ar[12,13]. Cu seed deposition was done following the TaN/Ta barrier sequence. All steps in the degas/barrier/seed sequence were performed using commercially-available ionized (PVD) Ta(N) and Cu sources clustered on a common mainframe.

The PVD Cu seedlayer chamber could achieve a high-ionization deposition condition by exploiting the high sputter-yield and relatively low ionization energy of Cu to achieve a plasma capable of continued operation with relatively low Ar process-gas flow[4]. This type of process condition is currently in common use on commercially-available Cu PVD systems in order to achieve a highly directional deposition, and results in a high degree of seedlayer coverage on the bottom surfaces of advanced BEOL interconnect structures. In addition, the chamber was also capable of generating an ionized Ar+ plasma which could be used in conjunction with the electrically biasable wafer chuck to induce a resputter enhancement to the seedlayer process, thereby enabling improved sidewall coverage on the lower sidewalls of the interconnect structures[5,6].

Following barrier/seedlayer deposition, the interconnect structures were filled with Cu using a commercially-available electroplating bath, annealed, and planarized using chemical-mechanical polishing (CMP).

The process performance was evaluated using top-surface scanning-electron microscopy (SEM) to assess Cu fill quality and voiding prevalence, as well as parametric electrical tests of line and via structures, and transmission-electron microscopy (TEM) of some selected splits

Preliminary process-window studies of Cu seedlayer thickness were done with conventional, low-pressure PVD Cu depositions ranging in thickness from 700 Angstroms down to 100 Angstroms. (All quoted thicknesses are calibrated according to the deposition on the wafer top-surface [field area] using X-ray fluorescence (XRF)). All Cu seed splits used a common TaN/Ta barrier sequence. Based on via-yield and inspection results, an optimal seedlayer thickness using the conventional Cu PVD process was identified which would serve as a benchmark control split for comparison against the Rf-plasma enhanced seedlayer process.

For the experiments discussed here, half the wafers received conventional, low-pressure seedlayer depositions with individual split thicknesses of 1.0x, 0.75x or 0.65x the optimal benchmark thickness determined from the initial screening experiments. For comparison, the other half received Cu seed splits with the same three deposition thicknesses, but which also employed the in-situ Ar+ Rf-plasma capability with bias sputtering for enhanced seedlayer coverage on the lower sidewalls of the features. The amount of Ar+ plasma processing was scaled in direct proportion to the deposition thickness in each split.

DISCUSSION

Top-surface SEM Inspection Results

Initial characterization of the results was done using top down, post-CMP SEM inspections of the single-damascene lines. The top-surface voids in the Cu fill material were categorized according to whether they occurred at the edges, corners (ends of circuit traces) or centers of the wiring structures. Representative pictures are shown in Figs 1(a) and 1(b).

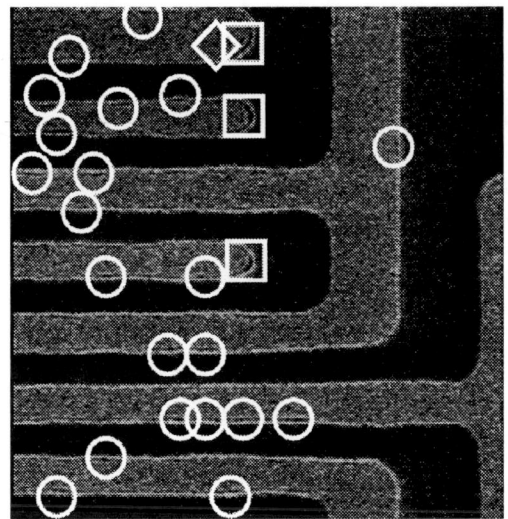 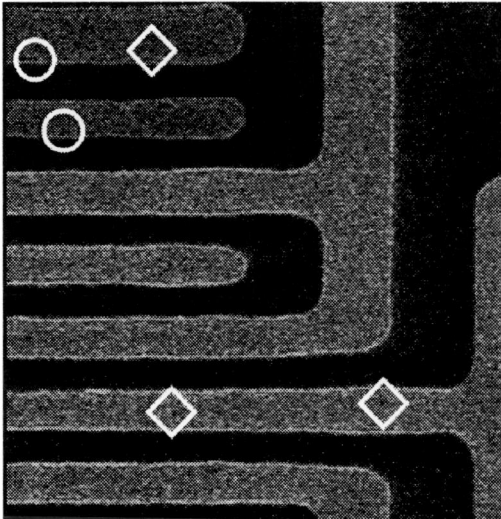

Fig.1(a) 0.65x Nominal Seed (conventional) Fig.1(b)0.65xNominal Seed (Rf-plasma enhanced) Top-surface, post-CMP SEM inspection defects of single damascene lines: Edge voids (circles), corner voids (squares) and center voids (diamonds).

Sampling of the defects was done at the center site and two edge sites on each wafer. The defect counts were then tabulated by split, as shown in Fig. 2, in order to gauge process extendibility.

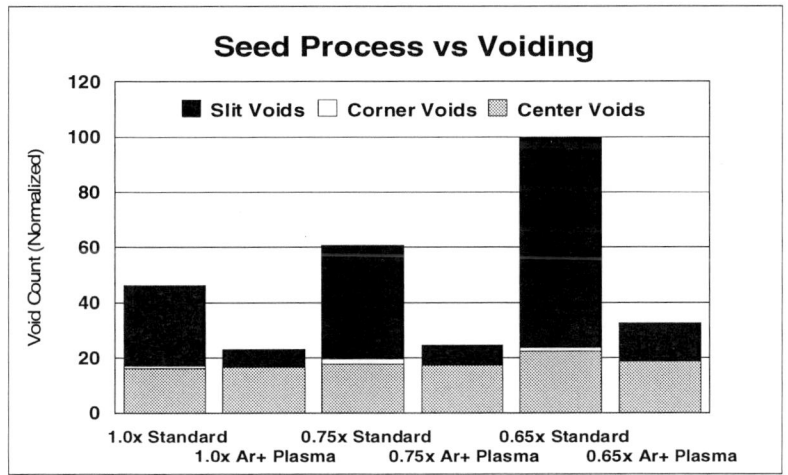

Fig. 2 : Top-surface void defect counts for conventional low-pressure and Ar+ plasma-enhanced sputter Cu seedlayer process for 45nm groundrule interconnects, showing reduction in slit voids for samples with Ar+ plasma enhanced coverage. Thicknesses and void counts are normalized.

For the seedlayer splits deposited at the 1.0x nominal thickness, the Rf-plasma enhanced seedlayer showed a 50% reduction in total defects relative to the conventional low-pressure seed.

As the seedlayer was scaled down in thickness, the enhancement from the Rf-Plasma capability become more pronounced: at 0.75x nominal thickness, a 60% defect reduction for the Ar+ enhanced seed vs. the conventional seed was observed, and at 0.65x nominal thickness a void-count reduction of 67% for the Ar+ enhanced seed vs. the conventional seed was measured. Over the range of seedlayer thicknesses studied, the 0.65x nominal Ar+ enhanced seed sample exhibited 33% fewer total defects than the 1.0x nominal conventional seed sample.

The defect reduction was most pronounced for the edge defects, a result that can be explained by the enhanced sidewall coverage afforded by the Ar+ Rf-plasma process. The relative insensitivity of the results to center void counts indicates the greater difficulty current Cu fill technologies have in resolving the pinch-off phenomena observed due to the high sticking coefficient of sputtered Cu. When combined with current Cu electroplating technologies, the process window in which it is possible to suppress center void formation can be very narrow.

Electrical Test Results

Corroboration of the inspection results was found in via-yield testing. Dual-damascene via-yield tests done at-level showed good yield on all splits, with both the conventional and Rf-plasma enhanced seedlayer. As a gauge of process robustness, via yields were compared further up in the process flow, after additional layers of BEOL interconnects had been built, with the associated temperature cycling. After five layers of BEOL processing, via yields at second-level metal for the conventional seeds show degradation of 30-40% at 0.75x nominal, and 80-90% at 0.65x nominal, whereas the Ar+ enhanced Cu seeds show consistently less yield degradation (0-20% at 0.65x nominal) from level-two to level-six, depending on the via structure.

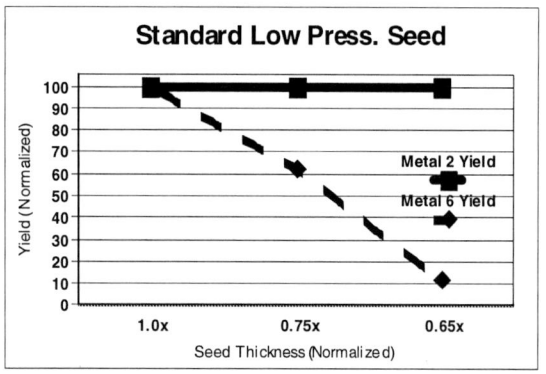 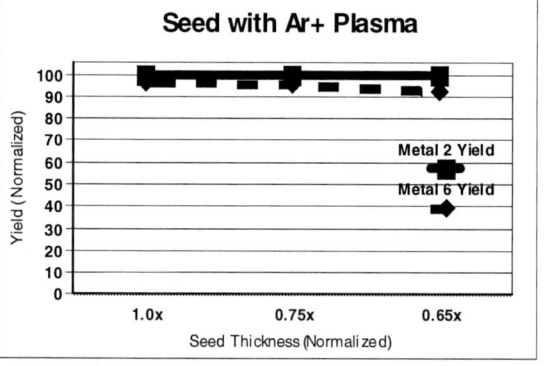

Fig. 3(a) Cu Std, Low pressure Dep Via Yields
The conventional seedlayer shows ~90% yield
During four levels of interconnect build.

Fig. 3(b) Cu seed with Ar+ plasma Via Yields
The via-yield degradation with Ar+plasma
Component shows negligible degradation.

TEM Analysis

In order to gain further insight into the discrepancies in Cu fill behavior between the conventional low-pressure and Ar+ Rf-plasma enhanced seed processes, TEM cross sections were performed on random vias on the experimental wafers with the thinnest seedlayer of each type (0.65x nominal).

Two cross-sections comparing the two processes are shown in Fig. 4(a) and 4(b).

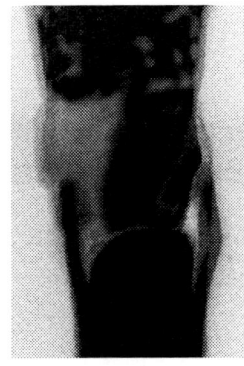 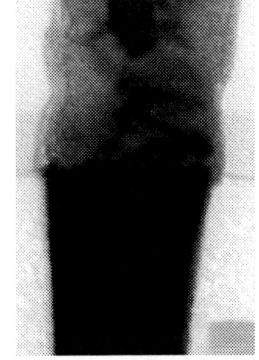

Fig. 4(a) TEM of via with low pressure seed   Fig. 4(b) TEM of via with Ar+ plasma
Note voiding on right sidewall.   Enhanced seed coverage

The TEM of the via with the conventional, low-pressure seed deposition (Fig. 4(a)) is shown magnified so that the interface with the single-damascene line below is visible. We see significant voiding at the right-side interface between the Cu fill material and the barrier sidewall, indicating marginal fill in this via chain. The presence of sidewall voiding is consistent with the known weakness of ionized PVD Cu seeds addressed by Rf-plasma enhancement: the highly directional deposition leads to thin Cu coverage on the lower sidewalls of the via. In contrast, the via with the Rf-Plasma enhancement (Fig.4(b)) shows robust Cu fill, with no voiding.

CONCLUSIONS

A benchmarking and extendibility study comparing conventional, low-pressure PVD Cu seedlayer and an enhanced version using Ar+ Rf-Plasma to induce a sputter component has been undertaken at the 45nm groundrule technology node. Characterization by means of SEM defect inspection, via-yield degradation and TEM cross-section all point to similar conclusions: the use of Ar+ Rf-plasma enhancement with advanced PVD Cu has the potential to widen process window by an amount on the order of 30% or more, owing to enhanced Cu sidewall coverage from the Ar+ sputter phenomenon in the lower sidewalls of high aspect-ratio structures.

ACKNOWLEDGMENTS

The authors would like to Xiaomeng Chen, Thomas R. Gow, Jr., Donna Cote and Thomas Ivers for management support, and Daniel C. Edelstein for helpful discussion and critical review of the manuscript. This work has been supported by the Independent Bulk CMOS and SOI Technology Development Alliance Projects at IBM Microelectronics Division, Semiconductor Research and Development Center, Hopewell Junction, NY 12533.

REFERENCES

1. D. Edelstein, *et al.*, Tech. Digest, Proc. IEEE Int'l Electron Devices Mtg. 1997, p. 773
2. A.H. Simon, , *et al.*, Advanced Metallization Conf. 2004, p.545

3. E. Klawuhn, *et al.*, J. Vac. Sci. Tech, v.18, p. 46 (2000)

4. J.H. Thompson, *et al.*, European Semicon. Design Production Assembly, v.23, p.97 (2001)

5. A.H. Simon, C.E. Uzoh, U.S. Patent # 5,933,753;

6. A.H. Simon, C.E. Uzoh, U.S. Patent # 6,768,203

7. R.M. Geffken, S.E. Luce, U.S. Patent # 5,985,762

8. G. B. Alers*, et al.*, Proc. IEEE 41st Int'l Reliability Physics Symposium 2003, p. 151

9. C.-C. Yang, *et al.*, U.S. Patent # 6,784,105

10. S.G. Malhotra, A.H. Simon, U.S. Patent # 6,949,461

11. C.-C. Yang, *et al.*, Proc. IEEE Int'l Interconnect Tech. Conf. 2005, p. 135

12. D. Edelstein, *et al.*, Proc. IEEE Int'l Interconnect Tech. Conf. 2001, p. 9

13. D. Edelstein, *et al.*, Advanced Metallization Conf. 2001, p.541

14. N. Kumar, , *et al.*, Advanced Metallization Conf. 2004, p.247

Mater. Res. Soc. Symp. Proc. Vol. 1079 © 2008 Materials Research Society

Self-formation of Ti-rich Layers at Cu(Ti)/low-k Interfaces

Kazuyuki Kohama[1], Kazuhiro Ito[1], Susumu Tsukimoto[1], Kenichi Mori[2], Kazuyoshi Maekawa[2], and Masanori Murakami[1]

[1]Department of Materials Science and Engineering, Kyoto University, Kyoto, 606-8501, Japan

[2]Process Technology Development Division, Renesas Technology Corp., Itami, 664-0005, Japan

ABSTRACT

In our previous studies, thin Ti-rich diffusion barrier layers were found to be formed at the interface between Cu(Ti) films and SiO_2/Si substrates after annealing at elevated temperatures. This technique was called "self-formation of the diffusion barrier," which is attractive for fabrication of ultra-large scale integrated (ULSI) interconnects. In the present study, we investigated the applicability of this technique to Cu(Ti) alloy films which were deposited on the four low dielectric constant (low-k) dielectric layers which are potential dielectric layers of future ULSI-Si devices. The microstructures were analyzed by transmission electron microscopy (TEM) and secondary ion mass spectrometry (SIMS), and correlated with the electrical properties of the Cu(Ti) films. It was concluded that the Ti-rich interface layers were formed in all the Cu(Ti)/dielectric-layer samples. The primary factor to control composition of the self-formed Ti-rich interface layers was the C concentration in the dielectric layers rather than the formation enthalpy of the Ti compounds (TiC and TiSi). Crystalline TiC was formed on the dielectric layers with a C concentration higher than 17 at.%.

INTRODUCTION

Although copper interconnects have been used extensively in ULSI devices for almost 10 years, there are still many materials-related issues that must be solved before Cu interconnects are used in deep sub-micron scale devices. In particular, large resistance-capacitance (RC) delay and poor device reliability are critical issues [1], and development of low dielectric interlayers and passivation materials is mandatory to reduce the RC delay [2,3]. One of the primary factors for the increase in electrical resistivity of nano-scale Cu wires is the existence of barrier layers which prevent the intermixing of Cu wires with the surrounding dielectric materials. The resistivity increases due to the barrier layers becoming significantly large with the reduction in line width of the Cu wires. Moriyama et al. [4] and Shimada et al. [5] concluded that a very thin barrier layer (< 5 nm) is required for the interconnects with a line-width of ~ 65 nm to achieve an effective interconnect resistivity of less than 4 μΩ-cm. Thus, a fabrication technique to prepare nano-scale Cu wires with ultra thin barrier layers should be developed.

To prepare such thin barrier layers in Cu alloy films, new fabrication techniques to form the barrier layers by annealing at elevated temperatures were extensively studied [6-11]. This technique is conventionally called "self-formation of the barrier layer," and application of this technique to the fabrication process of Cu interconnects is attractive. We recently succeeded in forming Ti-rich barrier layers using Cu(Ti) alloy films prepared on SiO_2/Si substrates after annealing at 400°C [12-14].

In the present study, we deposited Cu(Ti) alloy films on low-k(SiO_xC_y)/Si substrates, and investigated the possibility for the formation of Ti-rich barrier layers at the interface between alloy films and the dielectric substrates. The microstructures of the Ti-rich interface layers were analyzed by TEM and SIMS, and correlated with the electrical properties of the annealed Cu(Ti) alloy films prepared on the dielectric substrates. Based on our present results, an optimized fabrication process of the self-formation for the Ti-rich interface layers was proposed.

EXPERIMENTAL PROCEDURES

Low-k/Si substrates were used for deposition of Cu(Ti) alloy films. Four types of low-k/Si substrates (which are named here as low-k①-④) were prepared (Table 1). The dielectric layers were grown on (100)-oriented Si wafers using conventional plasma chemical vapor deposition methods. The dielectric constants and thicknesses are 2.6-3.0 and about 450 nm, respectively. The concentrations of C, O and Si are summarized in Table 1. Prior to film deposition, the substrates were ultrasonically cleaned with acetone and isopropyl alcohol. The Cu(Ti) alloy films were deposited onto these substrates in a radio frequency magnetron sputter system in which the base pressure prior to deposition was approximately 1×10^{-6} Pa. To prepare the Cu(Ti) films, small rectangular Ti plates were mounted on a Cu target. The purities of the Cu target and the Ti plates were 99.9999 % and 99.9 %, respectively. During deposition, the sputtering power and working pressure were kept at 300 W and about 1 Pa, respectively, and the substrate holder was placed 100 mm above the target.

The compositions of the Cu(Ti) films were found to be Cu(10 at.% Ti). Films with this Ti composition were reproduced by controlling the area percentage of the Ti plate coverage on the Cu target. The film thicknesses were measured using a stylus surface profiler and cross-sectional TEM: the typical thickness was controlled to approximately 450 nm. After deposition, the Cu(Ti) films were annealed isothermally at 500°C or 600°C in an Ar gas ambient for 2 hours and were cooled in the furnace. The electrical resistivities in the Cu(Ti) films were measured by a four-point probe method. The film microstructures were analyzed by X-ray diffraction (XRD) and TEM. The elemental depth profile (mapping) in the films was obtained by SIMS.

Table 1. The Si, O and C concentrations (at. %) of the low-k layers.

Dielectric layers	Low-k①	Low-k②	Low-k③	Low-k④
Si (at.%)	18.8	18.4	~18	~18
O (at.%)	24.9	26.6	~29	~29
C (at.%)	17.0	17.3	~14	~14

RESULTS AND DISCUSSION

Identification of self-formed Ti-rich interface layers formed on the low-k/Si substrates

The self-formed Ti-rich interface layers in the Cu(Ti)/SiO_2 samples consisted of Ti silicides and Ti oxides [12-14]. In order to identify the Ti-rich interface layers in the Cu(Ti)/low-k samples, cross-sectional TEM observation and selected area diffraction (SAD) analysis were

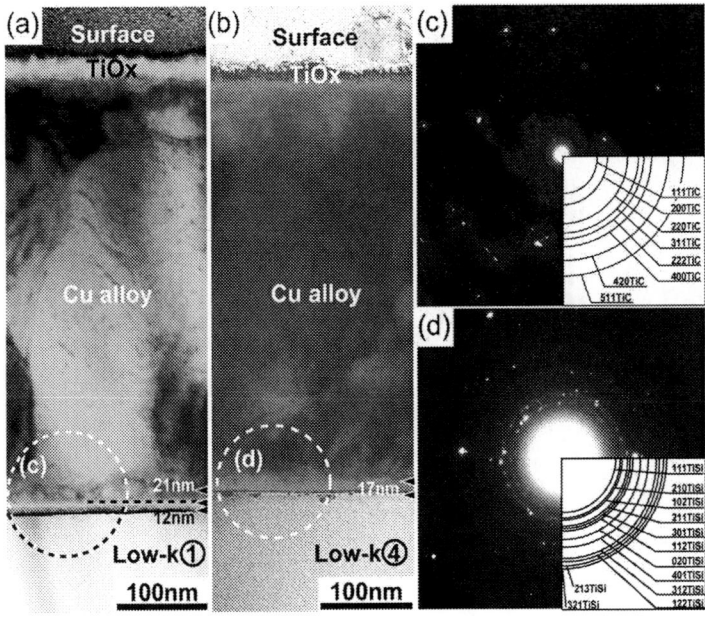

Figure 1. Cross-sectional TEM bright filed images of the (a) Cu(10 at.%Ti)/low-k① and (b) Cu(10 at.%Ti)/low-k④ samples after annealing at 600°C for 2 hrs. (c) and (d) SAD images taken from the area marked with a broken circle in (a) and (b), respectively.

carried out. Figures 1(a) and 1(b) show cross-sectional TEM bright field images of the Cu(Ti)/low-k① and Cu(Ti)/low-k④ samples after annealing at 600°C for 2 hours. Ti-rich interface layers of about 30 nm and 17 nm were observed to have formed on the low-k①/Si and low-k④/Si substrates. The SAD images obtained around the Ti-rich interface layer in the Cu(Ti)/low-k① and Cu(Ti)/low-k④ samples (marked with broken circles (c) in Fig. 1(a) and (d) in Fig. 1(b)) are shown in Figs. 1(c) and 1(d), respectively. Different diffraction-ring patterns were obtained from these two samples. The diffraction spots in Fig. 1(c) are in good agreement with the diffraction-ring pattern of TiC, but those in Fig. 1(d) are in good agreement with that of TiSi. This indicates that the TiC and TiSi polycrystalline grains were formed on the low-k①/Si and low-k④/Si substrates, respectively.

The SIMS depth profiles of the Cu(Ti)/low-k① and Cu(Ti)/low-k④ samples after annealing at 600°C are shown in Figs. 2(a) and 2(b). The Ti segregation was observed both at the surface of the Cu(Ti) alloy film and the Cu(Ti)/low-k interface. The Ti segregation at the interface was smaller than that at the surface. A weak reaction of Ti atoms with the low-k layer decreased Ti-atom diffusion to the Cu(Ti)/low-k interface and increased Ti-atom diffusion to the surface of the Cu(Ti) alloy films. Copper peaks beside the surface edges of the Cu alloy films were observed. This corresponds to the Cu_4Ti formation below the surface of the Cu alloy film, which was similarly identified by TEM/SAD analysis. The super-saturated Ti atoms segregated at the surface would provide the driving force for the Cu_4Ti formation. Figures 2(c) and 2(d) show the Ti-rich interface layer portions of the SIMS profiles as shown in Figs. 2(a) and 2(b), respectively. For comparison, the profiles of the four elements were equally distributed with respect to the vertical axis. The similar penetration of the O atoms at the Ti-rich interface layer was observed, suggesting Ti oxide formation at the interface layer. Since the Ti oxide formation

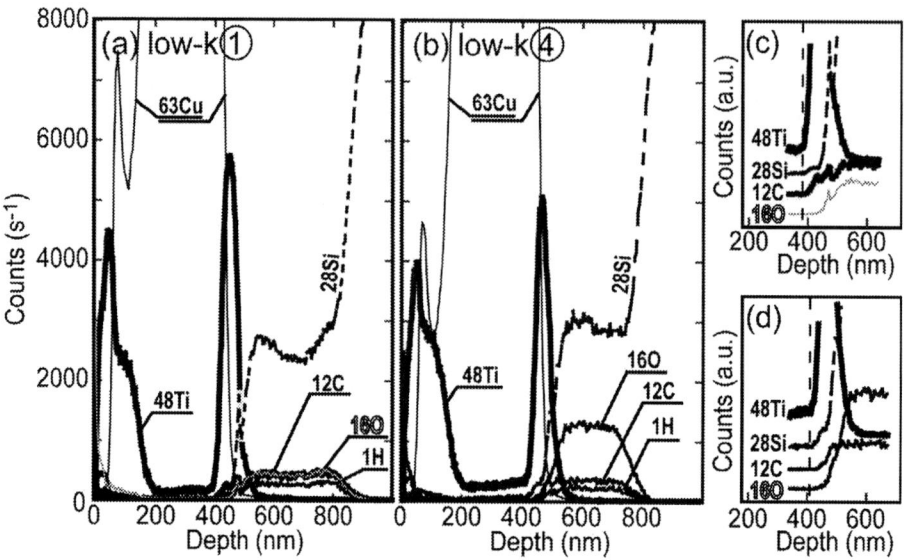

Figure 2. SIMS spectrum profiles of the (a) Cu(10 at.%Ti)/low-k① and (b) Cu(10 at.%Ti)/low-k④ samples after annealing at 600°C for 2hours. (c) and (d) Ti-rich interface layer portions of the profiles (a) and (b), respectively.

was not identified by TEM/SAD analysis, the amorphous Ti oxide must be formed. On the other hand, the penetration of the Si and C atoms was different. The penetration of the C atoms was larger in the Cu(Ti)/low-k① sample (Fig. 2(c)) than that in the Cu(Ti)/low-k④ sample (Fig. 2(d)) and the penetration of the Si atoms was smaller in the Cu(Ti)/low-k① sample. These results correspond to the TiC and TiSi formation in the Cu(Ti)/low-k① and Cu(Ti)/low-k④ samples, respectively. Thus, the Ti-rich interface layer that self-formed on the low-k①/Si and low-k④/Si substrates consisted of crystalline TiC and TiSi, respectively, in addition to amorphous Ti oxides.

Figure 3 shows schematic illustrations of the Cu(Ti)/low-k samples after annealed at 600°C for 2 hours. Crystalline TiC grains were formed in the Cu(Ti)/low-k① sample, but that crystalline TiSi grains were formed in the Cu(Ti)/low-k④ sample, in addition to the amorphous Ti oxides. The weak reaction of the Ti atoms with the low-k dielectric layers enhanced the Ti diffusion to the surface and thus, the Cu_4Ti polycrystalline grains were formed below the surface of Cu(Ti) alloy films. Since the formation enthalpy of TiC [15] is larger than that of TiSi [16], the TiC formation in the Cu(Ti)/low-k① sample can be explained by the formation enthalpy of

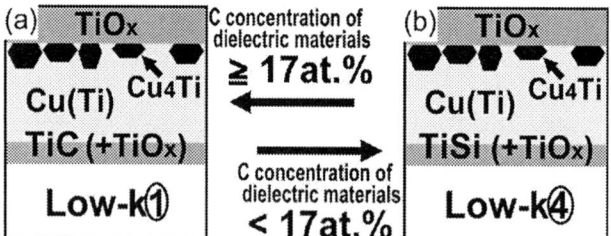

Figure 3. Schematic cross-sectional illustrations of (a) Cu(10 at.%Ti)/low-k① and (b) Cu(10 at.%Ti)/low-k④ samples after annealing at 600°C for 2hours.

the Ti compounds. However, the TiSi formation in the Cu(Ti)/low-k④ sample cannot be explained by the formation enthalpy of the Ti compounds. The Si, O and C concentrations of the low-k layers are different (Table 1). The C concentration of 14 at.% in the low-k④ layer is lower than those in the low-k① (17 at.%) layer, although the Si concentration is similar. Thus, the C concentration (rather than just the formation enthalpies) was concluded to play an important role in determining what Ti-rich compound (TiC or TiSi) would be present in the self-formed Ti-rich interface layer. Based on the C concentration in the dielectric layers as listed in Table 1, crystalline TiC would form on the dielectric layers with the C concentration higher than 17 at.%. This C-concentration rule was confirmed in the Cu(Ti)/SiCO, Cu(Ti)/SiCN and Cu(Ti)/SiO$_2$ samples [17].

The Ti-rich interface layers self-formed on the low-k①/Si substrate consisted of crystalline TiC and amorphous Ti oxides. The Ti oxide layers were demonstrated to exhibit a thermally stable diffusion barrier [9]. The TiC compounds with the high melting point of 3100°C are also expected to show thermal stability as barrier layers. The SIMS profiles of the Cu(Ti)/low-k① samples after annealing at 600°C for 2 hours (Fig. 2(a)) showed that the Ti-rich interface layers was sealed to prevent Cu penetration in the low-k①/Si substrates. Thus, the Ti-rich interface layers that self-formed on the low-k①/Si substrate are also expected to show good thermal stability as barrier layers.

Electrical resistivities in the annealed Cu(Ti)/low-k samples

The electrical resistivities of the Cu(Ti)/low-k samples annealed at 500°C or 600°C are shown in Fig. 4. The electrical resistivities of the as-deposited Cu(Ti) alloy films were reduced significantly after annealing, because the Ti atoms supersaturated in the Cu(Ti) alloy films segregated at the film surfaces and Cu(Ti)/low-k interfaces after annealing at 500 and 600°C. The resistivities of the Cu(Ti) alloy film are reduced to about 3.5-4 μΩ-cm upon annealing at 500°C as shown in Fig. 4. In the present experiment, the Ti concentration of the Cu(Ti) alloy films was higher than those used in the previous studies [12-14], and thus, the high resistivities are due to impurity scattering of electrons because the excess Ti atoms remained in the Cu alloy films. Also, the higher resistivities are explained partly by grain boundary scattering because the present Cu alloy films had fine bamboo structures. The resistivities decreased with increasing annealing temperature in the Cu(Ti)/low-k samples. The decrease of the resistivities with temperature can be explained by the decrease of the Ti atoms that remain in the alloy films.

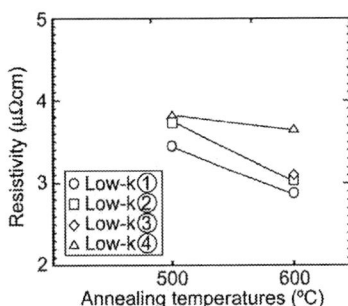

Figure 4. Annealing-temperature dependence of resistivities of Cu(10 at.%Ti)/low-k samples. The dielectric layers were low-k①, low-k②, low-k③, and low-k④.

CONCLUSIONS

Ti-rich interface layers were found to be formed in the Cu(Ti) alloy films prepared on the low-k①-④/Si substrates after annealing at 500 and 600°C. The Ti-rich interface layers formed in the samples consisted of crystalline TiC or TiSi in addition to amorphous Ti oxides. Crystalline TiC was formed in the Cu(Ti)/low-k① sample, and crystalline TiSi was formed in the Cu(Ti)/low-k④ sample. The compositions of the self-formed Ti compounds after annealing at 600°C is determined by the C concentration in the dielectric layers rather than the formation enthalpy of the Ti compounds. Crystalline TiC is formed when the C concentration in the dielectric layer is higher than 17 at.%.

ACKNOWLEDGMENTS

This work was supported by Grants-in-Aid for Scientific Research from The Ministry of Education, Culture, Sports, Science and Technology (18360324). The authors would like to thank Shorai Foundation for Science and Technology, and Iketani Science and Technology Foundation for financial supports.

REFERENCES

1. M.T. Bohr, and Y.A. El-Mansy, *IEEE Trans. Electron Devices* **45**, 620 (1998).
2. S.P. Murarka, *Mater. Sci. Technol.* **17**, 479 (2001).
3. Y. Morand, *Microelectron. Eng.* **50**, 391 (2000).
4. M. Moriyama, M. Shimada, H. Masuda, and Masanori Murakami, *Trans. Mater. Res. Soc. Jpn.* **29**, 51 (2004).
5. M. Shimada, M. Moriyama, K. Ito, S. Tsukimoto, and M. Murakami, *J. Vac. Sci. Technol.* **B 24**, 190 (2006).
6. P.J. Ding, W.A. Lanford, S. Hymes, and S.P. Murarka, *J. Appl. Phys.* **75**, 3627 (1994).
7. D. Adams, T.L. Alford, N.D. Theodore, S.W. Russell, R.L. Spreitzer, and J.W. Mayer, *Thin Solid Films* **262**, 199 (1995).
8. C.J. Liu, and J.S. Chen, *Appl. Phys. Lett.* **80**, 2678 (2002).
9. C.J. Liu, J.S. Jeng, J.S. Chen, and Y.K. Lin, *J. Vac. Sci. Technol.* **B 20**, 2361 (2002).
10. M.J. Frederick, R. Goswami, and G. Ramanath, *J. Appl. Phys.* **93**, 5966 (2003).
11. M.J. Frederick, and G. Ramanath, *J. Appl. Phys.* **95**, 3202 (2004).
12. S. Tsukimoto, T. Morita, M. Moriyama, K. Ito and M. Murakami, *J. Elec. Mater.* **34**, 592 (2005).
13. S. Tsukimoto, T. Kabe, K. Ito and M. Murakami, *J. Elec. Mater.* **36**,258 (2007).
14. K. Ito, S. Tsukimoto, T. Kabe, K. Tada and M. Murakami, *J. Elec. Mater.* **36**, 606 (2007).
15. Metal Databook 4th edition, in Japanese, (Japan Institute of Metal, 2004) p.101.
16. D.G. Archer, R.J. Kematick, C.E. Myers, S. Agarwal and E.J. Cotts, *J. Chem. Eng. Data* **44**, 167 (1999).
17. K. Kohama, K. Ito, S. Tsukimoto, K. Mori, K. Maekawa, and M. Murakami, *J. Elec. Mater.* in press.

Mater. Res. Soc. Symp. Proc. Vol. 1079 © 2008 Materials Research Society

Selective Oxidation and Resistivity Reduction of Cu-Mn Alloy Films for Self-forming Barrier Process

Jun Iijima, Yoshito Fujii, Koji Neishi, and Junich Koike

Department of Materials Science, Tohoku University, 6-6-11 Aoba, Aramaki, Aoba-ku, Sendai, 980-8579, Japan

Abstract

Optimum conditions of annealing atmosphere and temperature for the reduction of Mn content from the Cu-Mn alloy layer in Cu-Mn self-forming barrier process were investigated. Mn was selectively oxidized at the surface by annealing in Ar gas containing an impurity level of O_2 (<0.01ppm). Resistivity of the film was decreased to 2.0 $\mu\Omega$cm after annealing. On the other hand, internal oxidation of Cu-Mn alloy was observed with no external protective surface oxide layer in Ar containing more than 10 ppm of O_2. An optimum oxygen concentration is found to be in between 0.01 and 10 ppm in 1atm of Ar gas at 350 $^\circ$C.

Introduction

The formation of thin and conformal diffusion barrier layer has become increasingly difficult as the technology node is reduced to less than 45 nm. A possibility of self-forming barrier process has been investigated as an alternative method. In this process, an alloying element in Cu metallization is supposed to migrate and to react with a dielectric material, and the reacted layer can act as a barrier layer. Previous researchers investigated the possibility of self-forming barrier process, using Mg [1, 2], Ta [3], Zr [4], Ti [5-7], Al [8], W and Mo [9], as an alloying element in Cu. Koike et al. [10-12] reported the self-formation of the barrier layer using Cu-Mn alloy thin films. The barrier thickness could be controlled in the range of 2 to 8 nm by heat treatment at 250 to 450 $^\circ$C. The barrier layer was thermally stable at 450 $^\circ$C for 100 h and at 600 $^\circ$C for 10 h. Using this alloy as a seed layer of dual-damascene interconnect structure, Usui et al. [13] reported that a self-forming barrier layer was successfully fabricated without a barrier at via bottom. They also demonstrated excellent resistance against stress-migration and electro-migration. However, Mn atoms tend to remain in the film, resulting in higher resistivity than pure Cu. In the present work, we investigated the optimum conditions of annealing atmosphere and temperature to minimize the resistivity by selectively oxidizing the Mn atoms on the film surface.

Experimental

Alloy films of Cu-4 at. % Mn were deposited directly on SiO_2 substrates to a thickness of 160 nm by a sputtering method. Relative resistance change was measured in-situ during heat treatment in various gas atmosphere of pure Ar, Ar+10, 50, 100 and 1000 ppm O_2 and in vacuum at a base pressure of 1.0×10^{-4} Pa. The dynamic flow rate of Ar and Ar+O_2 was set at a rate of 200 cc/min. It is noted that approximately 0.01 ppm O_2 was detected as impurity in pure Ar gas at room temperature before annealing. With increasing temperature above 150 $^\circ$C, the O_2 pressure decreased to less than 0.01 ppm O_2 which is the lowest detection limit. This is probably because some of the impurity oxygen was adsorbed on the chamber surface. Annealing temperature was varied from 150 to 350 °C with duration of up to 5400 sec. Accurate resistivity values at room temperature were measured with a standard four-point probe apparatus after annealing. The cross section of the annealed samples was observed by TEM. Chemical composition of each layer was also measured using EDS attached to TEM.

Results

Figure 1 shows relative resistance change R/R_0 with annealing time in various atmosphere of Ar plus a ppm level of O_2 at 150 °C. Here R_0 is the resistance of as-deposited samples, and R is the resistance of annealed samples. The relative resistance decreases rapidly with annealing time in all atmospheres, and is found to almost saturate after 200 sec. This result suggests the reduction of the Mn content from the Cu-Mn alloy layer. It is noted that a gradual increase in resistivity change is due to temperature rise of the measurement system by keeping the whole set up at high temperatures for long time. Four-point-probe resistivity values of the annealed samples are shown at right-hand side of each curve in Fig. 1. In pure Ar, the lowest resistivity of 5.7 μΩcm was obtained. In contrast, notable surface discoloration indicated that strong oxidation took place in more than 100 ppm of O_2. Resistivity of the annealed samples in Ar+O_2 were all higher than the corresponding value of the sample annealed in pure Ar. Thus the trace amount of oxygen is needed for the selective oxidation of Mn in the Cu-Mn alloy film.

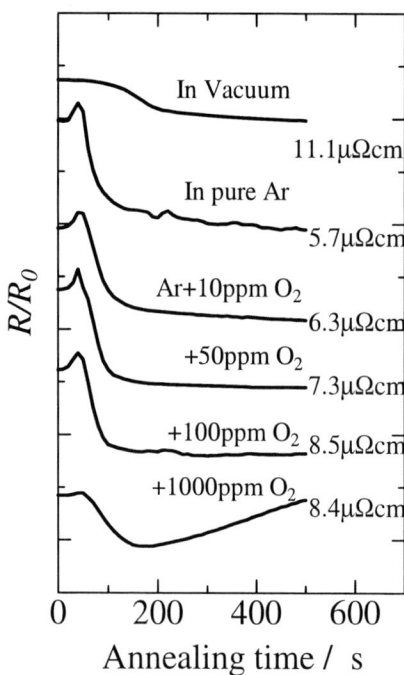

Figure 1. Relative resistance change during annealing at 150 °C in various atmospheres.

Figure 2 shows the relative resistance change and the four-point-probe resistivity of the films annealed in Ar+10ppm O_2 at 150 to 350 °C. The resistance change at 150 °C decreases slowly and saturates after 1500 s at a resistivity of 4.3 $\mu\Omega$cm. The resistance change at 300 °C decreases rapidly and is followed by a gradual increase. At 350 °C, the initial decrease of the resistance is almost the same as in the case of annealing at 300 °C. However, it increases rapidly by further annealing. Thus, the Ar+10ppm O_2 atmosphere gives rise to internal oxidation of the Cu-Mn alloy owing to substantial oxygen diffusion into the alloy film.

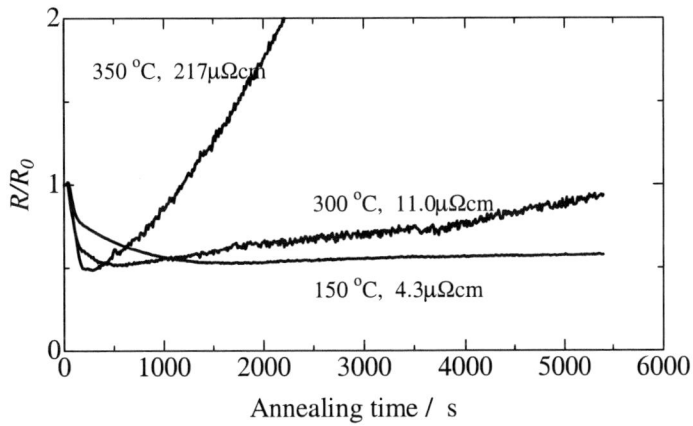

Figure 2. Relative resistance change during annealing in Ar + 10ppm O_2.

Figure 3 shows relative resistance change R/R_o in pure Ar atmosphere at 250, 300 and 350 °C and four-point-probe resistivity values measured after annealing. Relative resistance decreases rapidly with annealing time, and is followed by a slow decrease. Resistivity value at 250 and 300 °C are still high and are 5.3 and 3.1 $\mu\Omega$cm. The resistivity of 2.0 $\mu\Omega$cm was obtained after annealing at 350 °C after 5400 sec. This value is equivalent to a level of pure Cu film.

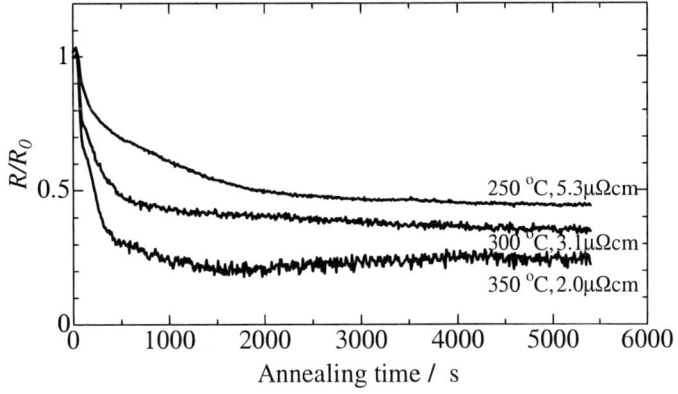

Figure 3. Relative resistance change during annealing in pure Ar.

Figure 4 shows the cross-sectional TEM image of the annealed sample in pure Ar at 350 °C for 1800 s. Large Cu grains are observed with micro twins at the center of the image. The sample surface shows the formation of a discontinuous oxide layer of 12.7 nm in average thickness. In addition, the interface with SiO_2 shows the formation of a thin diffusion barrier layer of 2 nm in average thickness.

Figure 4. TEM image after annealing in pure Ar at 350 °C for 1800 s.

Figure 5 shows EDS spectra taken from (a) the surface layer, (b) the film interior, and from (c) the SiO_2 layer. A weak Si peak appears in all the spectra, even in a spectrum from the bonding epoxy layer. This is due either to the excitation of the Si peak from the dead layer of the EDS detector or to the contamination with Si atoms produced during ion milling. Thus, a weak Si peak should be neglected. The EDS analysis shown in Fig. 5 (a) confirms that the surface layer consists of Mn and O, indicating the formation of the external Mn oxide layer. The spectrum from the film interior in Fig. 5 (b) shows no Mn and O, indicating that Mn atoms are expelled from the film and Cu film is protected from oxidization by the surface Mn oxide layer. Figure 5 (c) shows only Si and O peaks, indicating a good diffusion barrier property of the interface layer.

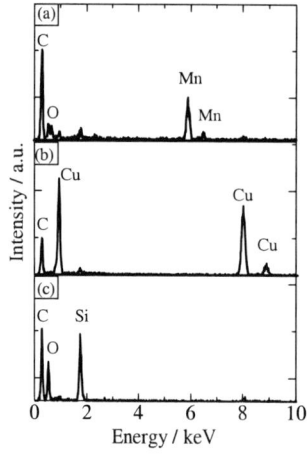

Figure 5. EDS spectra annealed in pure Ar at 350 °C for 1800 s from (a) the film surface, (b) the film interior and (c) the SiO_2 layer.

The above results suggest that Mn is selectively oxidized at the surface and external Mn oxide is formed by annealing in the ultra low partial pressure of oxygen in Ar gas containing an impurity level of less than 0.01 ppm O_2. On the contrary, with the oxygen concentration of more than 10 ppm, internal oxidation of the alloy film took place with no external protective surface oxide. Therefore, an optimum oxygen concentration is found to be in between 0.01 and 10 ppm in 1atm of Ar gas at 350 °C.

Discussion

Surface selective oxidation is classified into two categories of internal oxidation and external oxidation [15, 16]. An alloy consists of a base metal Cu with an alloying element Mn in solid solution. General internal and external oxidation is depicted schematically in Fig 6. Oxygen affinity is higher for Mn than for Cu. The alloy is in contact with an atmosphere with oxygen potential (p_{O2}). The alloying element Mn may form oxides by reacting with oxygen, either externally at the surface of the film or internally within the film. Whether an alloying element is oxidized internally or externally depends on the oxygen solubility in Cu matrix (N_O), the concentration of the alloying element in the alloy (N_{Mn}), the diffusivity of oxygen in the alloy (D_O), and the diffusivity of the alloying element in the alloy (D_{Mn}). The transition from internal to external oxidation is governed by the relative magnitude of the solubility–diffusivity product of oxygen ($N_O D_O$) and that of manganese ($N_{Mn} D_{Mn}$). The equation of $N_O D_O << N_{Mn} D_{Mn}$ (1) gives the criterion for external oxidation [15, 16]. The equation enables to predict how exposure condition affects the formation of external oxide layer. A low external oxygen partial pressure (p_{O2}) corresponds to a low value of N_O. The low p_{O2} will lead to the situation of $N_O D_O << N_{Mn} D_{Mn}$. According to our calculation, the value of $C_{Mn} D_{Mn}$ takes 6.8×10^{-23} $m^2 s^{-1}$ at 350 °C. On the other hand, the values of $C_O D_O$ take 2.7×10^{-25} $m^2 s^{-1}$ at 0.01 ppm O_2 and $2.7 \times 10^{-23.5}$ $m^2 s^{-1}$ at 10 ppm O_2. Therefore, annealing in Ar gas containing 0.01 ppm O_2 satisfies the condition of Eq. (1) for the external oxidation. Annealing in Ar+10 ppm O_2 gas cannot satisfy the condition of Eq. (1) and results in the internal oxidation. Therefore, an optimum oxygen concentration is found to be in between 0.01 and 10 ppm in 1atm of Ar gas at 350 °C, which is in a good agreement with experiment.

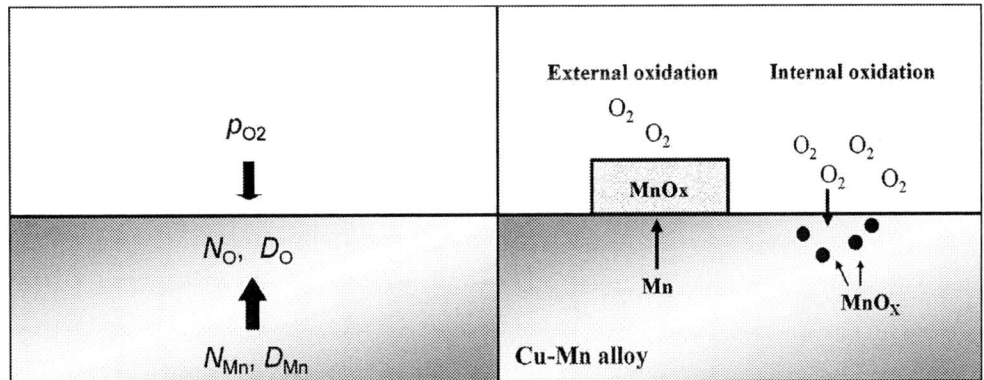

Figure 6. Schematic illustration of internal and external oxidation.

Conclusion

Optimum conditions were investigated to reduce resistivity of the Cu-Mn alloy film for self-forming barrier process. Resistivity of the film was decreased to 2.0 $\mu\Omega$cm by annealing in Ar gas containing an impurity level of O_2 (<0.01ppm) at 350 $^\circ$C. The low resistivity was achieved by external oxidation of Mn under a very low partial pressure of oxygen. In contrast, resistivity remained at a high level by annealing in Ar + O_2 of more than 10 ppm. The optimum condition was achieved when the Cu-Mn film was externally oxidized under a very low oxygen partial pressure (<0.01ppm) at 350 $^\circ$C. The obtained results could be explained consistently by the internal-external oxidation model.

References

1. W. Lee, H. Cho, B. Cho, J. Kim, Y. S. Kim, W. G. Jung, H. Kwon, J. Lee, P. J. Reucroft, C. ee, and J. Lee, J. Electrochem. Soc. 147, 3066 (2000).
2. M. J. Frederick, R. Goswami and G. Ramanath: J. Appl. Phys. 93 (2003) 5966.
3. C. J. Liu, J. S. Chen, and Y. K. Lin, J. Electrochem. Soc. 151, G18 (2004).
4. C. J. Liu and J. S. Chena J. Vac. Sci. Technol. B 23(1) 90 (2005)
5. C. J. Liu, J. S. Jeng, J. S. Chen, and Y. K. Lin, J. Vac. Sci. Technol. B 20, 2361 (2002).
6. S. Tsukimoto, T. Kabe, K. Ito, and M. Murakami   J. Electron. Mater. 36, 258 (2007).
7. K. Ito, S. Tsukimoto, T. Kabe, K. Tada and M. Murakami J. Electron. Mater. 36, 606 (2007).
8. K. Shepherd, C. Niu, D. Martini, J.A. Kelber Applied Surface Science 158, 1 (2000)
9. J.P. CHU and C.H. LIN J. Electron. Mater. 35, 1933 (2006).
10. J. Koike and M. Wada, Appl. Phys. Lett. **87**, 041911 (2005).
11. M. Haneda, J. Iijima, and J. Koike, Appl. Phys. Lett. **90**, 252107 (2007).
12. J. Koike, M. Haneda, J. Iijima, Y. Otsuka, H. Sako, K. Neishi, J. Appl. Phys. **102**, 043527 (2007).
14. T. Usui, H. Nasu, S. Takahashi, N. Shimizu, T. Nishikawa, M. Yoshimaru, H. Shibata, M. Wada, J. Koike, IEEE Trans. Electron Devices. **53**, 2492 (2006).
15. N. Birks and G.H. Meier, *Introduction to high temperature oxidation of metals,* E. Arnold, ondon. (1983).
16. Paul G. Shewmon, *Transformations in metals*, McGraw-Hil, New York (1969).

Al Diffusion in Polycrystalline Cu

Florian Gstrein, Harold Kennel, Andre Budrevich, Barbara Miner, John Plombon, and Ebrahim Andideh

Intel Corporation, Hillsboro, OR, 97214

ABSTRACT

The diffusion of aluminum (Al) from a source sandwiched between polycrystalline copper (Cu) thin films was investigated as a function of time and temperature through secondary ion mass spectroscopy (SIMS) and continuum simulations. Extracted diffusion coefficients for the bulk were in line with literature values. In order to simulate the experimentally derived diffusion profiles at temperatures where bulk diffusion is not the dominant diffusion mechanism (room temperature to 350 °C), it was necessary to explicitly include the re-distribution of Al as a result of Cu grain growth during anneal. Aluminum has the tendency to segregate to the Cu/liner and Cu/etch stop (ES) interface. The tendency of Al to segregate to the liner is ten times stronger for ruthenium (Ru) than for tantalum (Ta). In 100 nm wide dual damascene structures lined with Ru, this segregation behavior was responsible for the Al depletion in bulk Cu and for the Al depletion at the Cu/ES interface.

INTRODUCTION

To meet the increasing feature-density requirements of future technology nodes, the dimensions of copper (Cu) dual damascene interconnects must be continuously scaled. Since the current density increases with shrinking feature size, Cu lines become more and more prone to electromigration-induced failure. The electromigration resistance of Cu can be improved through doping with elements such as Ag, Sn, and Al [1-3]. Relatively little is known about the fundamental diffusion properties of these dopants in polycrystalline Cu thin films and about the dopant distribution in Cu dual damascene structures. In the present study, we report on the diffusion mechanism of aluminum (Al) in Cu thin films. To understand the experimentally derived Al diffusion profile in terms of grain boundary diffusion, bulk lattice diffusion and grain size evolution during anneal, secondary ion mass spectroscopy (SIMS) depth profiles of Al-doped Cu thin films were complemented with continuum simulations. Aluminum segregation to the Cu/ES interface and to two technologically relevant liner materials, tantalum/tantalum nitride (Ta(N)) and tantalum nitride/ruthenium (TaN/Ru), were first investigated for simple planar stacks. The results are then compared with the spatial distribution of Al in Cu-filled trenches.

EXPERIMENT

All samples were fabricated in state-of-the-art 300 mm barrier/seed deposition tools with integrated degas and pre-clean capability, an electroplating tool and a CVD etch stop deposition tool. Figure 1 (a) shows the blanket wafer film stack used in this study. The bottom interface consisted of a PVD Ta(N) based barrier or a TaN/Ru bi-layer, respectively.

A 200 Å thick PVD Cu-Al layer was sandwiched between two 1000 Å thick PVD Cu layers. The total Al dose in the stack was on the order of 7×10^{15} atoms/cm^2. Symmetrical diffusion stacks were capped either with a Ta or Ru cap, whereas non-symmetrical stacks were capped with a silicon carbide/nitride ES layer. Annealing experiments as a function of time were conducted in forming gas in a vertical diffusion furnace that was kept at the desired anneal temperature. Samples were introduced through a load lock. Figure 1 (b) shows a schematic of the corresponding dual damascene structures. Dual damascene structures were prepared using standard PVD Ta(N) barrier and seed deposition, and were subsequently plated with a state-of-the-art EP Cu process, chemically-mechanically polished and capped with a silicon carbide/nitride etch stop layer.

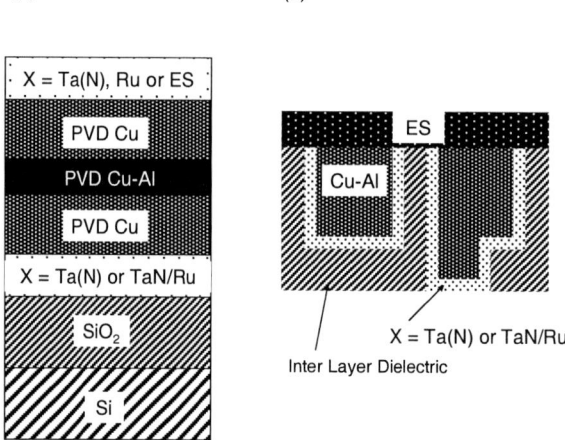

Figure 1. (a) Planar sample stack used for the determination of the Al diffusion profile in PVD as a function of temperature, liner material and cap material. **(b)** Dual damascene structures doped with Al.

SIMS measurements were done with a SC-Ultra SIMS instrument. A 1.5KeV Cs+ primary beam was used and positive secondary cluster ions (Cs+, Cs16O+, CsAl+ and Cs63Cu+) with an energy of 5keV were collected at low mass resolution (~300) [4]. For SIMS on patterned structures, large 1 x 0.5 mm test fields were used that consisted of 100 nm wide Al-doped Cu trenches of uniform pitch, width and orientation. Additionally, the Al concentration in Cu lines was measured with a field emission gun transmission electron microscope equipped with a detector for energy dispersive X-ray analysis (FE TEM/EDX). The spatial resolution is approx. 1nm and approx. 1% Al can be detected within the analyzed volume.

DISCUSSION

The evolution of Al depth profiles with time of symmetrical stacks with a Ta cap is shown in Fig. 2 for several characteristic temperatures 20°C, 150°C, 350°C and 450°C. At 450°C, Fig 2 (a), the Al source layer broadens symmetrically. When the anneal temperature is lowered to 350°C, Fig. 2 (b), little broadening of the source layer is observed, however, diffusion tails develop and saturate after approximately 5 minutes. At 250 °C (results are not shown) saturated diffusion tails form after 20 minutes of anneal, whereas at 150°C (Fig 2(c)) even after 80 minutes of anneal no saturation of the diffusion tails is reached. The concentration of the saturated tails is essentially independent of the anneal temperature and at saturation about half of the total Al dose is contained in the tail plateaus. Even at room temperature, Fig 2 (d), the Al source layer is far from being stable. Significant formation of diffusion tails is observed after 2 months. At all temperatures, Al segregation to the Ta liner material takes place. While the segregation of Al to the Ta liner (bottom) and Ta cap (top) is symmetrical for low anneal temperatures, the extent of segregation becomes increasingly asymmetrical when the anneal temperature is raised with more Al segregating to the top Ta interface than to the bottom Ta interface, see Fig 2 (a) - (d). The asymmetry of the Al segregation is not completely understood at this point.

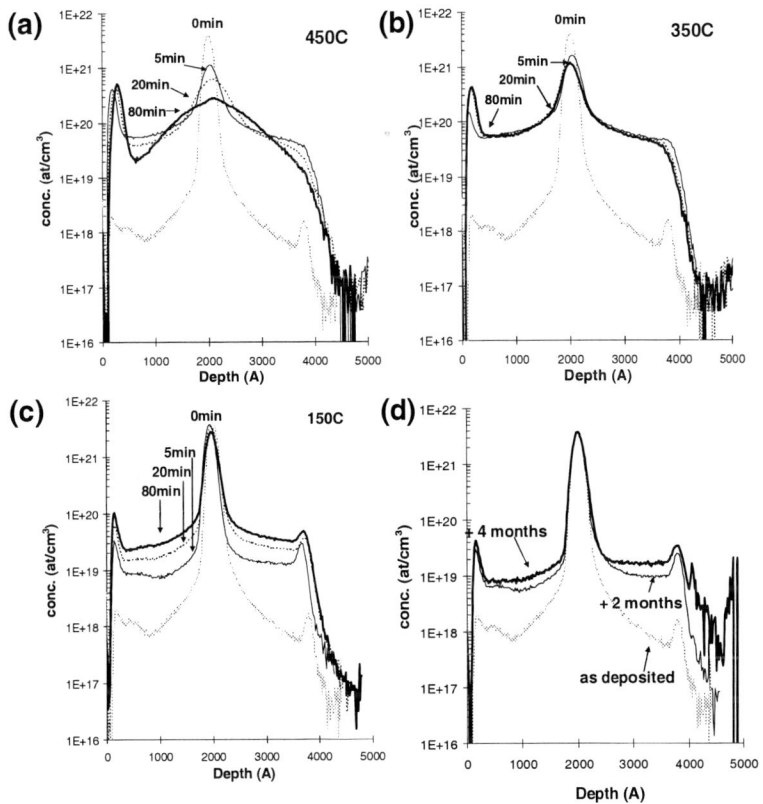

Figure 2. Evolution of SIMS depth profiles of Al in PVD Cu stack for various anneal temperatures and anneal times (0, 5, 20 and 80 minutes). The stack was annealed at (**a**) 450°C, (**b**) 350°C, and (**c**) 150 °C. (**d**) PVD Cu stack at room temperature as deposited and after 2 & 4 months of ageing.

To understand the re-distribution of Al from a source layer as a function of temperature and time, continuum simulations were employed that treat Al diffusion as transport inside grains (bulk diffusion) and along grain boundaries [5]. Values for the bulk and grain boundary diffusivities were taken from the literature [6-7]. Rapidly saturating grain growth during anneal was taken into account by including an empirical rate constant in the simulation. This rate constant was treated as a single exponential growth process with a characteristic time constant.

Simulated Al diffusion profiles at 350°C are shown in Figure 3. Since grain boundary diffusivity is orders of magnitude greater than bulk diffusivity [6-7] and grain boundary volume increases with decreasing grain size, the redistribution of Al is facilitated by small grain sizes. For a somewhat unrealistically small grain size of 50 Å, Fig. 3 (a), simulations predict, that the source layer diminishes quickly with time. Due to the small Cu grain size, the average diffusion path that Al experiences before it encounters a grain boundary is short and dopants are thus quickly transported away by high-diffusivity grain boundaries. A film with grain size of 5000 Å, which is on the same order of magnitude as the total film thickness used in this report, exhibits no saturation of the tail concentration and little loss of Al from the source layer is observed, see Fig 3(b). Although the diffusion along grain boundaries is still very rapid, the process is limited by the long diffusion path of Al inside the Cu bulk. Only when the grain size is allowed to evolve from a smaller grain size to 5000 Å in the early stages of the simulated anneal, does the simulation model fit the experimental data, see Fig. 3(c).

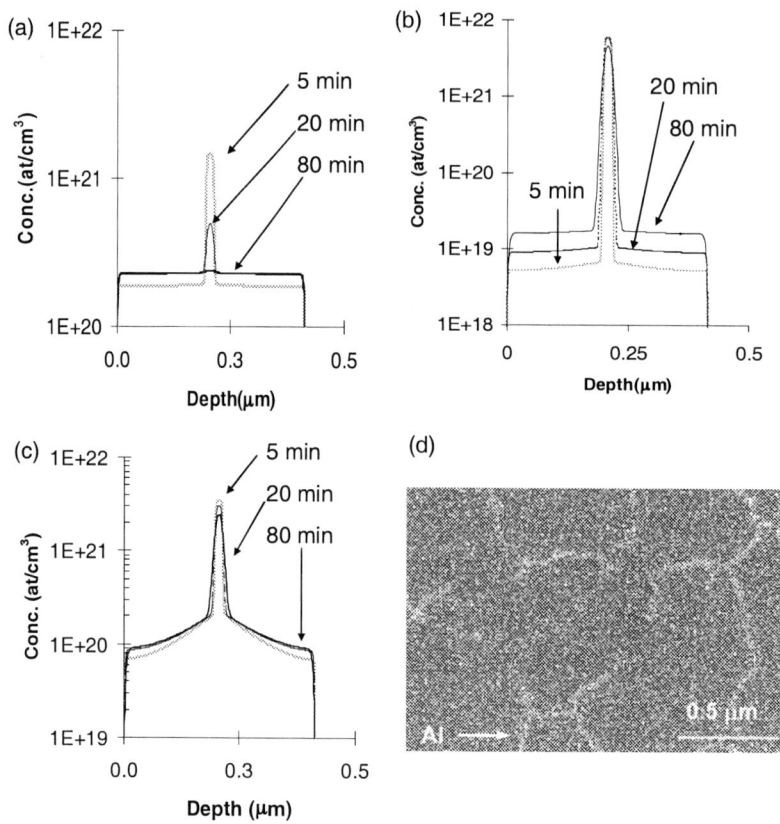

Figure 3. Simulation results for Al diffusion in Cu for annealing temperature of 350°C and annealing times between 5 and 80 minutes. **(a)** 50 Å grain size, **(b)** 5000 Å grain size, **(c)** 5000 Å grain size + grain evolution during anneal. **(d)** Auger map of Al in polycrystalline Cu annealed at 250 °C for 80 min. The grain size is on the order of 5000 Å. White dots are Al.

As Cu grains in the highly unstable as-deposited film grow, grain boundaries move through the Al-doped regions thus providing an effective path by which Al is captured inside the grain boundaries. The time needed to stabilize the grain size is represented by the time needed to reach saturation of the diffusion tails and this time decreases with temperature. Since Cu grain-growth at room temperature is a well known phenomenon [8], grain-growth induced re-distribution of Al at temperatures as low as room temperature is expected and was indeed observed, see Fig 2 (d). An Auger scan of a polycrystalline, Al-doped Cu thin film confirms the grain size is on the order of the film thickness and that more Al resides within the grains boundaries than in the Cu bulk, see Figure 3 (d).

The bulk diffusivity was determined by fitting the experimentally observed broadening of the Al source layer for 5 - 80 min of anneal. The bulk diffusivity values obtained for 450 °C and 350 °C were 1.8×10^{-15} cm^2/sec and 6.3×10^{-17} cm^2/sec, respectively. These values are in excellent agreement with the literature [6-7]. For 250 °C the extraction of diffusivity is complicated due to the limited broadening of the source layer and roughness-induced peak broadening during SIMS depth profiling. Since the Cu films in this report are thin and the barrier layer serves as a sink for Al, grain boundary diffusivities could not be extracted from the gradient of the diffusion tails.

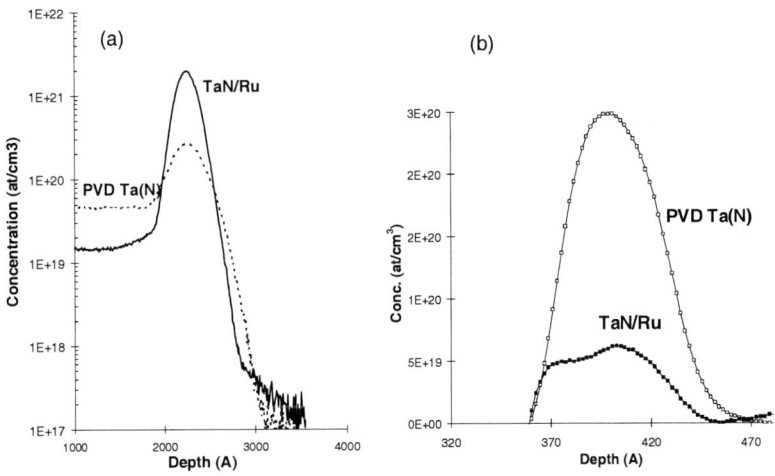

Figure 4. (a) Planar stack capped with ES material. According to SIMS depth profiling the Al segregation to the Cu/Ru interface is an order of magnitude larger than the segregation at Cu/Ta(N) interface. **(b)** Al concentration at the Cu/ES interface for a large assembly of patterned trenches as measured by SIMS.

Diffusion profiles are further influenced by the presence of the interface material. For a non-symmetrical PVD Cu stack capped with ES material, Fig. 4 (a) depicts the segregation of Al to a Ta(N)-based liner material and a Ru/TaN based liner material, respectively. For identical anneal conditions, we find that a stack with a Ru liner accumulates approximately two times more Al during doping than the Ta(N)-based stack. For Ru, most of the Al segregates to the Cu/Ru interface and the Al dose is ~10 times higher than that measured at the Cu/Ta(N) interface. In Fig 4 (a), the Al conc. at depth of 1000 A represents the bulk concentration of Al in the particular stack.. The bulk concentration in the Ta(N) stack is about 5 times larger than the Al bulk concentration in the Ru/TaN stack. For planar stacks both liner materials exhibited similar amounts of Al at the Cu/ES interface.

In order to understand the distribution of Al in real interconnect structures in the light of this segregation behavior, dynamic SIMS measurements were conducted on dual damascene structure which are depicted in Fig 1. (b). The diffusion profiles consist of an Al peak at the ES interface that quickly reaches a plateau as the sputter front moves away from the Cu/ES interface and into the trench. At the Cu/ES interface, SIMS detects the convoluted amount of Al at ES/Cu interface, Al residing in the bulk of the Cu lines and Al that has segregated to the sidewalls. Inside the metal trench, only Al in the bulk and Al residing at the sidewalls are detected. The Al concentration at the Cu/ES interface can thus be determined by subtracting the Al plateau region (Al in the bulk + Al residing at the sidewalls) from the Al depth profile, the result of which is shown in Fig. 4 (b). As a result of the strong segregation tendency of Al to the Ru sidewalls, see planar stack results in Fig. 4 (a), the Al concentration at Cu/ES interface is strongly depleted with respect to Al, while a healthy dose of Al is retained at the Cu/ES layer of trenches lined with Ta(N). These results have been independently confirmed by FE TEM/EDX line scans which detect within the accuracy of the measurement technique no Al at the Cu/ES interface but Al segregation to the Ru sidewalls. The data supports the model that the main pathway for Al to reach the ES/Cu interface is Al diffusion across the Cu/liner interface. Al atoms are more mobile across Cu/Ta(N) interface than Cu/Ru interface.

CONCLUSIONS

This report for the first time presents a model of Al diffusion in Cu thin films with explicit consideration of Cu grain growth during anneal and the segregation tendency of Al to the Cu/liner interface that is consistent with experimentally derived diffusion profiles. Simple planar stacks were used to predict the distribution of Al in dual damascene structures.

ACKNOWLEDGMENTS

The authors acknowledge Patrick Paluda for fab integration support of the experiments described in this report, Gregory Scott & Janet Forsyth for their support of the SIMS measurements, and Milt Jaehnig & John Richards for help with Auger spectroscopy.

REFERENCES

1. A. Isobayashi, Y. Enomoto, H. Yamada, S. Takahashi, S. Kadomura IEEE IEDM Technical Digest., 13-15 Dec. 2004 p 953 - 956
2. T. Tonegawa, M. Hiroi, K. Motoyama, K. Fujii, and H. Miyamoto, in Proc. IEEE IITC., 2003, pp. 216–218
3. Y. Matsubara, M. Komuro, T. Onodera, N. Ikarashi, Y. Hayashi, and M. Sekine, in VLSI Symp. Tech. Dig., 2003, pp. 127–128
4. Aluminum implanted in Cu was used as a standard to determine the relative sensitivity factor of Al in Cu. CsAl+ ion intensities were point-to-point normalized to the Cs+ signal.
5. S. Banerjee, "Modeling of Dopant Incorporation in Polysilicon Incorporating Effects of Grain Boundary Motion", M.S. Thesis, Boston University, 1996.
6. Kaur and Gust, in "Handbook of Grain and Interphase Boundary Diffusion Data", Vol. 1 (Ziegler Press, Stuttgart, 1989) p. 197
7. H. Oikawa, T. Obara, and S. Karashima, Metall. Trans. 1 (1970) 2969
8. M. A. Gribelyuk, S. G. Malhotra, P. S. Locke, P. DeHaven, J. Fluegel, C. Parks, A. H. Simon, R. Murphy, Proc. IEEE IITC, 2000, pp. 188 - 190

Mater. Res. Soc. Symp. Proc. Vol. 1079 © 2008 Materials Research Society   1079-N05-12

Sacrificial Passivation of Nanoscale Metal Powders for Transient Liquid Phase Bonding

Nick S Bosco[1], Beth Manhat[2], and Jolanta Janczak-Rusch[1]

[1]Laboratory for Interface and Joining Technology, Empa, Überlandstrasse 129, CH-8600 Dübendorf, Switzerland

[2]Department of Chemistry, Portland State University, Portland, OR, 97207

ABSTRACT

Differential scanning calorimetry (DSC) was used to evaluate the extent of liquid phase formation in particle compacts (compressed dry powders) comprised of organically capped nanoscale Ag particles and Sn. The Ag nanoparticles employed where synthesized with various organic molecules and procedures to produce particles with a range of capping thicknesses and decomposition temperatures (T_d), as measured by thermogravimetric analysis (TGA). A baseline sample containing commercially available un-capped micrometer scale Ag was also investigated for comparison. Results indicate that all of the Sn initially present formed a liquid phase when heated through its melting point when combined with Ag particles exhibiting a comparatively thick cap of low T_d. Slightly smaller fractions of Sn liquid were obtained when the Ag's cap was thin and of a high T_d while particles with thin-low T_d caps exhibited the highest levels of Sn consumption and similar to that observed with the un-capped micron-scale Ag particles. The reduction in the amount of Sn liquid formed is attributed to solid state reaction between the Ag and Sn particles resulting in the formation of a more refractory phase. The extent of the subsequent liquid phase reaction was also evaluated and is demonstrated to not necessarily be adversely effected by the presence of the organics. The significance of this work is the demonstration that organic molecules may be employed to prevent solid state reaction in particle compacts at elevated temperatures, yet allow the subsequent liquid phase reaction proceed uninhibited.

INTRODCUTION

The future commercialization of high-temperature wide band-gap semiconductors, such as SiC and GaN, for applications from laser diodes to power electronics will make current die-attach materials obsolete. Neither the solder alloys nor epoxies that are in wide use today could support the high-performance or high-reliability operational demands of these devices at temperatures beyond 250°C [1-4]. A die-attach material and process that is capable of producing microstructures tailored for their enhanced mechanical, thermal and electrical properties while obtainable through benign processing conditions is therefore required. Transient liquid phase (TLP) bonding possesses the potential to meet these needs [5]. The TLP process employs a melting point depressant (MPD), between the base metals to be joined and relies on interdiffusion for isothermal solidification at the bonding temperature. The integrity of the resulting bonds is enhanced by the presence of the liquid phase during bonding, yet the bonds are able to withstand operation at temperatures above the bonding temperature once solidification is complete.

In order to commercially benefit from the full potential of TLP bonding, however, it must be possible to tailor the joint's terminal microstructure and achieve that structure within a reasonable combination of bonding temperature and time. Previous work has demonstrated that the use of a homogenous MPD interlayer of requisite thickness is necessary to affect bonding and an extensive bonding time may be required to convert that layer into a targeted terminal microstructure [6]. The origin of this requisite thickness arises from a compromise between minimizing the distance over which diffusion must occur and accommodating solid state interdiffusion prior to melting the interlayer; the consequence of too thin an interlayer being its total conversion into a higher melting point phase prior to reaching its melting point and therefore its ability to form a liquid phase to initiate the bond. Considerable interest consequently lies in inhibiting solid-state interdiffusion between the base material and MPD in the early stages of bonding.

In the current paper the method of Sacrificial Passivation is presented as a route to preclude solid-state interdiffusion in the early stages of the TLP process. It is developed for a model Ag-Sn system and focused around the ability to synthesize organically capped nano-scale Ag particles as the base material. It is proposed that the organic cap precludes the reaction of (passivates) the Ag nanoparticles, the base material, with particles of Sn, the MPD, when in intimate contact at elevated temperatures. Furthermore, the timely decomposition of that cap provides for the subsequent, desirable liquid phase reaction to proceed uninhibited [7]. The concept of Sacrificial Passivation is illustrated in Fig. 1 through contrast with an un-passivated system: (a) during heating while below the melting point of Sn, T_mSn, solid-state reaction with Ag consumes the Sn in the non-passivated system into the more refractory Ag_3Sn phase, while the amount of Sn in the passivated system is preserved, (b) upon reaching T_mSn all of the Sn initially present in the passivated system melts; in the un-passivated system only a fraction is now available to form the liquid phase, (c) after further heating, isothermal solidification will occur in the un-passivated system and, if the passivating layer is removed, in the passivated system as well. One consequence of the reduced amount of liquid phase in the non-passivated system may be excessive porosity upon solidification [6, 8]. This illustrated process is analogous to transient liquid phase sintering where a lower melting point material is provided to aide in the rate and extent of densification. In this case the Sacrificially Passivated system would provide for a larger fraction of the sintering aide to form a liquid phase, consequently reducing the amount required for densification and therefore decreasing overall process time.

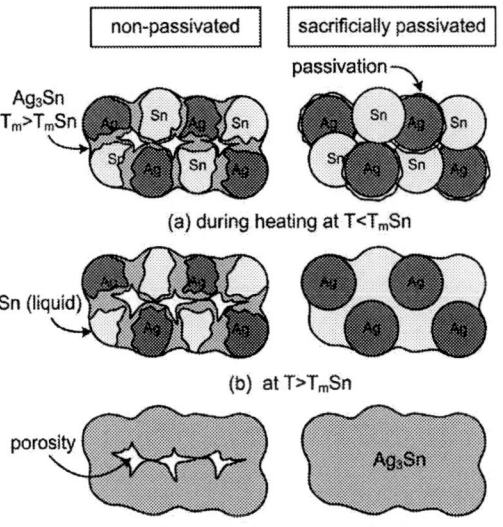

Figure 1. Schematic illustration of the Sacrificial Passivation effect, applied to the Ag-Sn system.

To address the current objective, the melting behavior of nanoparticle compacts comprised of organically capped Ag nanoparticles and micrometer scale Sn particles has been investigated. Since the characteristics of the organic cap are expected to dictate its effectiveness as a Sacrificial Passivation layer, this study is also constructed as a start to develop the processing-structure-property link between Ag nanoparticle synthesis and Sacrificial Passivation. To this end, the organically capped Ag nanoparticles were synthesized with various organics and procedures. These particles where then first characterized by thermogravimetric analysis (TGA) to determine the amount of cap present on the particle and its decomposition temperature, the two traits suspected most influential in a Sacrificial Passivation layer. Microstructural analysis was also performed to support these results.

EXPERIMENTAL METHODS

Silver nano-particle synthesis and characterization

The synthesis and isolation of the capped Ag nano-particles used were conducted during a concurrent study, therefore only salient details are presented here [9]. A chemical reduction method was employed to synthesize the particles of interest. In this method a solution of the metal salt ($AgNO_3$) is slowly added to a heated solvent, reductant and capping agent. Four procedural factors were varied in this concurrent investigation to elucidate their affect on the caps decomposition temperature and relative amount and are: (i) synthesis route, (ii) organic molecule, (iii) molar ratio of the precursor metal salt to organic molecule and (iv) addition rate of the metal salt to the reaction mixture. The Taguchi Method of experimental design and data analysis was employed to realize the influence of each factor over three levels with only nine experimen-

61

tal trials. The particles produced in these nine trials were used in the current study and are designated by their organic cap: Polyvinylpyrrolidone (PVP1-3), Poly(vinyl alcohol) (PVA1-3) or Sodium dodecyl sulfate (SDS1-3). Repetitions of particles with the same organic reflect samples produced with variations in the other synthesis factors.

The as-prepared capped nanoparticles were characterized for phase by powder x-ray diffraction (XRD) and size and character by transmission electron microscopy (TEM). The decomposition temperature and amount of the organic cap was measured via thermogravimetric analysis (TGA). For this measurement the resulting powders were weighed, placed in the TGA's alumina crucible and heated at 20 °C/ min under He to 900 °C as sample mass loss was monitored. The decomposition temperature of the organic molecules, themselves, was also evaluated.

Particle compact characterization

Ten sets of binary powder mixtures, corresponding to nine of the passivated Taguchi trials and a baseline sample, were prepared for evaluation by differential scanning calorimetry (DSC). The passivated sets were prepared with the synthesized capped Ag nano-scale particles and the non-passivated set with micron sized Ag particles (Alfa Aesar). The binary powder mixtures were made by dry milling the Ag particles in air with Sn powder (Alfa Aesar) in a ratio of Ag-27 wt% Sn. This composition was chosen for this investigation to correspond with the Ag-Sn ε-phase (Ag_3Sn), thereby ensuring a unique melting point once homogenization is complete. The powder mixtures were pressed into particle compacts under 2 GPa and approximately 10 mg of each placed directly within the DSC. The DSC temperature profile consisted of two identical, consecutive scans characterized by heating and cooling rates of 20 °C/ min and a maximum temperature of 300 °C.

Two sets of particle compacts, as fabricated above, were also prepared for the purpose of microstructural analysis. These consisted of the passivated Ag nano-scale particles SDS3 and the baseline sample. These compacts were heated in an oil bath at 20 °C/ min to a maximum temperature of 220 °C and quenched to room temperature; a profile intended to simulate the state of the DSC samples immediately prior to the melting of Sn during the first scan. The treated compacts were then sectioned, polished, etched and imaged by scanning electron microscopy (SEM).

RESUTLS AND DISCUSSION

Ag Nanoparticles

The synthesized particles were confirmed to be Ag by XRD and are successfully stabilized by the PVP, PVA and SDS caps. A representative XRD pattern of the PVP3 capped particles is presented in Fig. 2 and indicates lattice spacings at 2.35, 2.04, 1.47, 1.23, and 1.17 Å assigned to diffraction from the (111), (200), (220), (311) and (222) planes, respectively. The lattice constant calculated for this set of planes is 4.08 Å, a value in agreement with Ag [10]. XRD patterns of the other synthesized particles yielded similar results. Particle diameter for each synthesis is estimated to be 100-300 nm from SEM images similar to the one presented in Fig. 3 of the PVP1 capped particles. These images also illustrate that the particles are discrete, suggesting they are successfully stabilized and of various shapes.

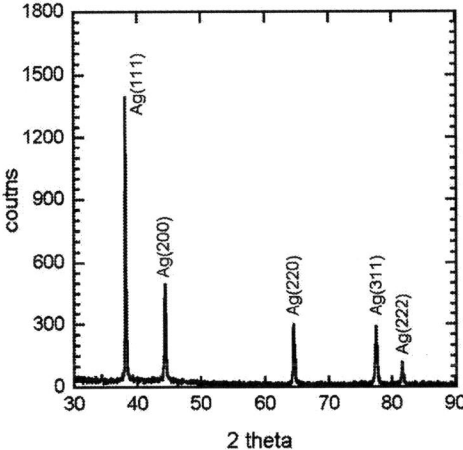

Figure 2. XRD pattern for the PVP3 capped Ag particles.

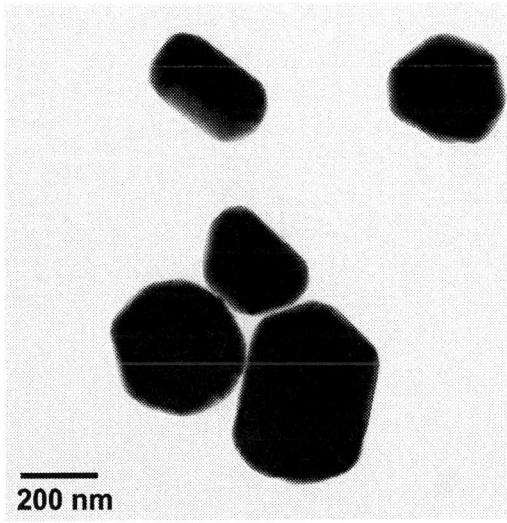

Figure 3. TEM image of the PVP1 particles.

A representative TGA scan of percent mass loss against temperature is presented in Fig. 4 for the PVA3 capped particles along with its first derivative, dM%/dT. For the purpose of the current evaluation, the amount of organic cap is defined as the total mass loss experienced by each sample when heated through 600 °C and the decomposition temperature the point of most rapid mass loss; represented as a minima in the dM%/dT curve. The amount of organic cap

(M%) and its decomposition temperature (T_d) as measured via TGA is reported in Table I for each sample.

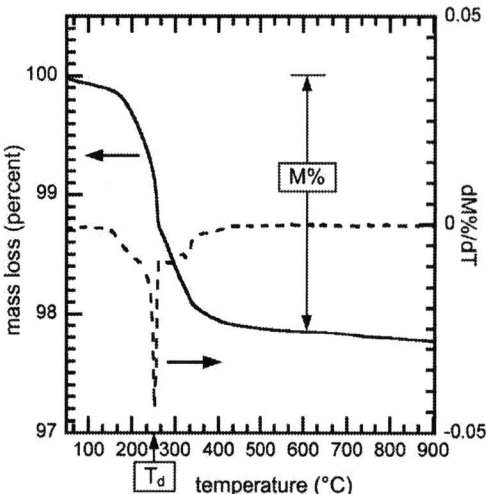

Figure 4. TGA plot for the PVA3 capped particles illustrating the cap's decomposition temperature (T_d) and amount (M%).

The T_d, similarly measured for each of the three organic molecules themselves (native values), are 485, 295 and 240 °C for the PVP, PVA and SDS molecules, respectively. While most of the decomposition temperatures measured for the organics on the particles fall in this regime, quite different behavior from the molecules themselves is observed. For instance the lowest decomposition temperature measured, 252 °C, is on the PVA3 particles while PVA's native T_d is 295 °C. Furthermore, the SDS1 capped particles (T_d SDS1 = 507 °C) demonstrated the highest decomposition temperature, which is also furthest from its organics native value (T_d SDS = 240 °C).

Table I. Tabulation of the nine synthesized Ag nanoparticles by their organic cap.

sample	decomposition temperature (°C)	amount (M%)	$\%\Delta H_{Sn}^{1}$	$\%\Delta H_{Sn}^{2}$
PVA1	504	0.60	64.6	0.1
PVA2	274	1.30	62.3	12.9
PVA3	252	2.42	65.5	17.9
PVP1	493	0.99	47.8	1.2
PVP2	506	0.60	78.3	18.1
PVP3	443	3.30	65.5	31.7
SDS1	507	0.60	87.0	18.0
SDS2	336	0.80	24.1	6.5
SDS3	265	2.80	99.2	3.5

Particle compacts

The set of DSC scans for the baseline (BL) particle compact is presented in Fig. 5. The endothermic peak is consistent with the melting of Sn at 232 °C. This set of scans is representative of all those obtained, where the Sn melting endotherm is prominent during the first scan and barely detectable when the sample is brought through the melting point of Sn during the second DSC scan.

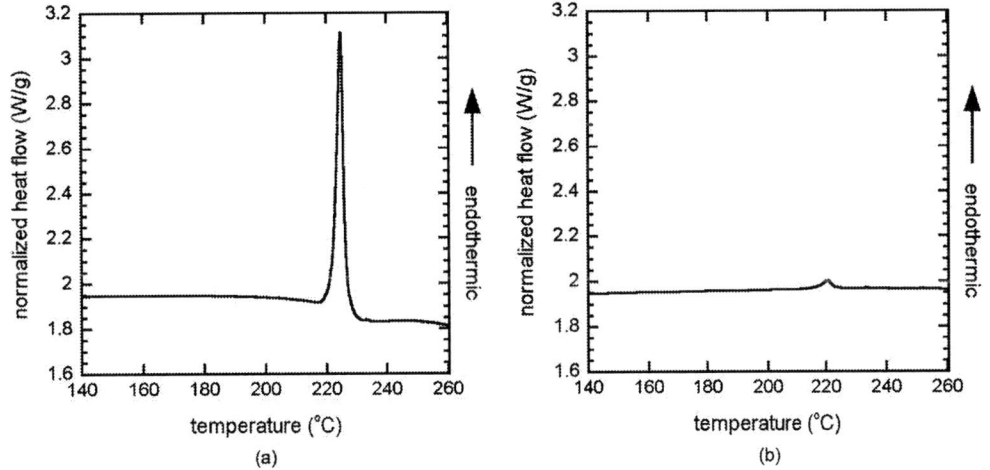

Figure 5. DSC scans of the baseline particle compact during the a) first and b) second heating.

The amount of liquid phase formed in the particle compacts during DSC scans may be quantified by calculating the area under the corresponding endothermic peak, ΔH_{actual}. This area represents the heat of formation of the liquid and can be compared to the theoretical value for Sn, ΔH_{Sn}. Considering the amount of Sn initially available to form a liquid phase is dictated by its weight percent in the particle compact, X_{Sn}, yields the following relation for the fraction of Sn liquid formed:

$$\%\Delta H_{Sn} = \frac{\Delta H_{actual}}{X_{Sn}\Delta H_{Sn}} \qquad (1)$$

Evaluating this expression for the first DSC scan of each particle compact quantifies the amount of Sn consumed by solid-state reaction prior to reaching its melting point, ΔH_{Sn}^1, Fig. 6 and Table I. In this case roughly 40 % of the initial Sn present is consumed into a higher melting point phase for the baseline sample. All of the samples synthesized with the PVA capped particles, despite their varied T_d and M%, showed comparable behavior to the baseline and therefore have been omitted from the remainder of the analysis. The compacts fabricated with the PVP and SDS capped particles, however, show varied behavior. While the PVP1 and SDS2 compacts demonstrate a larger initial consumption of Sn than the baseline, the balance of compacts suggests that the capping layer has at least partially passivated the Ag particles from the initial solid-state reaction. For the SDS3 particles, the theoretical value of the Sn melting endotherm is achieved suggesting that in this sample solid-state reaction has been completely prevented.

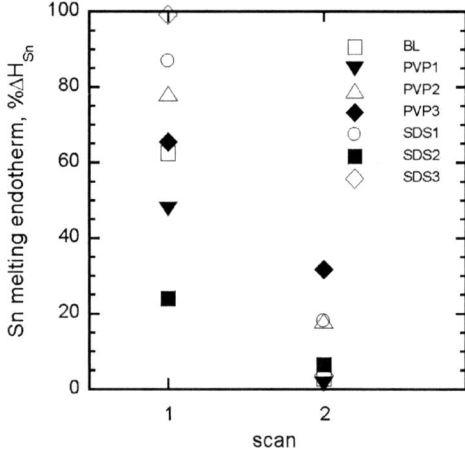

Figure 6. Percent theoretical Sn liquid formation for the first and second DSC scans.

Evaluating equation (1) for the second DSC scan produces a measure of the extent of subsequent liquid phase reaction, ΔH_{Sn}^2, Fig. 6 and Table I. In this case a very limited endothermic reaction is detected ($< 5\ \%\Delta H_{Sn}$) for the baseline, PVP1, SDS2, SDS2, and SDS3 coated particles, suggesting that nearly all of the Sn initially present has been converted into a higher melting point phase mid-way through the second scan. The compacts containing the PVP2, PVP3 and SDS1 coated particles, however, exhibit that over 18 % of the initial Sn present is unreacted and therefore available to form a liquid phase.

The two particle compacts fabricated for microstructural analysis are in agreement with the DSC results, Fig. 7. The SEM images of the polished and etched cross-sections of samples containing the baseline and SDS3 particles were analyzed to determine the area fraction of Sn that remained following their heat treatment. By assuming an isotropic microstructure and considering the relative densities of the phases present, the area fraction of Sn was converted to a weight percent then compared to the initially assigned value. This analysis yielded values of 59 and 92 wt% Sn for the baseline and SDS3 particle compacts, respectively. The result describes the amount of Sn not initially consumed through solid-state reaction, and therefore serves as a direct comparison to the interpretation of the DSC measurements.

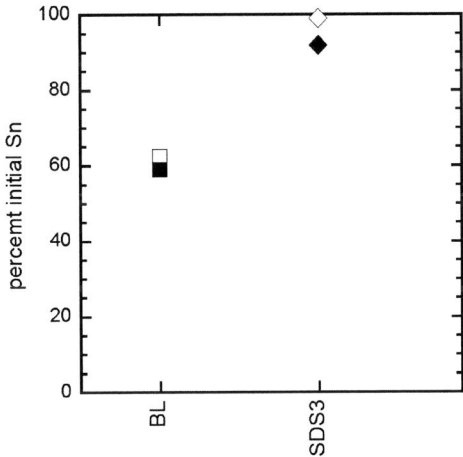

Figure 7. Comparison of the percent of initial Sn present as determined by DSC (solid) and microstructural analysis (open).

The overall performance of the particle compacts is summarized in Fig. 8. This plot is presented to compare both the ability to preclude the Ag-Sn solid-state reaction and allow the subsequent liquid phase reaction to occur uninhibited. It is constructed by considering the difference in Sn melting endotherms through the first, ΔH_{Sn}^{1}, and second, ΔH_{Sn}^{2}, DSC scans in comparison to the overall reaction of the baseline sample:

$$ \text{performance} = \frac{\%\Delta H_{Sn}^{1} - \%\Delta H_{Sn}^{2}}{100 - \Delta H_{Sn}^{BL2}} \qquad (2) $$

A value of unity represents the ideal case of Sacrificial Passivation. This may be achieved by obtaining 100 % of the theoretical Sn melting endotherm through the first scan and a value similar to the baseline measurement for the second, ΔH_{Sn}^{BL2}, suggesting a similar extent of the liquid phase reaction has occurred. A value that approaches zero represents non-ideal cases. These may arise by initially un-reactive particles remaining un-reactive through the second scan, or when solid-state consumption of Sn does occur.

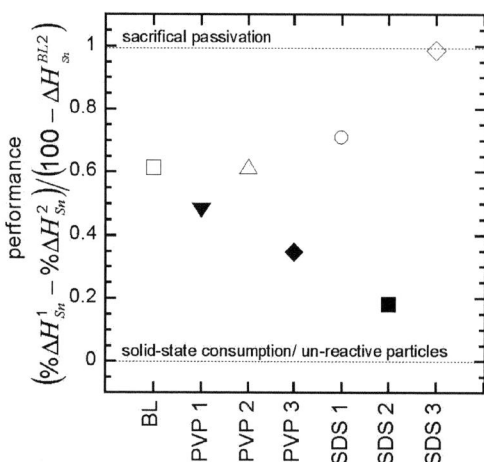

Figure 8. Overall Sacrificial Passivation performance of the particle compacts.

For the particle compacts investigated in the current study, the one comprised of the SDS3 capped Ag particles exhibits the ideal behavior for Sacrificial Passivation. The compacts comprised of PVP2 and SDS1 capped particles exhibits slightly better behavior than non-passivated baseline sample, while the compacts containing the PVP1, PVP3 and SDS2 capped particles exhibit the lowest ratings.

SUMMARY

The previous study on the synthesis and capping behavior of these particles has already established the link between synthesis conditions and trends in their caps' structure (T_d and M%) [9]. Comparing the current observations to the characteristics of the capped Ag nanoparticles now provides insight into the structure-property link for the Sacrificial Passivation. With the SDS capped particles, a thick cap (high M%) of low T_d promotes an effective Sacrificial Passivation layer (SDS3). The passivating effect is also exhibited for thinner caps (low M%) of high T_d, though to a lesser extent (SDS1). When the SDS cap is thin and exhibits a low T_d, however, its ability to passivate the particle decreases (SDS2). A similar interrelationship exits for the PVP capped particles. The best performance was observed for the PVP2 particles which have a thin cap of high T_d. When the thickness of the PVP cap increases for a similarly high T_d (PVP1 and PVP3) however, its ability to passivate the particle decreases. These observations suggest that either a "thick-weak" or "thin-strong" cap is best suited for Sacrificial Passivation.

REFERENCES

[1] D. R. Frear, S. N. Burchett, H. S. Morgan, J. H. Lau, The Mechanics of Solder Alloy Interconnects, Van Nostrand Reinhold, New York, 1994.

[2] Z. Mei, J. J. W. Morris, J. Elec. Mat. 21 (1992) 401-407.

[3] H. D. Solomon, J. Electronic Packaging 111 (1989) 75-82.

[4] S. Vaynman, M. E. Fine, D. A. Jeannotte, Metallurgical and Materials Transactions A (Physical Metallurgy and Materials Science) 19A (1988) 1051-1059.

[5] W. D. MacDonald, Eagar, T.W., The Metal Science of Joining (1992) 93-100.

[6] N. S. Bosco, F. W. Zok, Acta Materialia 52 (2004) 2965-2972.

[7] N. S. Bosco, B. A. Manhat, J. Janczak-Rusch, Scripta Materialia, Corrected Proof, In press (Available online 11 January 2008).

[8] N. S. Bosco, F. W. Zok, Acta Materialia 53 (2005) 2019-2027.

[9] N. S. Bosco, B. A. Manhat, J. Janczak-Rusch, submitted to Materials Chemistry and Physics (2008).

[10] JCPDS, file no. 04-0783, Joint Committee on Powder Diffraction Standards, Swarthmore, PA.

Nanoindentation of Lead Free Solders for Harsh Environments

Vitor Farinha Marques, Patrick Grant, and Colin Johnston
Department of Materials, University of Oxford, Oxford, OX5 1PF, United Kingdom

ABSTRACT

To better understand the factors governing the reliability of lead free solders during severe excursions in temperature, the hardness and elastic modulus of the micro-phases formed in a Sn-Ag-Cu/Cu solder joint were characterized using nanoindentation at temperatures from 25 to 175°C. The creep behaviour of the different micro-phases was also studied as function of temperature. The hardness and elastic modulus of Cu_6Sn_5, Cu_3Sn had a weak dependence on temperature, while primary Sn, eutectic regions and electroplated Cu hardness and modulus were sensitive to temperature. Creep studies indicated that intermetallic were more creep resistant than softer phases that readily underwent creep, the type and rate of which was shown to be strongly temperature dependent.

INTRODUCTION

Current environmental legislation restricts the use of lead containing solders in microelectronic packages for domestic use, and a comprehensive range of "drop-in" alternatives have been developed successfully [1]. The aerospace industry is currently exempt from this legislation on safety grounds. However, it is inevitable and widely accepted that the aerospace sector will become eventually subject to similar restrictions. At the current time, domestic lead free alternatives are not qualified for the aerospace sector and industrial studies suggest that the performance and failure behaviour of lead free solders in the harsher aerospace environment is different to reported domestic behaviour [1-2].

During service, microelectronic assemblies are subjected to coefficient of thermal mismatch induced strains because of the multiplicity of materials used, such as alumina or reinforced fibers PCBs, as well as the interconnect solder themselves. Resulting stresses are dependent upon the elastic modulus and the ability of the materials – primarily the solders – to relieve the thermally induced stresses by non-elastic processes including yield and/or creep [3].

Further, the microstructure, phases and their fraction, and joint geometry also tend to change with time and temperature. Consequently, development of successful lifing methodologies for interconnects is extremely challenging,

Room temperature nanoindentation has been explored for its potential to characterize the materials used in real microelectronic joints/interconnects, rather than studying the constituent material in bulk form [4-6]. However, there was been a relatively little work on the nanoindentation of evolving interconnect phases as function of temperature and time. The present study investigates the hardness, elastic modulus and creep behaviour of Cu_6Sn_5, Cu_3Sn, Ag_3Sn, Cu, primary Sn and eutectic regions in Sn-Ag-Cu/Cu joints as a function of temperature. Ultimately it is envisaged that this type of data will improve the understanding of lead free solder reliability under harsh environments, and be able to be used in improved models of interconnect strains, stresses and lifetimes.

EXPERIMENTAL

Small ball grid arrays (BGAs) of 3x3 600µm diameter 95.6Sn3.7Ag0.6Cu (Indium Corp.) balls were reflowed on Cu and Ag pads using a forced convection oven at maximum peak temperature of 245°C for 10s. One set of BGAs were then submitted to ageing at 225°C for 3.5 weeks to promote thick Cu_6Sn_5+Cu_3Sn and Ag_3Sn intermetallic compound (IMC) phases at the SAC/Cu and SAC/Ag interfaces respectively. Samples were polished in cross-section using standard polishing methods to a final finish of 0.04µm colloidal silica.

Nanondentation experiments were performed at a range of temperatures from 25 to 175°C using a Micromaterials Nanotest 600 machine equipped with a high temperature stage and a Berkovich indenter with 150nm radius [7]. Hardness and elastic modulus were analyzed using the Oliver and Phar method [8]. Mechanical characterization of primary Sn, eutectic regions and the electroplated Cu pad was performed on as-reflowed samples and IMCs in the aged samples. Similarly to Alex *et al*, a 5-10nm Pt to layer was evaporated onto the sample surface to avoid oxidation, and was assumed not to play a significant role in the subsequent indentation behaviour [9]. Indentation tests were performed at a maximum load of 2mN at a load/unload rate of 0.067 mN/s. The creep exponent n of a material is derived from the expression $\dot{\varepsilon} = K\sigma^n$ where $\dot{\varepsilon}$ is the strain rate, K is the material dependent constant and σ is the applied stress. The creep exponent can be obtained from a series of nanoindentation tests at a range of hold times at constant load [6]. The creep behaviour in this study is divided into two sets: (1) the creep behaviour of all phases at temperature T=25 °C using hold times of 30s at maximum load, and (2) the creep behaviour as function of temperature between 25 to 175°C for the softer Cu, primary Sn and eutectic regions, with hold times of 100s for Cu and 500s for primary Sn and eutectic regions.

RESULTS AND DISCUSSION

Figure 1 shows the microstructure of as-reflowed solder joints characterized by primary Sn dentrites, eutectic regions containing primarily Sn-rich regions and finely dispersed Ag_3Sn particles, and well-developed Cu_6Sn_5 intermetallics. After ageing, there was significant growth of $Cu_3Sn+Cu_6Sn_5$ and Ag_3Sn at the SAC/Cu and SAC/Ag interface respectively. As intended, it was possible to probe these IMCs using nanoindentation with a negligible influence of adjacent phases.

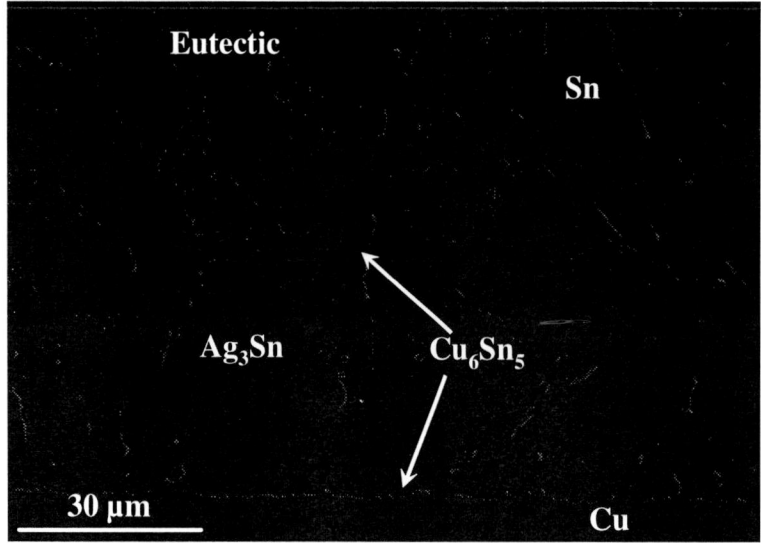

Figure 1. Scanning electron microscope image of the as-reflowed Sn-3.7Ag-0.6Cu/Cu interface region.

Figures 2(a) and (b) show the hardness and elastic modulus determined by nanoindentation for all the phases indentified in the SAC/Cu solder joints as function of temperature up to 175°C respectively. The eutectic regions were slightly harder than primary Sn because of the hardening effect of the contained low fraction and fine dispersion of Ag_3Sn. The Cu hardness of ~1.9 GPa was largely unaffected by temperature. As expected, the various IMCs were comparatively hard at room temperature and increasing temperature had little effect on hardness, with Ag_3Sn ~3 times softer than Cu_3Sn and Cu_6Sn_5.

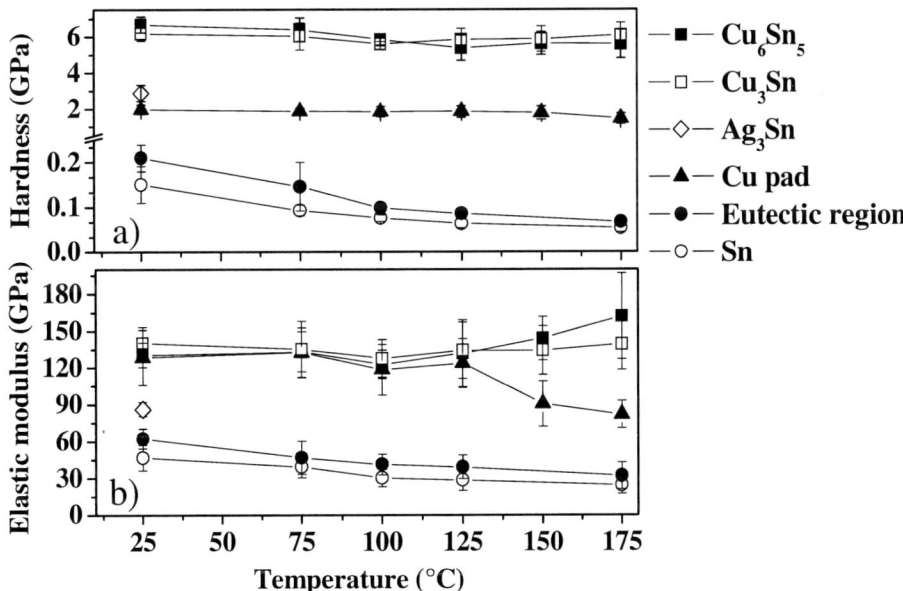

Figure 2. (a) Hardness and (b) elastic modulus as function of temperature the phases observed in a 95.6Sn3.7Ag0.6Cu/Cu couple.

The elastic modulus of the phases as a function of temperature showed the same general trend as hardness. However, at temperatures higher than 150°C, the elastic modulus of Cu_6Sn_5 increased from a near constant 130 to 160GPa. At room temperature, the stable form is Cu_5Sn_6-η' with an ordered long period superlattice (LPS) structure, while at temperatures above ~185°C, it has been suggested that Cu_5Sn_6-η with the NiAs-type structure is stable [10]. Calorimetry experiments have suggested that the η'→ η transformation occurred after 6 min at a lower temperature of 175°C or after longer periods of time for temperatures as low as 150°C: consequently it is surmised that kinetics of the η'→ η transformation are sluggish and the precise transition temperature ill-defined [10]. In the present work, nanoindentation experiments took ~20 hours at the target temperature. Similarly to [10], Figure 2 (b) suggested that the η'→ η transformation could occur at or close to 150°C during extended thermal excursions.

The elastic modulus of the electroplated Cu pad also showed two different regimes: up to 125°C there were no significant changes in elastic modulus or hardness, while at 175°C the modulus and hardness decreased to 80GPa and 1.2GPa respectively. Similar results have been reported for bulk and film Cu samples using conventional methods and nanoindentation, with the change in elastic modulus attributed to a combination of grain growth and annealing [9, 11].

Figures 3(a) and (b) show typical nanoindentation depth as function of time profiles for eutectic regions and Cu_6Sn_5 at a constant load of 2mN. The data in Figures 3(a) and (b) was fitted to the equation $h-h_0=A+ln(Bt+1)$ where A and B are fitting parameters, h_0 is the initial depth and h is the depth at time t. This procedure produced a smooth best-fit function from which the strain rate $(1/h.dh/dt)$ and stress $(F/(24.5.h^2))$ where F is the force were readily obtained. Figures 3(c) and (b) shows the corresponding log-log plots of strain rate versus stress where the gradient of a best-fit straight line to the data is the creep exponent n.

The eutectic region underwent significant plastic deformation over all the load dwell, while Cu_6Sn_5 showed an initial sharp increase in displacement followed by a plateaux of comparatively small displacement. Table 1 summarizes the measured creep exponents for all the phases at 25°C using hold times of 30s at constant load. The creep exponents for Cu, primary Sn and eutectic regions were in the range typical for soft metals and Pb–free solder alloys [12-15]; the IMC creep exponents were in agreement with previous works using nanoindentation, with the Ag_3Sn creep exponent significantly lower than that of Cu_3Sn and Cu_6Sn_5 [13,16]. Because the IMCs were shown to have very low creep rates, only the creep behaviour of Cu, primary Sn and eutectic regions as function of temperature are discussed further.

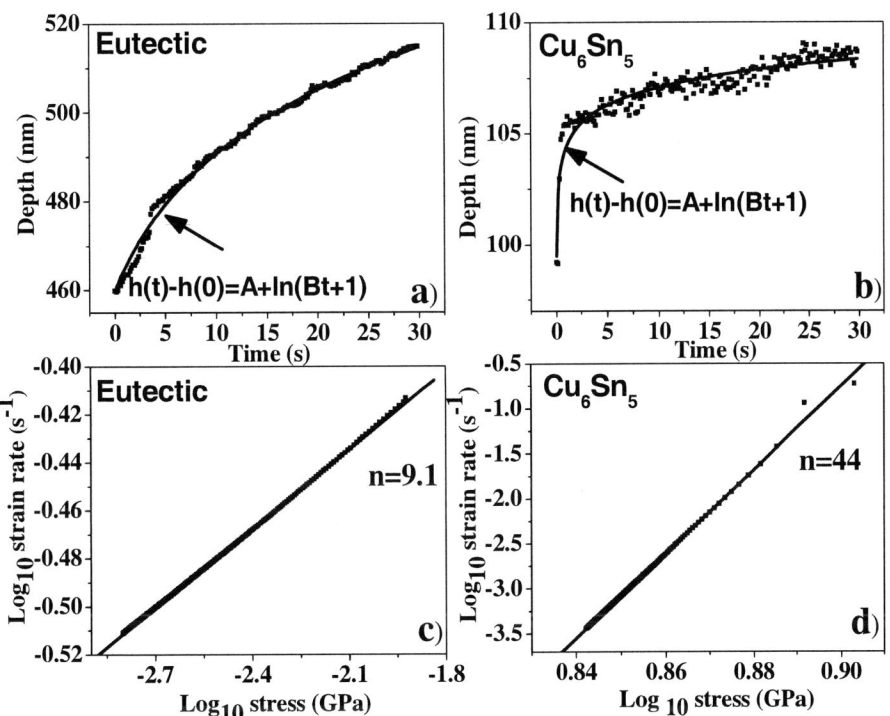

Figure 3. Examples of the (a) eutectic region and (b) Cu_6Sn_5 indentation depth as a function of time under a constant load of 2mN, and (c) and (d) the corresponding strain rates and stresses used to obtain the creep exponent n.

Figure 4 shows the creep exponent as function of temperature for primary Sn, eutectic regions and Cu. At $T=25°C$, the creep exponent of the eutectic region was slightly higher than that for primary Sn, similar to the trend in hardness and elastic modulus. As temperature increased, the creep behaviour of the eutectic region progressively approached that of the primary Sn, because the influence of the contained IMCs reduced as they coarsened and their volume fraction reduced. The creep exponent behaviour of primary Sn and eutectic region could be divided into two different regimes: relatively high values of n at lower temperatures where creep was controlled by lattice diffusion controlled climb; and relatively low values of n at higher temperatures where creep became dominated by dislocation core diffusion controlled climb [17]. Finally, the Cu creep exponent reduced from 12 to 9 as the temperature increased. However, given that this temperature change was a much smaller fractional change relative to the Cu melting point, it is postulated that the reduction in n is due to the previously annealing and grain growth (elimination of grain boundary area) effects rather than by changes in the creep mechanism.

Table 1. Measured creep exponents for the all micro-phases in Sn-3.7Ag-0.6Cu/Cu solder joint at T=25°C.

Phase	Cu_6Sn_5	Cu_3Sn	Ag_3Sn	Cu	Eutectic	Sn
Creep exponent	44.0±10.2	38.1±10.6	20.5±7.9	12.4±7.0	9.1 ±3.3	7.8±3.05

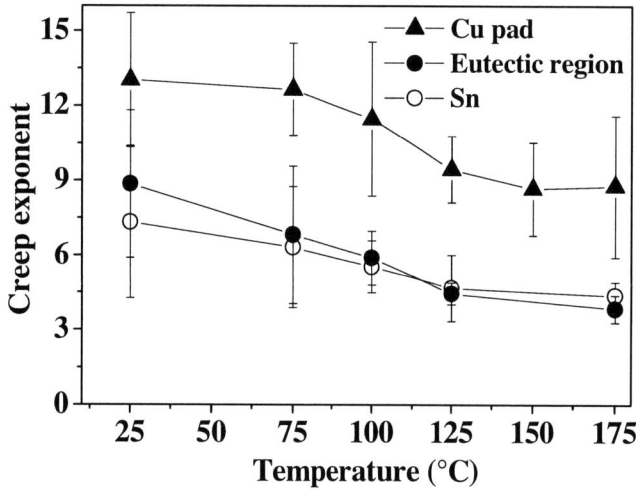

Figure 4. Creep exponents for Sn, eutectic and electroplated Cu pads as function of temperature.

CONCLUSIONS

Nanoindentation has been used to characterize some of the important mechanical aspects of the micro-phases that evolve in real solder joints. The variation in properties has also been obtained as a function of temperature, and all data was consistent with previous work using conventional bulk mechanical tests. The creep behaviour of the phases was also studied: IMCs showed very low creep rates at all temperatures and are therefore suggested to play a limited role in stress relaxation during thermal excursions (although they play a critical role in stress concentration at the interface). The softer and less creep resistant primary Sn, eutectic regions and the Cu pad readily underwent creep, the extent of which was shown to be strongly temperature dependent. This type of constitutive behaviour readily available by nanoindentation is a critical requirement for more realistic models currently under development for interconnect stresses, strain and lifetimes currently under development.

ACKNOWLEDGMENTS

The authors would like to thank the UK Engineering and Physical Sciences Research Council-UK for financial support.

REFERENCES

[1] R. Mahmudi, A.R. Geranmayeh and A. Rezaee-Bazzaz, Mater. Sci. Eng. A **448,** (2007) 287-293

[2]Andrew A. Shapiro, J. Kirk Bonner, Oladele A. Ogunseitan, Jean-Daniel M. Saphores, and Julie M. Schoenung, IEEE Transactions **29**, (2006) 60-70

[3] Tae-Sang Park, Soon-Bok Lee, J. Electron. Packag. **127**, (2005) 237-244

[4] R.R. Chromik, R.P. Vinci, S.L. Allen, M.R. Not, J. Mater. Res. **18**, (2003), 2551-2661

[5] X. Deng , N. Chawla , K.K. Chawla, M. Koopman, Acta Mater **52**, (2004) 4291-4303

[6] R. Goodall, T.W. Clyne, Acta Mater. **54,** (2006), 5489-5499

[7] B.D. Beake, J.F. Smith, Philos. Mag. A **82**, (2003) 2179-2186

[8] W.C. Oliver and G.M. Pharr, J. Mater. Res. **7** (1992).1564-1580

[9] Alex A. Neville R. Moody, William and W. Gerberich, J. Mater. Res. **19,** (2004) 2650-2657

[10] T. Laurila, V. Vuorinen, J.K. Kivilahti, Mater. Sci. Eng R **49**, (2005) 1-60

[11] O.D. Sherby: In Nature and Properties of Materials: An Atomistic Interpretation, edited by J. Pask (Wiley, New York, 1967) 376

[12] Shou-Yi Chang, Yu-Shuien Lee, Ting-Kui Chang, Mater Sci. Eng A **423** (2006) 52-56

[13] H. Rhee, J.P. Lucas, K.N. Subramanian, J. Mater. Sci. **13**, (2002) 477-484

[14] M.L. Huang, L. Wang, C.M.L. Wu, J. Mater. Res. **17**, (2002) 2897-2903

[15] I. Dutta, C. Park, S. Choi, Mater Sci. Eng A **379,** (2004) 401-410

[16] R.R. Chromik, D-N. Wang, A. Shugar, L. Limata, M.R. Notis, R.P. Vinci, J. Mater. Res. **20,** 8, (2005)2161-2172

[17] M. Kerr, N. Chawla, Acta Mater. **52,** (2004) 4527-4535

Mater. Res. Soc. Symp. Proc. Vol. 1079 © 2008 Materials Research Society

Effects of Pulse Duration and Polarity on the Electromigration Behavior of Copper Interconnects under Pulsed Current Stress

Meng Keong Lim[1,2], Chee Lip Gan[1], Yong Chiang Ee[2], Chee Mang Ng[2], Bei Chao Zhang[2], and Juan Boon Tan[2]

[1]School of Materials Science and Engineering, Nanyang Technological University, Block N4.1, 50 Nanyang Avenue, Singapore, 639798, Singapore
[2]Chartered Semiconductor Manufacturing Ltd, 60 Woodlands Industrial Park D, Street 2, Singapore, 738406, Singapore

ABSTRACT

Direct current (D.C.) is usually employed to characterize the electromigration reliability of interconnects. However, D.C. characterization techniques might not reflect the actual reliability of interconnects that are carrying pulsed D.C. or A.C. (alternating current) signals during operation. This study investigates the effects of unipolar and bipolar pulsed currents on the electromigration lifetime of copper (Cu) interconnects. A series of long period pulsed current (i.e. 2, 16, 32 and 48 hours) were applied to Cu interconnects. Lifetime enhancement is observed when the half-period of pulsed current is shorter than the median-time-to-failure (t_{50}) of D.C. stressed samples. Relatively minor increase in resistance occurring in-between pulses for unipolar pulsed current stressed samples, and occurrence of damage healing in bipolar pulsed current stressed samples are reasons attributed for the observed enhanced lifetime. We obtained longer t_{50} when the period of the pulsed current is shorter.

INTRODUCTION

Electromigration, a persistent reliability issue that affects both aluminum (Al) interconnects in the past and advanced copper (Cu) interconnects in present integrated circuits, is usually characterized by applying direct current (D.C.) when conducting reliability studies [1-3]. However, most interconnects carry time varying electrical signals during operation conditions. A mismatch between the actual operation conditions versus the accelerated testing conditions applied in reliability studies is therefore present, and the failure behaviors extrapolated from these accelerated testing conditions might be different from those of operation conditions.

Limited electromigration reliability studies that were conducted under pulsed D.C. or A.C. (alternating current) conditions observed that Al interconnects exhibit longer electromigration lifetime as compared to D.C. conditions [4,5]. Although Cu interconnects are expected to exhibit enhanced lifetime under pulsed D.C. or A.C. conditions too, however, substantial work in this area is either lacking or limited at this moment. If the electromigration behavior of Cu interconnects carrying time varying current is well characterized, then the information acquired may be applied to relax the progressively stringent current density design rules and interconnect technology challenges beyond the 22 nm technology node [6].

This paper intents to look at the effects of unipolar and bipolar pulsed currents on the electromigration lifetime and behavior of Cu/low-κ interconnects through their resistance evolution. Very long period pulsed currents (i.e. 2, 16, 32 and 48 hours) were applied in this preliminary study on Cu electromigration under pulsed D.C. stressing, which is aimed to serve as a link for transiting from the well characterized D.C. conditions to low and/or high frequency pulsed D.C. conditions, where substantial work is lacking at this moment.

EXPERIMENTAL DETAILS

Samples were fabricated using a 65 nm CMOS process with single damascene process at metal-1 (M1) and dual damascene process at metal-2 (M2) and metal-3 (M3). The Cu test line, which is 200 μm long and 0.3 μm wide, is in the M2 level and is connected to M1 at one end and M3 at the other. Figure 1 shows the schematic layout of the test structure.

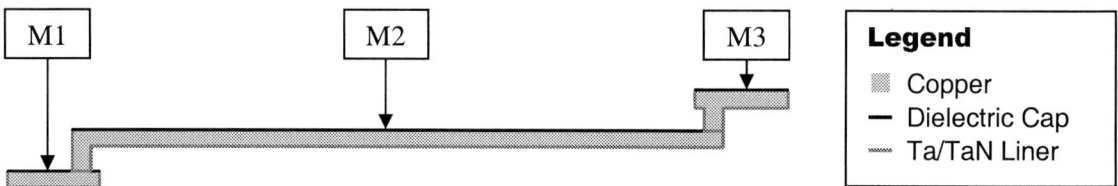

Figure 1. Schematic diagram of Cu interconnect test structure for electromigration study.

Electromigration stressing was carried out on ceramic packaged samples using an Xpeqt electromigration test system. This test system measures the lifetime of the samples by means of resistance monitoring using Kelvin connection. A range of very long period unipolar pulsed current and bipolar pulsed current (i.e. 2, 16, 32 and 48 hours), which have a duty cycle of 50% and a maximum current density of 2.0 MA/cm^2, were applied to the Cu lines during this study. Unipolar pulsed current alternated between 2.0 MA/cm^2 and 0.057 MA/cm^2 periodically, while bipolar pulsed current alternated between 2.0 MA/cm^2 and -2.0 MA/cm^2 periodically. Electron current always flows upstream (i.e. M1 to M2 to M3) for unipolar pulsed current test, while bipolar pulsed current test always begins with electron current flowing upstream. As the lowest amount of current the test system can supply is non-zero, the unipolar pulsed current applied in this study has the lowest possible current density of 0.057 MA/cm^2 during the lower binary state. Due to this reason, continuous resistance monitoring was enabled for unipolar pulsed current test. Lifetime of samples that were subjected to a 2.0 MA/cm^2 D.C. stress was also measured and treated as a baseline. The temperature applied throughout this study is 350°C. Failure is defined as a 10% increase in resistance.

RESULTS & DISCUSSION

(A) Unipolar Pulsed Current Test

Figure 2 shows the typical resistance evolution when a Cu interconnect was subjected to unipolar pulsed current with 32 hours period. Only the first 200 hours of resistance development is shown as similar trend is observed for extended stress duration. Typical resistance developments consisting of near constant resistance then followed by a sharp increase in resistance are observed during the initial stage of the test [7]. The near constant resistance signifies an incubation period which consists of void nucleation and then followed by void growth. While the sharp increase in resistance indicates the instance when the void has grown beyond the edge of a via. Beyond the point where resistance raised sharply, gradual increase in resistance and near constant resistance are observed when the stress current density applied are 2.0 MA/cm^2 and 0.057 MA/cm^2, respectively. These observations are characteristics of the *on-time* model [8], where electromigration damage is incurred and accumulated when current flows while neither damage nor healing takes place during the time between pulses. The gradual increase in resistance when 2.0 MA/cm^2 of current is flowing can be attributed to a growing full-spanning void which results in current having to flow through a longer length of Ta/TaN liner, where the resistivity of the liner is higher than that of Cu [9]. We ascribed the absence of electromigration damage healing when 0.057 MA/cm^2 of current is flowing to weak back-stress [10]. Instead of observing constant resistance during the time between pulses as described by the *on-time* model, relatively minor increase in resistance is observed since a low current density of 0.057 MA/cm^2 was applied. This implies that electromigration damage was incurred even at a low current density and line length product (jL) of 1140 A/cm. This was not unexpected since Cu/low-κ samples were reported to fail even at an extremely low jL value of 375 A/cm [11].

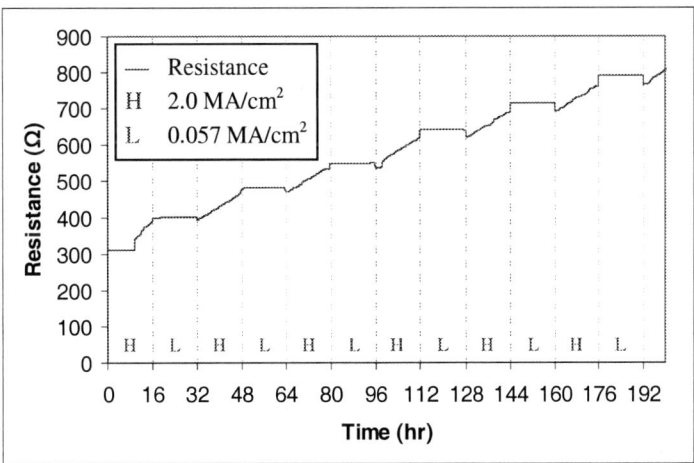

Figure 2. Resistance evolution exhibited by a Cu interconnect that was subjected to a unipolar pulsed current of 32 hours period.

Figure 3 shows the cumulative failure probability of Cu interconnects subjected to unipolar pulsed current of different periods in a log-normal plot. Longer median-time-to-failure (t_{50}) is observed when unipolar pulsed current of shorter period is applied to Cu interconnects. Based on resistance evolution that resembles the *on-time* model, we infer that unipolar pulsed current with shorter period is associated with longer time to accumulate electromigration damage that is sufficient to cause a 10% increase in resistance. Exemplifying this point, Cu interconnects that are subjected to unipolar pulsed current of 2 hours, 16 hours, 32 hours and 48 hours period for a test duration of 50 hours experienced 2.0 MA/cm^2 of current stressing for a cumulative duration of 25 hours, 26 hours, 34 hours and 48 hours, respectively. Comparable t_{50} are obtained for samples that were subjected to D.C. and unipolar pulsed current with 48 hours period, while samples that were subjected to shorter periods showed longer t_{50}. This suggests the existence of a certain threshold period for unipolar pulsed current, where lifetime enhancement may only be attained when the period is shorter than the threshold period. This threshold period is assumed to be twice the t_{50} of D.C. stressed samples (where $t_{50,D.C.}$ is 15.8 hour), since more that half the population of samples would have failed if the pulse duration (i.e. half-period) is longer than the t_{50} of samples subjected to D.C. condition.

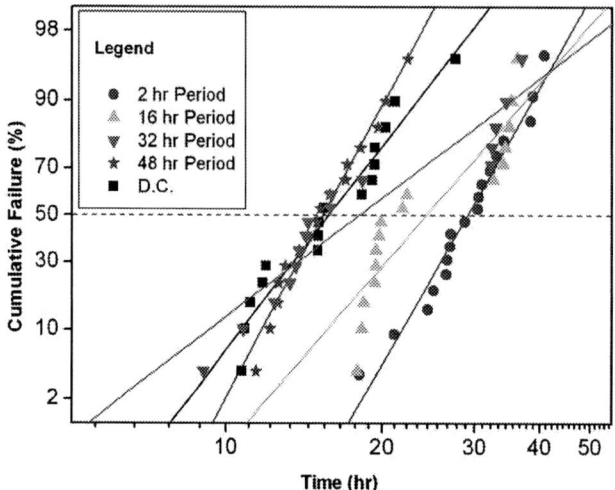

Figure 3. Cumulative failure probability versus log(time-to-failure) of Cu interconnects that were subjected to different periods of unipolar pulsed current.

(B) Bipolar Pulsed Current Test

The typical resistance evolution when a Cu interconnect was subjected to bipolar pulsed current with 16 hours period is shown in Figure 4. Near constant resistance was observed for the initial 8 hours when electron current is flowing upstream (i.e. M1 to M2 to M3), while a sharp increase in resistance was observed near the end of the first period (0th – 16th hour) when electron current is flowing downstream (i.e. M3 to M2 to M1) during the latter 8 hours. Near constant resistance either indicates a nucleating void or a growing void that has not grown beyond the edge of a via [7]. A sharp increase in resistance, on the other hand, signifies that the void has grown beyond the edge of via. The void is expected to be located below via 2 (V2), which is connecting M2 and M3, since the end of M2 below V2 is the cathode during the 8th – 16th hour. A sharp decrease in resistance was observed shortly after the commencement of the second period (16th – 32nd hour). Thereafter, the second period displayed similar resistance evolution as seen during the first period. The decrease in resistance to a value that is similar to the initial resistance at the start of the test indicates that the void below V2 has shrunk, and the metal/void interface is within the edge of V2. This suggests that electromigration damage can be healed when current changes direction. The extent of healing cannot be easily inferred by monitoring the development of resistance, although complete healing is not expected. Progressing from the second period (16th – 32nd hour) to the third period (32nd – 48th hour) and then to the fourth period (48th – 64th hour), the resistance evolution in Figure 4 shows that longer time is required to heal and shrink the void below V2 to a size that is within the edge of V2 from the instance when current flow changes from downstream to upstream; whereas a shorter time is required for the void to grow beyond the edge of V2 from the instance when current flow changes from upstream to downstream. This observation implies that the void never heals completely and that the remaining void is getting larger at the end of every successive period. Beyond the fourth period, the absence of a sharp decrease in resistance indicates that the void never shrinks to a size that is within the edge of V2.

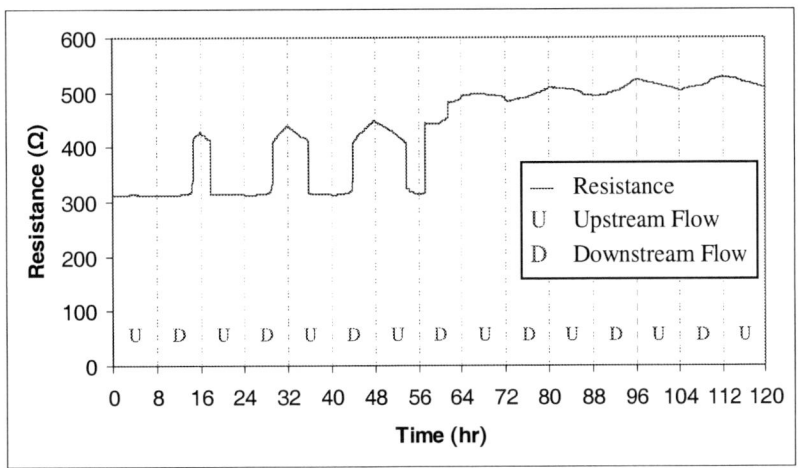

Figure 4. Resistance evolution of a Cu interconnect that was subjected to a bipolar pulsed current of 16 hours period.

The log-normal cumulative failure probability plots of samples that were subjected to different periods of bipolar pulsed current are shown in Figure 5. Longer t_{50} is observed when samples were subjected to bipolar pulsed current, provided that the half-period is shorter than the t_{50} of D.C. stressed samples. Near similar lifetime was obtained for samples that were subjected to D.C. and bipolar pulsed current, whose half-period is longer than the t_{50} of D.C. stressed samples. These two observations suggest that insufficient electromigration damage was induced in the interconnect to cause failure during the first half-period when the half-period is shorter than the t_{50} of D.C. stressed samples.

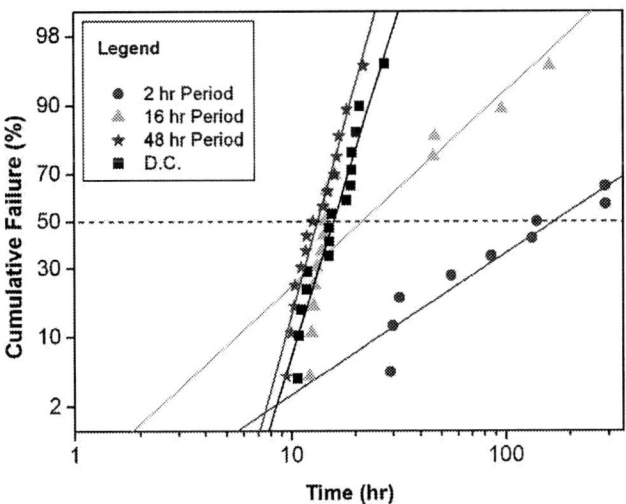

Figure 5. Cumulative failure probability versus log(time-to-failure) of Cu interconnects that were subjected to different periods of bipolar pulsed current.

The log-normal cumulative failure probability plot also shows that samples that were subjected to shorter pulse period exhibit longer t_{50}, which was observed in Al interconnects too [5]. We postulate that enhanced lifetime occurs due to the following two factors, namely, shorter half-period and therefore shorter duration to incur and accumulate the required magnitude of electromigration damage needed for failure to occur, and increased frequency of damage healing. The results obtained from this test shows that complete immortality was not achieved even when the bipolar pulsed current was symmetrical (i.e. 50% duty cycle) and the half-period was shorter than the shortest time-to-failure of D.C. stressed samples.

CONCLUSIONS

In this paper, we reported that Cu interconnects show enhanced electromigration lifetime when subjected to pulsed current, as compared to D.C. stress condition. Cu interconnects that were subjected to unipolar pulsed current exhibited enhanced lifetime because the increase in resistance during the time between pulses is relatively minor. Damage healing occurs in samples that were subjected to bipolar pulsed current, therefore longer stress duration is needed to incur and accumulate sufficient damage to cause failure. Samples that were subjected to shorter period of unipolar pulsed current and bipolar pulsed current showed longer t_{50}. However, enhanced lifetime will not be observed if the period of the pulsed current is above a threshold period of 31.6 hours, which is twice the t_{50} of D.C. stressed sample.

ACKNOWLEDGMENTS

The authors would like to thank Ms. Tan Tam Lyn for experimental assistance. One of the author, M. K. Lim, is receiving a scholarship from the Joint Industry Postgraduate Program, which is sponsored jointly by Chartered Semiconductor Manufacturing Limited and Singapore Economic Development Board.

REFERENCES

1. C.-K. Hu, L. Gignac and R. Rosenberg, *Microelectronics Reliability* **46**, pp. 213-231 (2006).
2. D. G. Pierce and P. G. Brusius, *Microelectron. Reliab.* **37**, pp. 1053-1072 (1997).
3. J. R. Black, *Proc. of 6th Ann. Reliability Physics Symp.*, pp. 148-159 (1967).
4. R. E. Hummel and H. H. Hoang, *J. Appl. Phys.* **65**, pp. 1929-1931 (1989).
5. J. Tao, J. F. Chen, N. W. Cheung and C. Hu, *Proc. of 34th Ann. Reliability Physics Symp.*, pp. 180-187 (1996).
6. International Technology Roadmap for Semiconductors (2007); see website http://www.itrs.net/
7. P.-C. Wang and R. G. Filippi, *Appl. Phys. Lett.* **78**, pp. 3598-3600 (2001).
8. D. W. Malone and R. E. Hummel, *J. Appl. Phys.* **83**, pp. 5750-5760 (1998).
9. C.-K. Hu, L. Gignac, S. G. Malhotra, R. Rosenberg and S. Boettcher, *Appl. Phys. Lett.* **78**, pp. 904-906 (2001).
10. I. A. Blech and C. Herring, *Appl. Phys. Lett.* **29**, pp. 131-133 (1976).
11. C. S. Hau-Riege, S. P. Hau-Riege and A. P. Marathe, *J. Appl. Phys.* **96**, pp. 5792-5796 (2004).

Mater. Res. Soc. Symp. Proc. Vol. 1079 © 2008 Materials Research Society

Pd Segregation at (001) B2-NiSi/Si Epitaxial Interface Studied by Density Functional Theory

Dae-Hee Kim[1], Hwa-Il Seo[2], and Yeong-Cheol Kim[1]

[1]Dept. Materials Engineering, Korea University of Technology and Education, 307 Gajeonri Byungchunnmyun, Cheonan, 330-708, Korea, Republic of

[2]School of Information Technology, Korea University of Technology and Education, 307 Gajeonri Byungchunnmyun, Cheonan, 330-708, Korea, Republic of

ABSTRACT

Pd segregation at (001) *B2-NiSi/Si* epitaxial interface was studied by using density functional theory (DFT). An epitaxial interface between 2x2x4 (001) *B2-NiSi* supercell and 1x1x2 (001) *Si* supercell was first constructed by adjusting the lattice parameters of *B2-NiSi* structure to be matched with those of *Si* structure. We chose *Ni* atoms as terminating layer of the *B2-NiSi*, and an equilibrium gap between the *B2-NiSi* and Si was calculated to be 1.1 Å. The *Ni* atoms in the structure moved away from the original positions along *z*-direction in a systematic way during the energy minimization. Two different Ni sites were identified at the interface and the bulk, respectively. The *Ni* sites at the interface farther away from the interface were more favorable for *Pd* substitution.

INTRODUCTION

Nickel monosilicide (*NiSi*) is an important material for contact applications in the semiconductor industry as it transitions from the 65 technology node to 45 nm node.[1] *NiSi* has many advantages over conventionally-used titanium silicide (*TiSi₂*) and cobalt silicide (*CoSi₂*) which include a lower temperature of formation, lower resistivity in narrow dimensions, reduced *Si* consumption, and formation kinetics that are controlled by diffusion of *Ni*, which leads to significantly smoother heterophase interface. Additionally, a one step annealing can be used for the self-aligned silicide process.[2]

NiSi, however, is less stable at high temperature than *TiSi₂* and *CoSi₂*. It agglomerates or transforms into more resistive *NiSi₂* phase at high temperature, causing overall resistivity increase. Recent studies demonstrate that adding elements such as *Pd*, *Pt*, or *Rh* reduces the agglomeration of *NiSi* and increases the formation temperature of *NiSi₂*. Laser-assisted local-electrode atom-probe (LEAP) tomography was employed to map the atomic-scale distribution of *Pd* in annealed $Ni_{1-x}Pd_x/Si$(001) structure.[3] The *Pd* was distributed nearly uniformly in the *NiSi* film and was segregated at the *NiSi/Si* (001) interface.

We employed a first principles calculation to construct a *NiSi/Si* (001) interface and to study the effect of *Pd* substitution into *Ni* sites that are located near interface and away from the interface. B2-NiSi structure was chosen for *NiSi/Si* structure, as the stable orthorhombic structure of *NiSi* as a bulk phase has higher interface energy with *Si* (001) surface.

CALCULATION

Density functional theory (DFT) calculations were performed using the Vienna *ab-initio* Simulation Package (VASP) code with the projector augmented wave (PAW) potentials and the local density approximation (LDA).[4-8] The residual minimization scheme direct inversion in the iterative subspace (RMM-DIIS) was used for calculating the ground state of electrons.[9,10] A cutoff energy was 500 eV for the plane wave expansion of the wave functions and Monk-horst pack k-point mesh of 4x4x4 produced well converged results.

Unitcells of orthorhombic and *B2* structures in *NiSi* were determined starting from literature values of lattice parameters. Lattice parameters along *x*-direction and *y*-direction in *B2-NiSi* were changed to match with the lattice parameters of *Si* for epitaxial contact. An optimum gap between *B2-NiSi* and *Si* (001) superstructure was calculated to be 1.1 Å, and a more detailed explanation will be shown elsewhere.[11] *Pd* substitution for *Ni* sites was calculated to understand the *Pd* segregation at the interface.

DISCUSSION

Fig. 1 shows unitcell structures of (a) *orthorhombic-NiSi* and (b) *B2-NiSi*. The *orthorhombic-NiSi* structure is pseudohexagonal and can be thought of as a distorted *NiSi* structure.[12] Both *Ni* and *Si* occupy position 4c (site symmetry *m*) with atomic positional parameters $(x, 0.25, z)$ and $(x, 0.75, z)$, respectively. The *B2-NiSi* is simple cubic with *Si* in the body center and *Ni* in the corner of the cubic.

Table 1 shows lattice parameters and energies of both *orthorhombic* and *B2-NiSi* from literature and calculation. The calculated lattice parameters are a little smaller than the experimentally determined ones. As the calculation is conducted at absolute zero temperature, the difference is reasonable. The calculated energy of *B2-NiSi* is a little high compared with that of *orthorhombic-NiSi*, as the latter phase is stable in bulk.

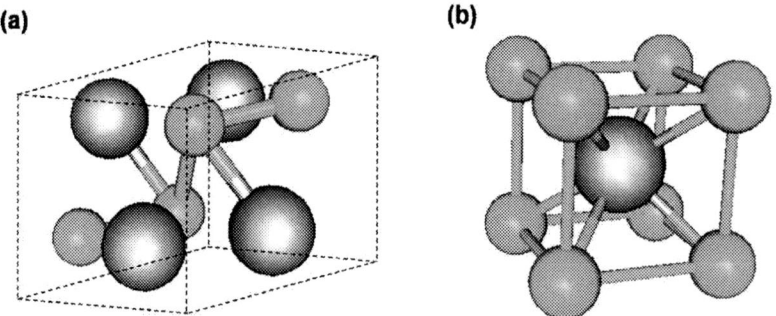

Fig.1. Unitcell structures of (a) *orthorhombic-NiSi* and (b) *B2-NiSi*. Dark (gray in color) spheres are *Si* atoms and light (green in color) spheres are *Ni* atoms.

Table 1. Lattice parameters and energies of *NiSi* from literature and calculation

Material	NiSi			
Structure	B2	Orthorhombic		
	a	a	b	c
Literature (Å)[12,13]	2.85	5.18	3.33	5.61
Calculation (Å)	2.82	5.10	3.31	5.48
Difference (%)	-0.92	-1.53	-0.75	-2.34
Energy (eV)/Molecule	-13.22	-13.75		

B2-NiSi structure is chosen to construct the *NiSi/Si* interface, as the *B2-NiSi* shows lower interface energy than the *orthorhombic-NiSi*. Fig. 2 (a) shows *B2-NiSi/Si* (001) structure constructed by *B2-NiSi* (001) 2x2x4 supercell and *Si* (001) 1x1x2 supercell. Four *Ni* atoms are shown on the plan-view of the interface shown in separate box. Fig. 2 (b) shows displacement of the *Ni* atoms along z-directions. There are four *Ni* atoms on (001) lattice plane in 2x2x4 *B2-NiSi* supercell, and two *Ni* atoms in diagonal position are equivalent in respect to the z-direction displacement. The *Ni* atoms in the structure move away from the original positions along z-direction in a systematic way during the energy minimization. Ni_{ij} indicates j-th *Ni* atom on the (001) lattice plane located in i-th layer. The 1st *Ni* atomic plane is equivalent to the 5th one, as the two layers form interface with Si.

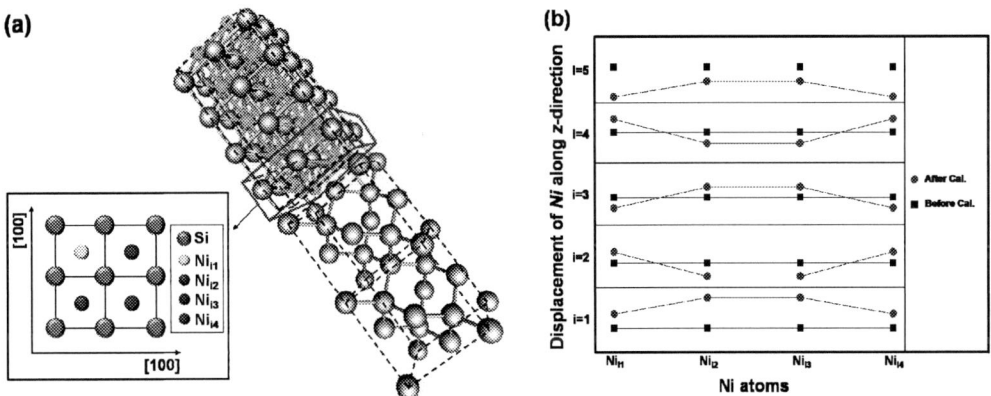

Fig. 2. (a) *B2-NiSi/Si* (001) structure constructed by combining *B2-NiSi* (001) 2x2x4 supercell and *Si* (001) 1x1x2 supercell. Inset shows the four *Ni* sites at the interface. (b) displacement of *Ni* atoms along z-directions. There are four *Ni* atoms on (001) lattice plane in 2x2x4 *B2-NiSi* supercell, and two *Ni* atoms in diagonal position are equivalent in respect to the z-direction displacement. Ni_{ij} indicates j-th *Ni* atom on the (001) lattice plane located in i-th layer.

Fig. 3 shows energy variation of the system when *Pd* substitutes *Ni* atom. The energy variation, $\Delta E_{Pd \to Ni}$, is calculated by the following equation.

$$\Delta E_{Pd \to Ni} = (E_{Ni_{1-x}Pd_xSi} + n_{Ni}\mu_{Ni}) - (E_{NiSi} + n_{Pd}\mu_{Pd})$$

,where $E_{Ni_{1-x}Pd_xSi}$ is the energy of $Ni_{1-x}Pd_xSi$, n_{Ni} the number of *Ni* atoms, μ_{Ni} the electrochemical energy of *Ni*, E_{NiSi} the energy of *NiSi*, n_{Pd} the number of *Pd* atoms, and μ_{Pd} the electrochemical energy of *Pd*, respectively. The *Pd* substitution at the 1st layer (i.e. near interface) into two different *Ni* sites shows a big energy difference. The Ni_{11} site is even less favorable than the sites in the bulk. The *Ni* site at the interface farther away from the interface (named Ni_{12} and indicated by a red circle in Fig. 3) is the most favorable for *Pd* substitution among the six *Ni* sites. We, therefore, understand that *Pd* segregates at the interface by substituting the energetically favorable *Ni* site between the two interface *Ni* sites.

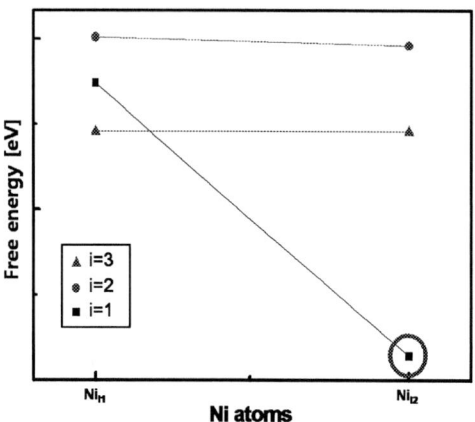

Fig. 3. Energy variation of the system with *Pd* substitution into *Ni* sites.

CONCLUSIONS

(001) *B2-NiSi/Si* epitaxial interface structure is constructed using DFT calculation. There are four *Ni* atoms on (001) lattice plane in 2x2x4 *B2-NiSi* superstructure, and two *Ni* atoms in diagonal position are equivalent in respect to the *z*-direction displacement. The *Ni* atoms in the structure move away from the original positions along *z*-direction in a systematic way during the energy minimization. The *Ni* site at the interface farther away from the interface is the most favorable for *Pd* substitution among the six *Ni* sites. We, therefore, understand that *Pd* segregates at the interface by substituting the energetically favorable *Ni* site between the two interface *Ni* sites.

REFERENCES

1. Front End Processes, International Technology Roadmap for Semiconductors, 2005.
2. T. Morimoto *et. al.*, IEEE Trans. Electron Devices, V.42, p.915 (1995).
3. Y. -C. Kim *et. al.*, Appl. Phys. Lett., V.91, p.113106 (2007).
4. G. Kresse *et. al.*, Phys. Rev. B, V.47, p.558 (1993); ibid. V.49, p.14251 (1994).
5. G. Kresse *et. al.*, Comput. Mat. Sci. V.6, p.15 (1996).
6. G. Kresse *et. al.*, Phys. Rev. B, V.54, p.11169 (1996).
7. G. Kresse *et. al.*, Phys. Rev. V.59, p.1758 (1999).
8. D. Vanderbilt, Phys. Rev. B, V.41, p.R7892 (1990).
9. D. M. Wood *et. al.*, J. Phys. A, V.18, p.1343 (1985).
10. P. Pulay, Chem. Phys. Lett., V.73, p.393 (1980).
11. D. –H. Kim *et. al.*, in preparation (2008).
12. M. Kh. Rabadanov *et. al.*, Inorg. Mat., V.38(2), p.120 (2002).
13. G. Profeta *et. al.*, Phys. Rev. B., V.70, p.235338 (2004).

Mater. Res. Soc. Symp. Proc. Vol. 1079 © 2008 Materials Research Society 1079-N09-02

Electromigration of Cu Interconnect Lines Prepared by a Plasma-based Etch Process

Guojun Liu, and Yue Kuo

Texas A&M University, Thin Film Nano & Microelectronics Research Laboratory, 235 J. E. Brown Engineering Bldg., MS 3122, College Station, TX, 77843-3122

ABSTRACT

The electromigration performance of Cu lines patterned by a Cl_2 plasma-based etch process has been studied with the accelerated isothermal lifetime test. An electromigration activation energy of 0.6 eV and a current density acceleration exponent of 2.7 were obtained. Both the copper-silicon nitride cap layer interface and the copper grain boundary were active diffusion paths. The applied mechanical bending stress changed the electromigration void distribution in the film, which leaded to the shorter lifetime and lower activation energy.

INTRODUCTION

Electromigration (EM) is one of the most critical reliability issues for modern ultra large scale integrated (ULSI) circuits due to the aggressive decrease of interconnect dimension and the subsequent very large current density [1]. Copper (Cu) interconnects are typically patterned by a damascene process, using the chemical mechanical polishing method to remove the Cu film outside of the dielectric trench. Recently, Kuo and Lee reported a new plasma-based Cu etch process [2, 3]. Instead of vaporizing the plasma/Cu reaction product formed in Cl- or Br-based plasma exposure, a dilute hydrogen chloride solution was used to dissolve the reaction product of $CuCl_x$ or $CuBr_x$. EM studies on the damascene Cu line showed that it had a lifetime one or two orders of magnitude longer than that of the aluminum (Al) line.[1] However, the EM performance of the Cu lines patterned by the plasma-based etch process has not been explored.

The EM induced failure occurs due to the flux divergence. Microscopic divergence can happen at a triple junction point of the grain-boundary network or any other location where the inhomogenous mass transport occurs [4]. Macroscopic divergence may occur at the bottom of via for the two-level testing structure where the via or the barrier layer blocks the EM flux, or the interface between Cu and the cap layer where the interfacial diffusion is the primary transport mechanism [5]. It is necessary to identify the diffusion path and to determine the activation energy (E_a) of the EM failure process before the Cu line's reliability can be understood. Previously, a wide range of E_a values were reported due to different diffusion mechanisms. For example, the lattice diffusion corresponded to an E_a value of 2.1 eV and the grain boundary diffusion showed an E_a value of 1.2 eV [6]. For Cu metallization, the surface or interface diffusion is often the primary failure mechanism. Proost reported the E_a value of 1.06 eV for the Cu-metal interface diffusion [7]. An E_a value between 0.6 and 0.88 eV was reported on the Cu-dielectric interface diffusion mechanism [8].

The EM performance depends not only on the film structure, but also the patterning process. Fillipi et al. compared the EM tests of the AlCu lines fabricated by damascene process and the reactive ion etching (RIE) process. The damascene lines showed a longer lifetime than the RIE lines. Although both processes showed a similar value of E_a, the current density acceleration exponent was close to 2 for the damascene structure and close to 1 for the RIE'd structure [9]. Most Cu lines reported in literature were fabricated by the damascene process. In this paper, authors reported the EM performance of the Cu lines patterned by the Cl_2 plasma-based etch process.

EXPERIMENTAL

Single level test lines were prepared on the multilayer Cu (800 nm)/Ta (10 nm)/SiN_x/Si film, using a Cl_2 plasma-based etch process. The Ta layer was used as a diffusion barrier for the Cu film. The EM line was 800 um long and 7 um wide resembling NIST straight line with 4 Kelvin contact pads. The Cu film was patterned the photoresist, followed by a Cl_2-based plasma etch process. The detailed Cu film etch process was described in [2]. A silicon nitride layer was deposited by a plasma enhanced chemical vapor deposition (PECVD) process at 300°C to cover the patterned Cu lines and prevent Cu oxidation during the EM tests. The silicon nitride layer at the probe contact pad area was etched off with a dilute buffered oxide etch solution.

The EM test system was built around a probe station (Signatone s-1160) with a hot chuck with the temperature control capability. The isothermal EM test was conducted with a feedback control loop that adjusted the current to the Cu line so that the effective temperature of the test line was kept within a narrow error band of the test temperature, T_{test}. The algorithm followed the principles of JESD61. The temperature coefficient of resistance (TCR), obtained prior to the EM test, was used to estimate the temperature of the test line. The current was applied from a programmable power source and the voltage drop along the test line was measured by a digital multimeter. The EM test stopped when the resistance increased more than 5%. The Cu line after EM was pictured with scanning electron microscopy (SEM). Planar views were obtained after the passivation SiN_x cap layer was removed by a dilute HF solution. The SEM observations provided qualitative information on the microstructure of the tested lines.

RESULTS AND DISCUSSIONS

Statistical EM Performance

The activation energy (E_a) and the current density acceleration factor (n) are usually expressed by the Black equation.

$$MTTF = \frac{A}{J^n} \exp(\frac{E_a}{kT}) \qquad (1)$$

where MTTF, J, and k are median time to failure, current density and Boltzmann's constant, respectively. A is a normalization time depending on the geometry and microstructure of the test line. From equation (1), we obtain

$$\ln(MTTF) = \ln(A) - n\ln(J) + E_a / kT \qquad (2)$$

The value of n can be obtained by the linear regression of ln(MTTF) vs. ln(J) at the same test temperature. The value of E_a can be obtained by the linear regression of [ln(MTTF)+nln(J)] vs. 1/kT once the value of n is known.

Since the temperature and the current density were coupled in the isothermal EM test, the EM tests could be conducted at the same Cu temperature of 350°C but different chuck temperatures, e.g., 20°C, 46°C, and 74°C, respectively. In such cases, the current density varied for the same T_{test}. The n value of 2.7 was obtained. It is generally considered that the n value should lie between 1 and 2. For void-growth limited failure, the n value was reported to be close to 1 while for nucleation limited failure it should be close to 2 [1]. For an n value greater than 2, it may exist a thermal gradient during the EM test. For the Cu lines in this test, the SEM observations showed that the voids were randomly distributed along the length. However, during void growth, the temperature increases locally in the vicinity of the void. From the Black equation, the underestimated temperature leads to an overestimation of the n value. Similarly, the n value increased with the current density due to the increase of Joule heating [10]. Therefore, the isothermal EM here resulted in a larger n value than the conventional EM test.

Figure 1 shows the cumulative failure distribution of Cu lines at the test temperatures from 320°C to 380°C. The times to failure (TTF) follow a log-normal distribution with the standard deviation σ between 0.19 and 0.30. Although these σ values are larger than those obtained from two-level test structure [11], they are comparable to the value obtained from the single level damascene test lines [12]. For the two-level structures, failure occurs in or adjacent to vias which are the weakest regions. For the single level structure, failure is more sensitive to Cu film microstructure and interface. The value of σ becomes large when more than one failure mechanisms exist in the test structure.

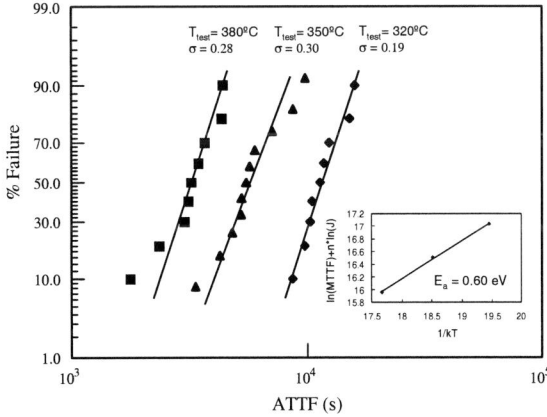

Figure 1. Cumulative failure distribution of Cu lines. The inlet is the linear regression of [ln(MTTF) + n*ln(J)] vs. 1/kT, using n =2.7.

The inset of Fig. 1 shows the linear regression of [ln(MTTF) + n ln(J)] vs. 1/kT using $n = 2.7$. The E_a value of 0.6 eV was obtained. There are four possible diffusion paths in the Cu structure: lattice, grain boundary, Cu/Ta interface, and Cu/SiN$_x$ interface. As discussed in the Introduction section, the lattice and the grain boundary diffusion generally show a high E_a value, i.e., larger than 1.2 eV. The Cu-barrier metal interface diffusion shows a higher E_a value than that of the Cu-dielectric interface diffusion because the former has a stronger adhesion force than the latter. The strength of adhesion affects the void nucleation and growth rate induced by EM. The PECVD SiN$_x$ cap layer usually adheres weakly to the Cu layer. In addition, the large difference of the thermal expansion coefficients between Cu and the dielectric film weakens the interface in the EM test at high temperature [13]. E_a value of 0.65 eV was reported for Cu line with PECVD SiN$_x$ cap layer [14]. The E_a value of 0.6 eV in this study indicated the Cu-SiN$_x$ cap layer interfacial diffusion may be the primary transport mechanism.

Failure analysis

It is well established that the basic requirement for EM to occur in an interconnect structure is the existence of flux divergence of metal atoms due to the electronic driving force. The atomic flux Γ at i-grain boundary can be expressed as follow.

$$\Gamma_i = \frac{N_i D_i}{kT_i} Z * qE \cos\phi_i \qquad (3)$$

where $N, D, Z * q$ are the atomic concentration, diffusion coefficient, and effective charge of the ions, ϕ_i is the inclination angle of the i-th grain boundary with respect to the electron flow. The flux divergence at a triple junction point is $\sum \Gamma_i$, which is zero only when the relative angle between two adjacent grain boundaries is 120° [15]. However, due to the variation of the grain size, this particular condition is hardly met. A non-zero divergence at the triple junction point leads to an accumulation or void. As illustrated in Figure 2(a), if $\Gamma_1 < \Gamma_2 + \Gamma_3$, the void is formed by the divergent flux along the grain boundaries. In such a case, the void at the early stage should have a triangle shape. Fig. 2(b) and (c) show the typical voids formed on the Cu line after EM at 350°C for 150 and 350 seconds, respectively. The voids were not due to the removal of the passivation layer, because no similar voids on the adjacent dummy lines were observed. Although Fig. 2(b) and (c) were from different lines, the voids presented at the similar location, i.e., the triple junction point of grain boundaries. Therefore, the void initiated from the grain boundary junction point and grew larger as the EM continued.

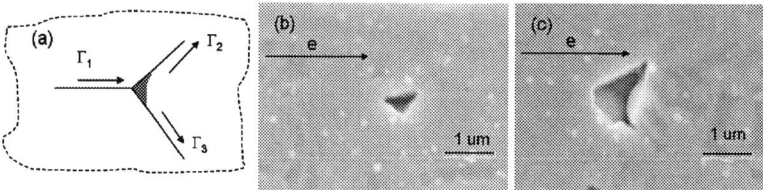

Figure 2. Illustration (a) and SEM images of void formation after EM at 350°C for (b) 150 seconds; (c)350 seconds.

Mass transport occurs not only along the grain boundary, but also along the interface. Cu atoms that are migrating along the upper Cu surface or Cu/dielectric interface are fed by two mechanisms: the Cu diffusion from inside of the metal line to the top surface along the grain boundaries or the grain thinning, as illustrated in Fig. 3(a). When the Cu atoms are depleted from the grain boundary or the inner free surface of the void that intersects the top Cu surface, the void tends to grow in a "v" shape. In such a case, the void grows wider and deeper with EM time. For grain thinning, the surface atoms on the upstream grain diffuse to the void boundary, and then move up the step to feed the atoms that flow at the Cu/dielectric interface [4]. In this case the void leads to the flat upstream surface. Fig. 3(b) shows the top view of the Cu line after EM at 350°C for 2400 seconds and the removal of SiN_x. The recess areas in Fig. 3(b) correspond to the voids at the Cu-SiN_x interface, which were due to the interfacial diffusion. Both grain boundary depletion voids and grain thinning voids were observed. The voids caused by grain thinning should have less impact on the line resistance change than the voids caused by grain boundary depletion because the lower-stream grains are still connected in the former situation. However, since both kinds of voids exist at the same time, it is hard to differentiate them from the resistance-time plot.

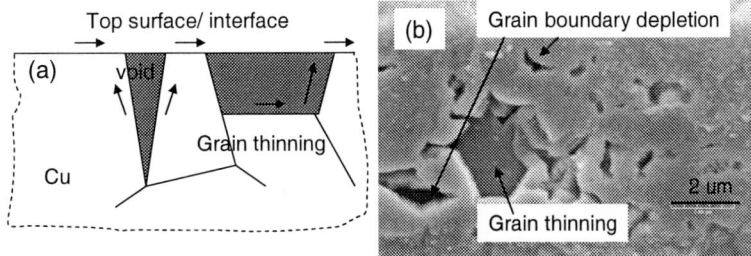

Figure 3. Illustration and SEM image of voids by grain thinning and grain boundary depletion at 350°C.

External bending stress effect on EM

To investigate the mechanical stress effect on EM, a series of EM tests were performed on the mechanically bended Cu lines. The patterned Cu sample was cut into a 2 x 2 cm plate, which was placed on the three-point bending Al holder on the hot chuck. The stress was caused by the mechanical bending and could be roughly estimated. Here, it is bout 85MPa, which is far more than the internal stress without mechanical bending. The MTTFs of the bended lines are shorter than those of the non-bended lines at 320°C and 350°C, i.e., 8852s and 4944s vs. 11711s and 6000s, respectively. At 380°C, the MTTFs of the bended and non-bended lines are comparable, i.e., 3320s vs. 3278, because the initial bending stress effect is less significant due to the increased thermal stress. Using n= 2.7, the E_a value of 0.5 eV was obtained, which is smaller than the E_a without bending stress, i.e., 0.6 eV. Similar activation energy shift was reported on Cu damascene lines with compressive stress being applied [16]. In the EM test, the resistance change is a complex collective behavior of the void growth and nucleation. After EM test, numerous voids were observed along the length. There is noticeable difference in the voids distribution between the bended and non-bended lines. For example, in the non-bended line, the consecutive voids tended to line up along the current flow, as shown in Figure 4 (a). However, in the bended lines, the consecutive voids tended to line up perpendicular to the current flow, as shown in Fig. 4(b).

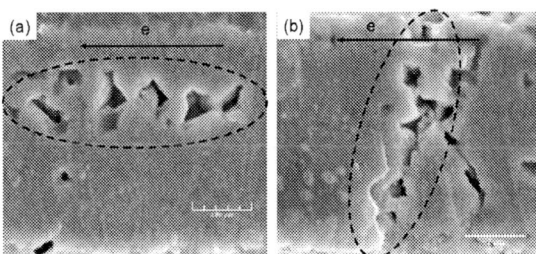

Figure 4. Typical void distribution in (a) non-bended line, (b) bended line of EM at 350°C.

In a simplified mode, assuming the total volume of the voids are comparable for the bended and non-bended lines, the resistance change in a conductor of overall length L and width W is given by the following equation:

$$\frac{\Delta R}{R_0} = \frac{l(t)}{L} \times (\frac{W}{W - w(t)} - 1) \qquad (4)$$

$l(t)$ and $w(t)$ are equivalent overall void length and width, respectively. From equation (4), the voids aligned perpendicular to the current flow play a larger impact on the resistance change than the equaxied voids do. Therefore, the line with the void distribution in Fig. 4(b) would have a shorter lifetime. Although equation (4) was based on a simplified model, more detail study and simulation on the void morphology showed the similar result [17]. Therefore, the applied bending stress affected the void distribution and shortened the EM lifetime.

CONCLUSION

The isothermal electromigration tests were performed on the single level Cu interconnect lines prepared from a Cl_2 plasma based etch process. An activation energy of 0.6 eV and a current density exponent of 2.7 were obtained. The SEM pictures showed that voids were initiated from the grain boundary junction points and grew preferentially along grain boundaries. Both the Cu-SiN$_x$ interface and the Cu grain boundary diffusion were critical to the EM failure. The Cu lines with the bended compressive stress showed a shorter lifetime and lower activation energy than the unbended lines due to the change of void distribution.

ACKNOWLEDGEMENTS

The authors would like to thank Helinda Nominanda for the SiN$_x$ cap layer deposition and Yu Lei for the technical discussion during the photolithography mask design.

REFERENCES:

[1] C. S. Hau-Riege, *Microelectronics Reliability*, **44**, 195-205 (2004).
[2] S. Lee, Y. Kuo, *J. Electrochem. Soc.*, **148**, G524-529 (2001).
[3] Y. Kuo, S. Lee, *Appl. Phys. Lett.*, **78**, 1002 (2001).
[4] C.-K. Hu, L. Gignac, R. Rosenberg, *Microelectronics Reliability*, **46**, 213-231 (2006).
[5] J. R. Lloyd, J. J. Clement, *Thin Solid Films*, **262**, 135-141 (1995).
[6] J. R. Lloyd, J. Clemens, R. Snede, *Microelectronics Reliability* **39**, 1595-1602 (1999).
[7] J. Proost, T. Hirato, T. Furuhara, K. Maex, J. P. Celis, *J. Appl. Phys.*, **87**, 2792-2802 (2000).
[8] A. V. Vairagar, S. G. Mhaisalkar, A. Krishnamoorthy, *Microelectronics Reliability*, **44**, 747-754 (2004).
[9] R. G. Filippi, M. A. Gribelyuk, T. Joseph, etc, *Thin Solid Films*, **388**, 303-314 (2001).
[10] S. Yokogawa, N. Okada, Y. Kakuhara, H. Takizawa, *Microelectronics Reliability*, **41**, 1409-1416 (2001).
[11] C.-K. Hu, *Thin Solid Films*, **260**, 124-134 (1995).
[12] C.-K. Hu, J. M. E. Harper, *Mater. Chem. Phys.*, **52**, 5-16 (1998).
[13] T. C. Lee, M. Ruprecht, D. Tibel, T. D. Sullivan, S. Wen, *IEEE International Reliability physics symposium proceedings*, **40**, 327-335 (2002).
[14] M. H. Lin, Y. L. Lin, K. P. Chang, K. C. Su, T. Wang, *Microelectronics Reliability*, **45**, 1061-1078 (2005).
[15] G. L. Baldini, I. D. Munari, A. Scorzoni, F. Fantini, *Microelectronics Reliability*, **33**, 1779-1805 (1993).
[16] G. Reimbold, O. Sicardy, L. Arnaud, F. Fillot, J. Torres, *IEDM*, 745-748 (2002).
[17] J. E. Sanchez, Jr., L. T. McKnelly, J. W. Morris, Jr., *J. Electron. Mater.*, **19**, 1213-1220 (1990).

Mater. Res. Soc. Symp. Proc. Vol. 1079 © 2008 Materials Research Society 1079-N09-09

Numerical Analysis of Packaging-Induced Failures in Cu/Low-k Interconnects

Aditya Pradeep Karmarkar[1], Xiaopeng Xu[2], Xiao Lin[2], Greg Rollins[2], Victor Moroz[2], and Xi-Wei Lin[2]

[1]Synopsys (India) Private Limited, My Home Tycoon, 4th Floor, Block A, Begumpet, Hyderabad, 500016, India
[2]Synopsys, Inc., 700, East Middlefield Road, Mountain View, CA, 94043

ABSTRACT

With decreasing feature sizes for every technology node, multi-level metallization schemes that employ copper interconnects and low-k dielectrics are required to achieve the requisite circuit performance. Here, the effects of the mechanical stresses originating from the packaging process on Cu/Low-k interconnects are assessed. The impact of package defects on interconnect reliability is also analyzed. It is seen that the package reliability varies with underfill mechanical properties. The packaging process introduces global level stresses that propagate to the local, i.e. interconnect, level. Moreover, the package defects also have an adverse impact on the mechanical stresses in the metallization structure. The package defects alter the mechanical stresses in the metal lines and affect the reliability. The complex interaction between packaging process induced stresses, package level defects and mechanical properties of various materials is analyzed in order to create robust interconnect designs.

INTRODUCTION

The current trends in the microelectronics industry demand decreasing feature sizes and increasing integration densities at each technology node. For deep sub-micron devices, multi-level metallization schemes that employ copper interconnects and low-k dielectrics are required to achieve the requisite circuit performance and yield. However, Cu/low-k interconnects are susceptible to mechanical stress induced failures due to their low mechanical strength [1]. Mechanical stresses originating from a number of sources affect the Cu/low-k interconnect reliability. Mechanical stresses generated in the Cu/low-k structures during packaging pose significant reliability challenges due to chip-package interaction. The packaging process generates mechanical stresses at the global level; which permeate to the local level, i.e. interconnect level, and are responsible for reliability failures and yield loss [2] - [4].

In this paper, the impact of chip-package interaction on Cu/low-k reliability is examined. Here, an advanced FEM based 3D simulator is used to examine interface delamination in the package and mechanical failures in interconnects, along with the interaction between these phenomena [5]. The simulator generates the 3D structures by performing fabrication process steps as described in [5], and solves the partial differential equations following the standard FEM procedure. Various sources of mechanical stress are considered during the process steps. The simulator can also account for the viscoelastic behavior of various materials at the processing conditions, which results in improved simulation accuracy. The J-integral method is used to determine the effects of various global and local factors affecting the formation and propagation of interface defects and cracks [6]. The multi-level multi-scale submodeling technique used here allows for the assessment of the stress propagation from the package to interconnect.

95

RESULTS AND DISCUSSION

Package Delamination

A global package model comprising of an FR4 substrate, silicon chip, Pb63-Sn37 solder bumps and epoxy underfill is created to assess the delamination behavior. An interfacial layer, or smear material, is simulated between the chip and the underfill to account for the mechanical properties of the interconnect structure. The mechanical properties of the smear material are calculated from the volume fractions of its constituents, i.e. copper, silicon dioxide, low-k dielectric and silicon nitride. Figure 1 shows a schematic cross-section of the model along an x-normal plane situated at the centre of the geometry [2], [7]. The thickness of the package along the x-direction is two µm. The boundary conditions are set to allow movements at the maxima along the y and z directions but constrain displacement at the minima and maxima along the x direction, representing an elongated structure along the x-direction. A 150 µm initial crack is introduced along the interface between the underfill and the smear material near the chip edge. The package structure is exposed to a thermal ramp from 125 °C to 25 °C to examine the evolution of mechanical stresses in the presence of interfacial defects and thermal cycling. The crack simulation is based on the cohesive FEM method as described in [8]. The J-integral is evaluated using the domain integral method originally proposed by Shih, *et al* [6]. The J-integral calculated at the crack tip corresponds to the strain energy release rate. The strain energy release rate is determined for various values of Young's modulus and coefficient of thermal expansion (CTE) for the underfill material.

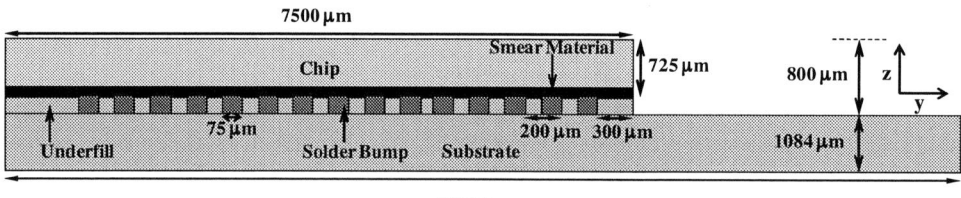

Figure 1: A schematic cross-section of the 3D global model along an x-normal plane

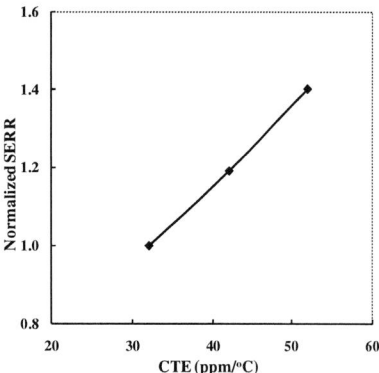

Figure 2: Normalized strain energy release rate with respect to underfill CTE

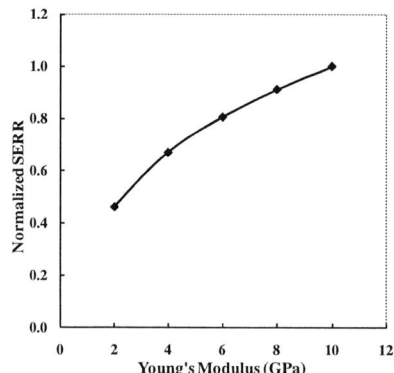

Figure 3: Normalized strain energy release rate with respect to underfill Young's modulus

Figure 2 and Figure 3 show the variation of the strain energy release rate with respect to the underfill CTE and the underfill Young's modulus, respectively. The strain energy release rate in both the figures is normalized to the value obtained for an underfill with a Young's modulus of 10 GPa and CTE of 32 ppm/°C. The strain energy release rates are normalized in order to clearly illustrate the variation in the strain energy release rate with underfill properties. Here, it can be observed that the strain energy release rate for an edge crack in a package increases with increasing underfill modulus and CTE. These results indicate that the amount of energy required to propagate the initial crack increases with increasing underfill Young's modulus and CTE.

Interconnect Stress

A multi-level, multi-scale submodeling technique is used to assess the effects of the packaging process and package defects on interconnect mechanical stress and reliability. Here, an interconnect structure containing five levels of copper lines embedded in low-k dielectric is included in the global model. The metal line width and thickness are 140 nm and 310 nm respectively, representing 90-nm technology. The horizontal and vertical spacing between the metal lines is 140 nm and 240 nm, respectively. This structure also consists of 50 nm thick barrier layers of silicon nitride. The entire structure is designed to mimic a flip-chip package. A three-dimensional submodel geometry and boundary conditions are extracted from the global model to study the propagation of mechanical stress from the global to the local level in more detail. Figure 4 (a) shows the submodeling technique. Figure 4 (b) is a schematic illustration showing the simulated z-displacements for the global model and the submodel, where the submodel boundary conditions are extracted from the global model.

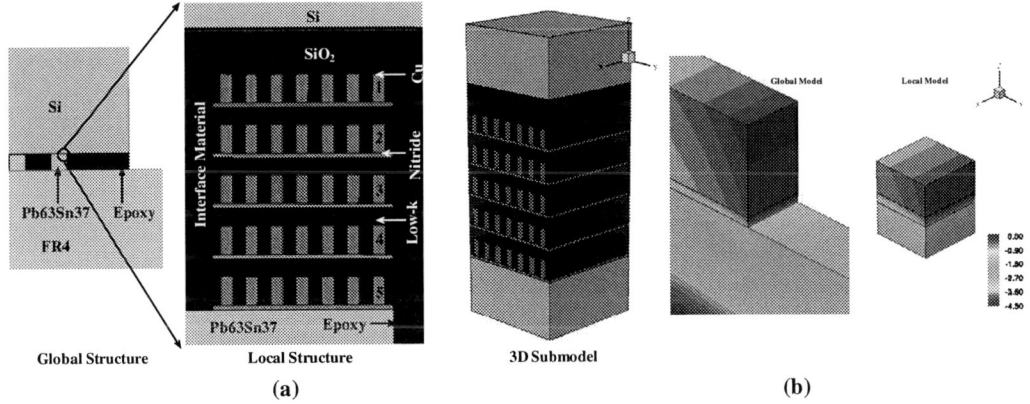

Figure 4 (a): Illustration of multi-level multi-scale submodeling technique; **(b)** Schematic illustration of z-displacements (μm) in the global model and the submodel

In order to analyze the behavior of the interconnect structure, various cases are studied. First, a global model and a submodel are simulated without any crack in the global model. These structures are exposed to a thermal ramp from 125 °C to 25 °C to simulate the thermal cycling during the packaging process. Figure 5 and Figure 6 show the distributions of the hydrostatic stress in the metal lines in the global model and the submodel, respectively. Here, the stress distributions in the submodel are determined using the viscoelastic models to improve the simulation accuracy.

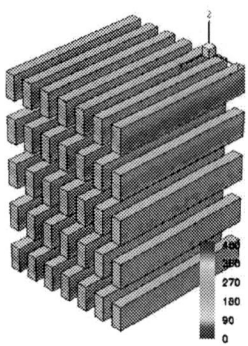

Figure 5: Hydrostatic stress (MPa) in global model without a global crack

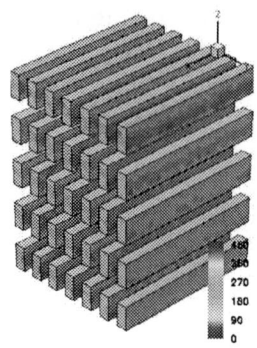

Figure 6: Hydrostatic stress (MPa) in submodel without a global crack

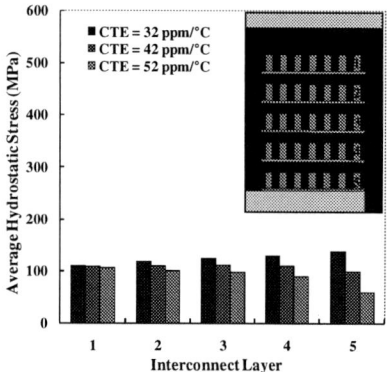

Figure 7: Average hydrostatic stress in metal lines for different underfill CTE

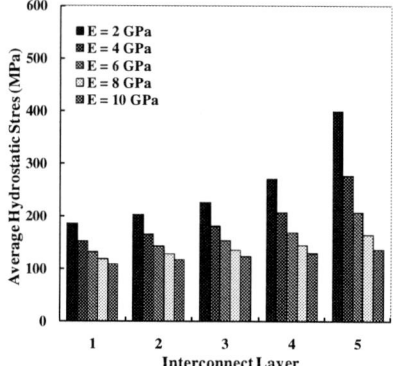

Figure 8: Average hydrostatic stress in metal lines for different underfill Young's modulus

Moreover, the average hydrostatic stress in one metal line in each metallization layer is determined to assess the impact of the packaging process on the mechanical stress. Here, the metal lines closest to the solder bump edge are selected, as shown in Figure 4. The average hydrostatic stress in the metal line is determined for various values of Young's modulus (E) and coefficient of thermal expansion (CTE) for the underfill material. In this case, simple linear elastic models are used to determine the stress distributions and the average stress values are extracted from the submodel. Figure 7 shows the average hydrostatic stress in a metal line in each layer for different values of underfill CTE and an underfill Young's modulus of 10 GPa. The inset in Figure 7 shows the relative position of each metal line. Similarly, Figure 8 shows the average hydrostatic stress for various values of underfill Young's modulus and an underfill CTE of 32 ppm/°C. From these figures, it is observed that the global stresses introduced during the packaging process propagate to the local interconnect level. Moreover, it is clearly seen that the mechanical stress in the metal lines decreases with increasing underfill CTE and Young's modulus. These results also indicate that the final stress distributions depend on the location of the metal line in the metallization structure.

An initial crack is introduced in the global model at the interface between the smear material and the underfill, such that the crack tip is 7 μm away from the metallization structure. The overall length of the crack is 293 μm. The global and the submodel are then exposed to a thermal ramp from 125 °C to 25 °C. Figure 9 and Figure 10 show the hydrostatic stress in the metallization structure in the global and the submodel in the presence of a global crack. Here, the stress in the submodel is determined using viscoelastic models to improve the simulation accuracy. These figures show that the presence of a package crack results in significantly higher stresses in the metal lines after thermal cycling.

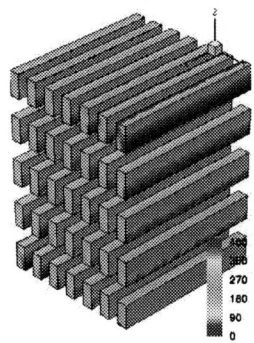

Figure 9: Hydrostatic stress (MPa) in the global model with a global crack.

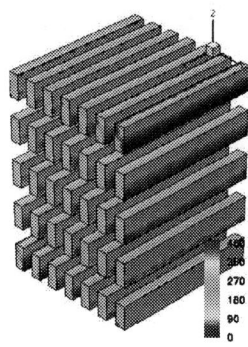

Figure 10: Hydrostatic stress (MPa) in the submodel with a global crack.

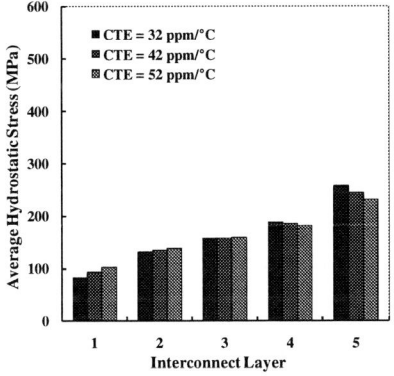

Figure 11: Average hydrostatic stress in metal lines for different values of underfill CTE and a global crack.

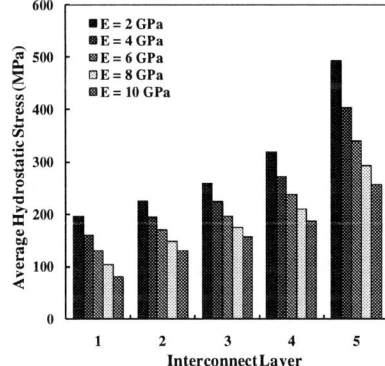

Figure 12: Average hydrostatic stress in metal lines for different values of underfill Young's modulus and a global crack

The average hydrostatic stress in one metal line is determined after exposing the global model and the submodel to a temperature ramp form 125 °C to 25 °C, as discussed previously. In this case, the global model consists of a 293 μm edge crack. Here, simple linear elastic models are used to determine the stress distributions and the average stress values are extracted from the submodel. Figure 11 shows the average hydrostatic stress in one metal line for various values of underfill CTE and an underfill modulus of 10 GPa. Similarly, Figure 12 shows the average hydrostatic stress for various values of underfill Young's modulus and an underfill CTE of 32 ppm/°C. In case of these figures, only the global model consists of an edge crack. From Figure 11 and Figure 12 it can be observed that the mechanical stress in interconnects shows significant increase after the introduction of a global crack. Moreover, it is also seen that the package crack alters the behavior of the mechanical stress with respect to the underfill mechanical properties. These results demonstrate the impact of a global crack on the packaging process induced stress in the interconnect structure.

The results discussed so far show that the packaging process has a large effect on the mechanical stress distribution in the interconnect structure. The package stress and reliability are affected by the mechanical properties of the underfill material. The existence of interface defects at the package level affects reliability of the global package and also impacts the mechanical stresses in the local interconnect structure. The defect location plays an important part in determining the final stress distributions, and hence reliability, in metallization structures. Moreover, interaction between the global and local level defects leads to changes in the mechanical stress at the local level and affects the interconnect reliability. The mechanical reliability in the Cu/Low-k structures depends on a number of global and local factors, which need to be considered to improve the overall yield.

SUMMARY

The effects of the packaging process, packaging defects and underfill material properties on the mechanical stresses in interconnect structures are studied here. It is observed that the global package reliability is affected by the variation of underfill mechanical properties. Moreover, the mechanical stresses introduced by a global packaging process propagate to the local interconnect structure. The mechanical stress distributions in local interconnect structures are significantly altered by the existence of package defects and cracks. All these factors impact the interconnect reliability and need to be considered in order to design robust metallization schemes and improve the overall yield.

REFERENCES

[1] Cherault, N., Besson, J., Goldberg, C., Casanova, N. and Berger, M.-H. "Finite element simulation of thermomechanical stress evolution in Cu/low-k interconnects during manufacturing and subsequent thermal cycling," *Proceedings of 35th European Solid-State Device Research Conference* (ESSDERC 2005), pp. 493-496, Sept. 2005.

[2] Zhai, C. J., Ozkan, U., Dubey, A., Sidharth, Blish, R. C. (II) and Master, R. N., "Investigation of Cu/low-k film delamination in flip chip packages," *Proceedings of 56th Electronic Components and Technology Conference*, 2006, pp. 709-717, May-Jun. 2006.

[3] Mercado, L. L. and Sarihan, V., "Evaluation of die edge cracking in flip-chip PBGA packages," *IEEE Transactions on Components and Packaging Technologies*, vol. 26, no. 4, pp. 719-723, Dec. 2003.

[4] Mercado L. L., Sarihan V. and Hauck, T., "An analysis of interface delamination in flip-chip package," *Proceedings of 50th Electronic Components and Technology Conference,* 2000, pp.1332-1337, May 2000.

[5] Fammos-TX Manuals, *Synopsys, Inc.*, March 2008.

[6] Shih, C. F., Moran, B., and Nakamura, T., "Energy release rate along a three-dimensional crack front in a thermally stressed body," *International Journal of Fracture*, vol. 30, pp. 79-102, 1986.

[7] Tsai, M.-Y., Lin, Y.-C, Huang, C.-Y. and Wu, J.-D., "Thermal deformations and stresses of flip-Chip BGA packages with low- and high-T_g underfills," *IEEE Transactions on Electronics Packaging Manufacturing*, vol. 28, no. 4, pp. 328-337, Oct. 2005.

[8] Xu, X.-P., and Needleman, A., "Numerical simulations of fast crack growth in brittle solids," *Journal of the Mechanics and Physics of Solids*, vol. 42, pp. 1397-1434, 1994.

Mater. Res. Soc. Symp. Proc. Vol. 1079 © 2008 Materials Research Society 1079-N03-05

Off-Angular Deposition Compensation for PVD Selective Re-sputtering Process

Hsien-Lung Yang, Fuhong Zhang, Kim Nelson, Jennifer M. Tseng, John Forster, Arvind
Sunddarrajan, Ajay Bhatnagar, Niranjan Kumar, and Prabu Gopalraja
Metal Deposition Product, Applied Materials Inc., Santa Clara, CA, 95054

ABSTRACT

In Copper back-end-of-line (BEOL), the "punchthru[TM] process" – removal of barrier material from via bottom during etch/re-sputter step, and gouging into the underlying Copper line - has been increasingly used in 65nm production for its superior reliability. However, with the adoption of porous low-k dielectric at 45nm node and beyond, the conventional punchthru process can cause physical damage to the porous dielectric, such as roughening of the trench bottom in dual damascene structures, micro-trenching in the bottom of single trenches, which may have reliability implications. This paper reported on the use of off-angular Tantalum neutral flux during the re-sputter process to improve the selectivity between the via and trench bottom in order to protect the trench bottom and via bevel, while still allowing sufficient gouging into the underlying Copper line. In addition, the plasma density and ion energy are adjusted to further optimize selectivity, and to eliminate any micro-trenching. Therefore, this paper demonstrated PVD high deposit/etch selectivity process based on transmission-electron microscopy (TEM) and studies of electrical test result. This approach has extended the PVD Tantalum barrier process to at least 32nm node.

INTRODUCTION

During the barrier portion of Tantalum barrier [1-2]/Copper seed deposition, the micro-electronics industry has increasingly used the "punchthru[TM] process" [3-5] – removal of barrier material from the via bottom during etch/re-sputter step, and gouging into the underlying Copper line. This process has shown superior reliability compared to non-punch through processes.[6] At 65nm generation and below, this process has been adopted by the industry as a "best known method", as a robust process with good electrical parameters (line resistance/via resistance) and reliability (stress- and electro-migration). However, with the introduction of porous low-k dielectric at more advanced technology nodes, the conventional punch through process can cause physical damage to the dielectric, e.g. roughening of the trench bottom in dual damascene structures, micro-trenching in the bottom of single trenches; this damage may have reliability implications. Integration of low-k dielectrics into microelectronic manufacturing will require the development of a barrier process with high deposit/ etch selectivity between the via and trench bottom in order to protect the trench bottom and via bevel, while still allowing sufficient gouging into the underlying Copper line.

EXPERIMENT

The structures used in this study were etched in low-k dielectric (K=~2.7). The structures contained dual-damascene via-chain with 2:1 aspect ratio and single trench line with approximately 2:1aspect ratio. The structures were deposited with TaN/Ta bi-layer initially, following with Ar+ sputtering on the bi-layer and addition barrier layer deposition step in the end. The PVD barrier chamber used could achieve a high-ionization deposition and in-situ Ar+ sputtering capability. The process performance was evaluated step coverage by using cross section of transmission-electron microscopy (TEM) and reliability by using internal electrical test wafers.

DISCUSSION

As we knew, selectivity depends on the difference in aspect ratio between trenches and vias. During the deposition step, source DC power from the target and AC bias power on the pedestal, are optimized to maximize the selectivity. Increasing the DC source power and while keeping AC bias power low, will increase the non-directional neutral flux, and will minimize the directionality of the ion flux at the wafer. This will lead to higher bottom coverage in the trenches than that in the vias, improving the deposition selectivity. High re-sputtering selectivity is more difficult to achieve since Ar ions arriving normal to the substrate will affect the bottom of trenches and vias equally, regardless of aspect ratio.

Increasing the DC power on the Tantalum coil which is attached on the inner chamber sidewall can generate more off-angular neutral flux (Fig. 1). The coil voltage will also increase proportionally, but this has no deleterious effect on the process (Fig. 2). The DC coil power is the key parameter to increase the re-sputtering selectivity.

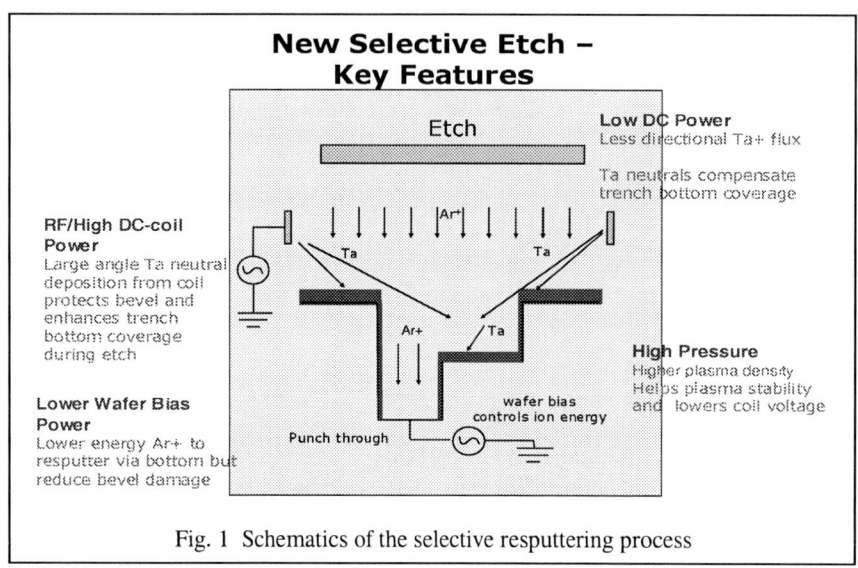

Fig. 1 Schematics of the selective resputtering process

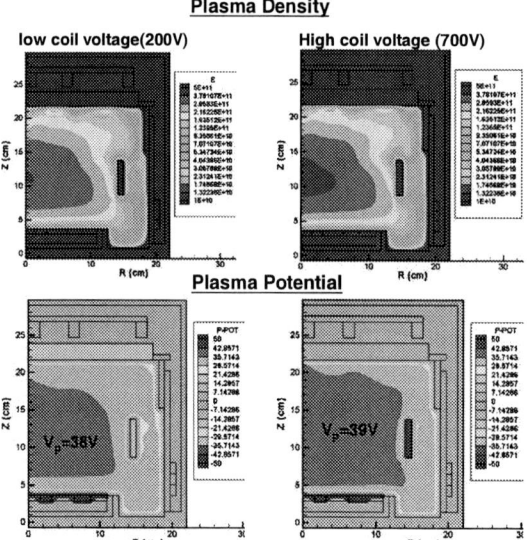

Fig.2 Simulation of plasma density and plasma potential vs. coil voltage

In addition, the DC source power needs to be optimized to maintain a stable plasma and to keep the Ta ion fraction as low as possible during re-sputtering process.If we defined the net deposition rate during sputtering step is the rate of Ta neutral/ion deposition minus the absolute rate of Ar/Ta ions sputtering away. This is always a minus number. Low Ta ion fraction during sputtering minimizes net deposition effect on via bottom, while high neutral fraction leads to higher Ta net deposition rate on trench bottom compared to via bottom. Low Ta ion fraction also leads to lower line resistance (Fig. 3).

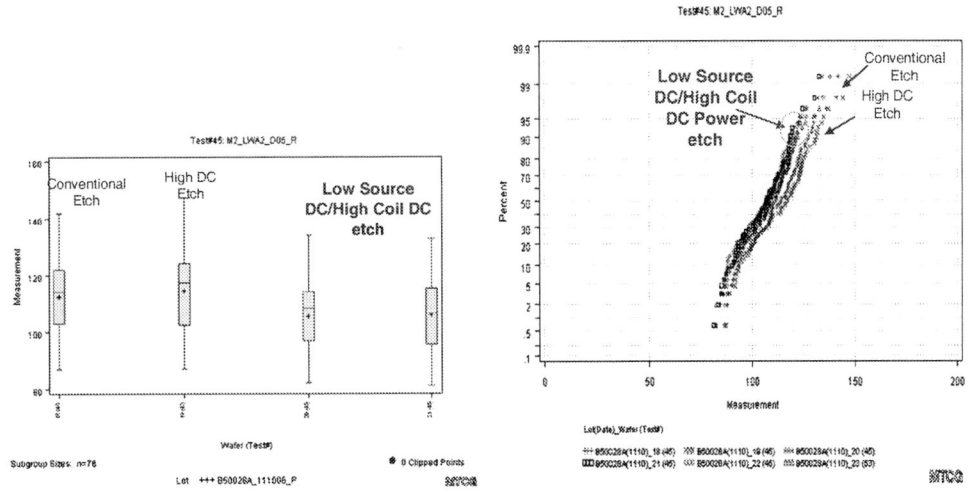

Fig. 3 Electrical result shows low Ta ion fraction resputtering process has lower line resistance and good yield

With higher DC coil power, more off-angular Ta flux from the coil strikes the wafer, possibly increasing overhang formation. Therefore both target DC power and coil DC power need to be optimized for proper trade-off between the overhang and the selectivity. Increasing the RF coil power improves the plasma stability and decreases the re-sputtering energy (Fig. 4). Proper adjustment of the RF coil power and AC bias power can keep the re-sputter/deposition ratio close to 1 (net deposition rate during sputtering is close to 0), thereby maximizing the re-sputtering selectivity. Also, increasing Ar flow improves plasma stability and lowers the coil DC voltage, which is beneficial for hardware stability.

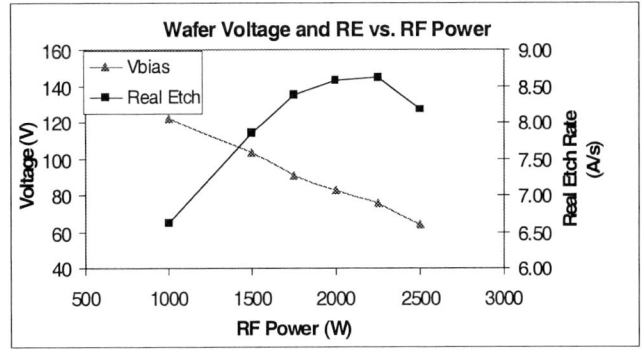

Fig. 4 RF power vs. Ion Energy (bias Voltage * eV) and real etch rate

103

CONCLUSIONS

The TEM analysis shows no micro-trenching damage in the single trench bottom (Fig. 5). The dual damascene bevel/trench bottom is also damage free with sufficient Cu gouging depth into the via (Fig. 4) after re-sputtering. This new PVD Tantalum barrier process has shown good manufacturability during an extended run (>1900KWH), achieving good process stability and good defect performance.

Several micro-electronic manufacturers have integrated this new barrier process into their low-k technology node, and have shown good reliability (both SM and EM) performance. This approach has extended the PVD Tantalum barrier process to at least the 32nm node.

Old BKM Etch   **Selective Etch**

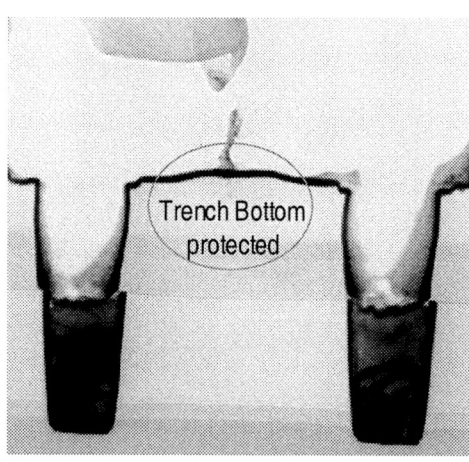

Fig.5 TEM cross-section of a dual damascene via on low-K dielectric showing trench bottom damage free while punching through via bottom from selective resputtering process

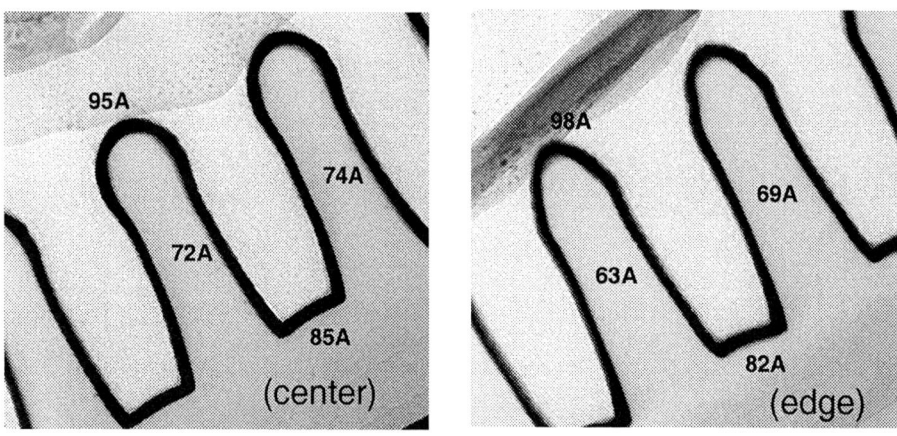

Fig.4 TEM crossection of a trench on low-K dielectric showing no mino-micro trenching damage in the bottom

ACKNOWLEDGEMENTS

The authors would like to thanks to Philip Wang for TEM support and Tza-Jing Gung for technical and management support. This work has been supported by the internal Mayden Technology Center Group

REFERENCES

[1] K. Ino, et al. J. Vac. Sci. Technol. A 15(5), 1997, 2627
[2] P. Catania, et al. J. Vac. Sci. Technol. A 10, 1992, 3318
[3] C.-C Yang, et al. Proc. IEEE Int' Interconnect Tech. Conf. 2005, p.135
[4] D. Edelstein, et al. Proc. IEEE Int' Interconnect Tech. Conf. 2001, p.9
[5] D. Edelstein, et al. Advanced Metallization Conf. 2001, p.541
[6] N. Kumar, et al. Advanced Metallization Conf. 2004, p.247

TVS Measurements of Metal Ions in Low-k Dielectrics: Effect of H₂O Uptake

Ivan Ciofi, Zsolt Tökei, Giovanni Mangraviti, and Gerald Beyer
IMEC, Kapeldreef 75, Leuven, B-3001, Belgium

ABSTRACT

Triangular Voltage Sweep (TVS) and Capacitance-Voltage (CV) measurements are gaining popularity in Back-End-Of-Line (BEOL) as techniques for studying drift of metal ions in low-k dielectrics. Recently, many works have been published on the topic. Although the experimental results that were presented are similar, the interpretations that were given are controversial.

In order to gain better insight, we systematically investigated Metal-Insulator-Silicon (MIS) planar capacitors with different gate materials: Al, Cu, Ti, Ta, Ru and Pt, used as a reference. The insulator was an SiOC:H low-k film with 7% porosity and k value of 3.0. Besides, we also fabricated planar capacitors with amorphous silicon (a-Si) gate, which represents the case of metal-free capacitors. TVS and CV measurements were performed on the samples and the results we obtained for the different gate materials were very similar, including Pt and a-Si. We conclude that under low stress condition (190°C, 1MV/cm) neither of the quoted metals, except Cu, can drift into low-k materials. The TVS peaks and CV shifts observed on MIS capacitors are related to H₂O uptake in low-k materials and are probably due to protons, generated during the measurement itself by electrical decomposition of H₂O in the low-k film.

INTRODUCTION

In order to meet the strict performance requirements for next generation interconnects, different solutions are being considered for reducing interconnect RC delay without compromising reliability. New dielectrics with increased porosity and different metal barriers with higher conductivity are investigated to further reduce line-to-line capacitance and resistance of Cu/low-k interconnects. Considering the huge effort that is required to optimize an integration scheme for a new technology generation, it is important to identify possible issues with new materials in an early phase of development. In this respect, MIS capacitors are very suitable as a test vehicle. They allow fast electrical characterizations of dielectric films and metal barriers that are chosen as insulator and top electrode (gate) of the capacitor structure. The dielectric constant is derived from the MIS capacitance with the Si substrate in accumulation, which can be obtained through high frequency CV measurements. Dielectric contamination by metal ions injected from the metal gate under BTS is evaluated from the induced CV shifts. MIS capacitors are also suitable to evaluate the effect on low-k films of aggressive integration steps, such as plasma etch and strip. An interesting alternative to CV measurements for characterizing MIS structures is TVS measurements. They are basically current-voltage measurements, obtained by applying a triangular voltage sweep to the capacitor under investigation. TVS can be also regarded as a quasi-static CV and, as such, be used for capacitance measurements ($C=I/\alpha$, where α is the sweep rate). However, TVS measurements are mainly known in BEOL for the capability to detect metal ions in the dielectric film. The measurement procedure is simple. Samples are placed on a hot chuck and, once at high temperature, BTS is performed in order to promote drift of ions from the metal gate into the dielectric and accumulate them at the silicon interface. After BTS, a voltage sweep is initiated. When the applied voltage changes polarity, possible metal ions cause peaks superimposed upon the capacitor's displacement current. The related concentration is calculated from the area under the peaks. Of course, mobile ions of different species can be present in the dielectric as contaminants. In this case, while CV measurements do not allow discrimination between the different species, TVS measurements are capable to differentiate ionic species since the related peaks are expected to occur at different values of the ramped voltage.

In a previous work, we reported about Cu drift in OSG low-k materials [1]. After BTS, the TVS trace of Cu dots on OSG films showed an ionic peak. Since such peak was not detected on Al dots, used as a reference, we could easily identify that as being due to Cu ions. However, on both Cu and Al dots we detected a different ionic peak that we could relate to water uptake in the low-k film: the peak disappeared after N_2-bake and reappeared after dipping the samples in deionized H_2O. TVS peaks of water-related ions were detected under low stress conditions and even without BTS. Water-related ions were also found to be responsible for CV shift in the direction of positive ions. Full correlation was also verified between water-related TVS peaks and CV shifts. The concentration of water-related ions evaluated by the two techniques was very similar for the same BTS, which indicates that CV shift and TVS peak had the same origin. In agreement with N. Lifshitz et al. [2], water-related ions are probably protons, generated during the measurement itself by electrical decomposition of water in the low-k film. N. Lifshitz et al. had indeed already observed water-related peaks on low deposition temperature dielectrics, such as SOX (Spin-On Oxide), Accuglass and PE-PTEOS, that were characterized by Al dots for possible applications in Al multilevel metallization. In our previous work [1], we also showed that water-related peaks have peculiar features that allow to distinguish them from Cu peaks (figure 1). Besides, these features do not seem to depend on the specific dielectric. Therefore, TVS measurements of Cu ions in low-k materials are still possible in presence of water as long as a careful analysis of the TVS trace is carried out.

In recent literature, TVS peaks and CV shifts under low stress conditions are also reported for Al, Cu, Ta and Ru dots on low-k materials, such as HOSP (hybrid organosiloxane polymer), MSQ (porous methyl silsesquioxane) and CDO (nano-porous carbon-doped oxide) [3, 4, 5, 6]. Correlation between TVS and CV results is often verified. In these papers the physical mechanisms that are proposed are in contrast with common knowledge in the field. In particular, Al, Ta and Ru are reported to drift in low-k films at quite low stress conditions (150°C, 0.5 MV/cm). On the other hand, Al, Ta and Ru are known to be not prone to drift in low-k materials. Al is commonly chosen as a reference material for low-k characterization. Ta and Ru are used as

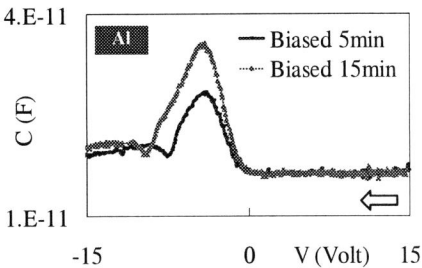

Figure 1. TVS peaks at 190°C on Al/OSG MIS, ascribed to water-related ions

Cu diffusion barrier in low-k damascene interconnects. Figure 1 shows TVS traces we measured on our Al/OSG MIS capacitors. The TVS traces reported by A. Mallikarjunan et al. [3] on Al/HOSP MIS capacitors, that can be considered representative for TVS peaks shown in recent literature for different metal/low-k combinations, are very similar. This indicates that the same ionic species is involved. However, in recent literature such TVS peaks are ascribed to metal ions injected from the metal gate under BTS and in some cases water is even excluded to play any role. Based on [3, 4, 5, 6], Al, Cu, Ta and Ru can drift into low-k materials at 150°C and 0.5 MV/cm, which is very close to IC operating conditions. This interpretation would have significant implications. The use of Al dots for low-k film characterization would be questionable, since Al ions would contaminate the dielectric film at low stress conditions. Moreover, Ta and Ru as metal barriers in Cu metallization would not be recommended because, while preventing Cu drift, they would contaminate themselves the intra-metal dielectric.

In order to refute or confirm the interpretation given in recent literature, we performed an in-depth investigation on low-k planar capacitors with different gate materials: Al, Ta, Ru, Ti, Cu, Pt and a-Si. In the following, we present and analyze the TVS traces that we measured on the planar capacitors that we fabricated for our study. CV measurements were also performed and a good correlation with TVS results was found. Since there is general agreement on this in literature, CV results are not included in this paper.

EXPERIMENTS

The planar capacitors we fabricated for our study are listed in Table 1. The substrate was n-type silicon with a resistivity of 12 Ohm·cm. The low-k dielectric was an SiOC:H material with 7% porosity and k-value of 3.0, which was deposited on top of 2 nm thick dry thermal oxide (ThOx), grown on the wafers to stabilize the silicon interface. The low-k film was deposited up to a thickness of 300 nm and for a case study it was exposed to N_2/H_2 plasma to reproduce intergration conditions for resist strip. On some wafers the low-k film was capped with oxide (Ox). Al, Ta, Ru, Ti, Cu and Pt were deposited through a shadow mask in order to define square metal dots of different areas: 1 mm^2 (Small), 4 mm^2 (Medium) and 9 mm^2 (Large). Samples with oxide cap represent the situation where the metal is not in direct contact with the low-k film. Al, Cu, Ti, Ta and Ru were sputtered up to a thickness of 100 nm, while Pt was e-beam evaporated up to a thickness of 50 nm. Pt is a noble metal and is not expected to drift into dielectrics. Therefore, Pt dots were used as a reference for our study. In order to exclude any interaction of the low-k material with any metal, planar capacitors with undoped amorphous silicon gate were also fabricated. A 500 nm thick layer of a-Si was deposited at 300°C on a capped low-k film and then patterned by litho and etch steps to define circular dots of different areas: 1.13 mm^2 (Small), 3.14 mm^2 (Medium). The etching bath was prepared by mixing water, buffered hydrofluoric acid and nitric acid in a ratio of 16 H_2O:1 Buf HF:40 HNO_3. Since drift of Si ions from the a-Si gate

Table I Description of the samples

Dielectric Stack	Gate Material	Deposition	Purpose
2nmThOx/300nmSiOC:H + N_2/H_2 Plasma	Al,Ta,Ru,Ti,Cu	Sputtering	Metal ions detection
	Pt	Evaporation	Reference
2nmThOx/300nmSiOC:H + N_2/H_2 Plasma/ 30nmOx	Al,Ta,Ru,Ti,Cu	Sputtering	Metal ions detection
	Pt	Evaporation	Reference
2nmThOx/300nmSiOC:H + N_2/H_2 Plasma/ 60nmOx	Al,Ta,Ru,Ti,Cu	Sputtering	Metal ions detection
	Pt	Evaporation	Reference
	a-Si	CVD	Reference

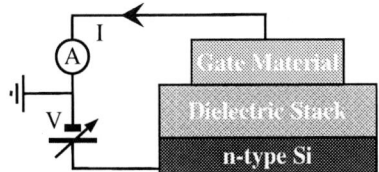

Gate Material: Al, Ta, Ru, Ti, Cu, Pt and a-Si

Dielectric Stack: 2nmThOx/300nmSiOC:H + N$_2$/H$_2$ Plasma
2nmThOx/300nmSiOC:H + N$_2$/H$_2$ Plasma/30nmOx
2nmThOx/300nmSiOC:H + N$_2$/H$_2$ Plasma/60nmOx

Figure 2. Schematic cross-section of the test structures and measurement configuration

into the dielectric through the Ox cap can be also excluded, a-Si dots represent a further reference for our study. For comparison, samples with square metal dots on the same dielectric stack were also fabricated. The schematic cross-section of the test structures and the measurement configuration are reported in figure 2. TVS measurements were carried out at 190°C on a hot chuck probe station by an HP4140B pA meter/DC voltage source. The voltage sweep rate was always 1V/sec and, unless otherwise specified, BTS was not applied before initiating the voltage sweep.

RESULTS AND DISCUSSION

For an ideal dielectric, the TVS current (I_{TVS}) consists of the capacitor's displacement current (I_d), which is the product of the low frequency capacitance (C) and the voltage sweep rate (dV/dt). Possible mobile ions in the dielectric contribute to current peaks (I_M) superimposed upon the capacitor's displacement current. Finally, dielectric leakage current (I_L) results in a further component that distorts the TVS trace.

$$I_{TVS} = C\frac{dV}{dt} + I_M + I_L \qquad (1)$$

In the following, we report and discuss TVS results obtained for the different samples. In the figures, TVS traces are normalized with respect to dot area and the sweep direction is indicated by the arrow.

Al, Ta, Ru, Ti, Cu and Pt dots on uncapped SiOC:H low-k film

Figure 3 shows the TVS traces we measured on the different metal dots. Ionic peaks were detected on all the samples, although for Pt and Cu dots peaks were much smaller than for the other metals (figure 3h). By increasing voltage sweep range and applying BTS, we could detect peaks of equivalent magnitude also on Pt and Cu dots. Figure 3g shows the case of Cu dots.

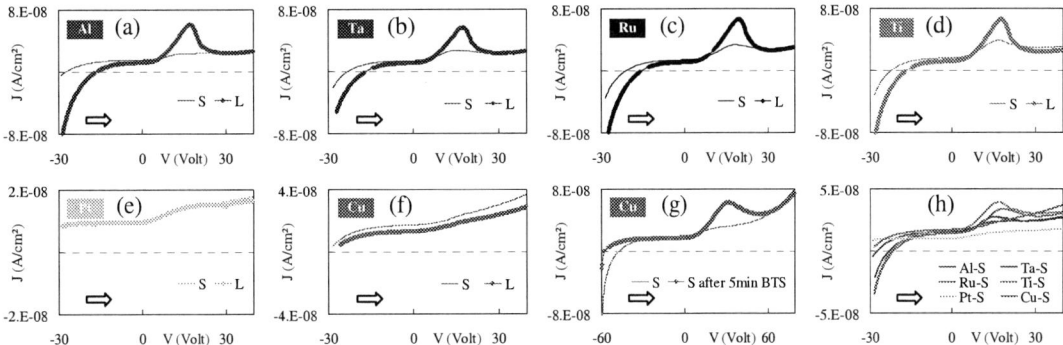

Figure 3. TVS traces for Al, Ta, Ru, Ti, Cu and Pt dots on uncapped low-k film

109

Furthermore, on Al, Ta, Ru and Ti dots there is a clear dependency of the peak magnitude from the dot area: smaller peaks are detected on smaller dots.

Al, Ta, Ru, Ti, Cu and Pt dots on 30nmOx capped SiOC:H low-k film

Figure 4 shows the TVS traces we measured on the metal dots. Similar ionic peaks were detected on all the samples, including Pt and Cu dots (figure 4g). Peaks are very similar for different dot areas.

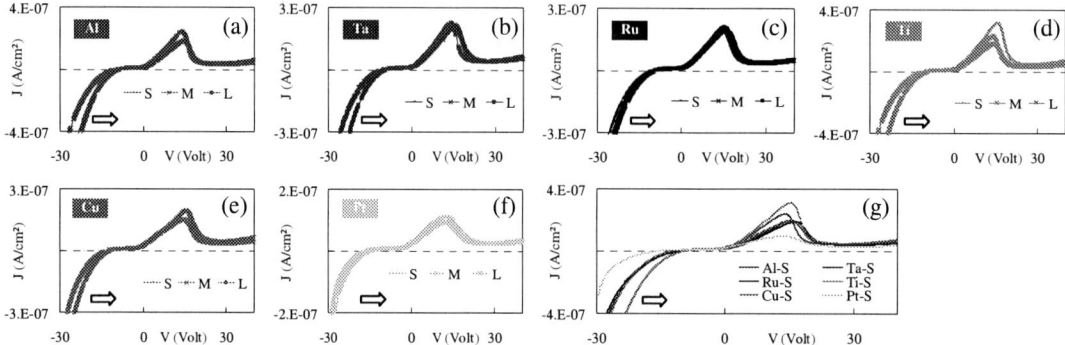

Figure 4. TVS traces for Al, Ta, Ru, Ti, Cu and Pt dots on capped low-k film

Al, Ta, Ru, Ti, Cu, Pt and a-Si dots on 60nm Ox capped SiOC:H low-k film

Figure 5a shows the TVS traces that we measured on planar capacitors with a-Si gate for two different dot areas. Apart from the magnitude, the peaks detected on a-Si were similar to those obtained for the different metal dots on the same dielectric stack (figure 5b).

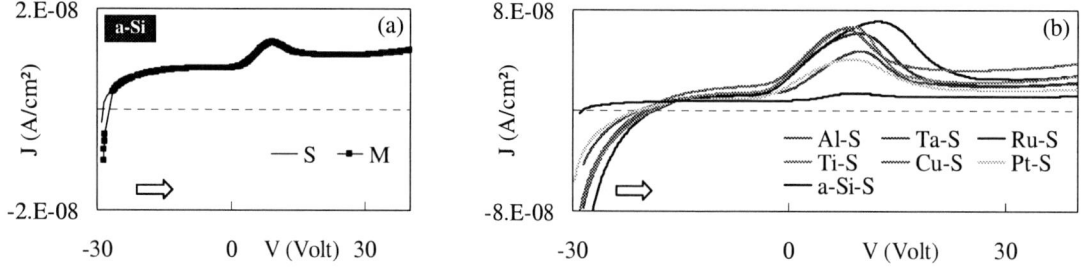

Figure 5. TVS for Al, Ta, Ru, Ti, Cu, Pt and a-Si dots on capped low-k film

Discussion

The ionic peaks we detected on our samples are comparable to those reported in literature [3] for metal dots such as Al, Cu and Ta. This clearly indicates that the same ionic species is involved. In addition, we show that such peaks can be detected on both Pt and a-Si reference dots. Therefore, we can certainly exclude that drift of ions from the gate material is involved in the generation mechanism. Water-related ions and different amount of water in the low-k film can explain the experimental results. At 190°C uncapped low-k film can release water. Since dots behave like cap for the film, small dots dry faster than large ones, which explain the peak magnitude dependency from the dot area. A longer resident time on the hot chuck at 190°C before starting the actual measurements can explain the smaller peaks on Pt and Cu dots. For dots on capped low-k films a good uniformity was indeed found for the different dot areas and for the different metal gates. The smaller peak detected on a-Si dots is instead probably due to the high resistivity of a-Si with respect to metals, which resulted in a high contact resistance. However, we noticed that the 60nm Ox cap was also removed during the etching bath we used to pattern the a-Si layer. Therefore, the smaller peak on a-Si dots could be also due to lower water content in the low-k film. We conclude that TVS peaks are not due to drift of ions from the gate material but most likely to protons, generated during the measurement itself by electrical decomposition of H_2O in the low-k film.

CONCLUSIONS

We investigated the origin of TVS peaks and CV shifts that were detected under low stress conditions on low-k planar capacitors with different gate materials. We demonstrate that drift of ions from the gate material is certainly not involved. In particular, our experimental results lead us to exclude Al, Ta, Ru and Ti drift into low-k materials under low stress conditions (190°C, 1MV/cm). TVS peaks and CV shifts are caused by positive water-related ions, most likely protons, generated during the measurement itself by electrical decomposition of water. The use of Al as a reference material for low-k characterization and of Ta, Ru and Ti as Cu diffusion barriers in low-k damascene interconnects is validated.

Since water uptake is expected to increase for next generation highly porous low-k materials, the effect of water in TVS and CV measurements must be taken into account for the correct interpretation of the measurement results.

REFERENCES
1. I. Ciofi, Zs. Tőkei, et al., *MRS Proceedings* **914**, 0914-F02-02 (2006)
2. N. Lifshitz and G. Smolinsky, *Appl. Phys. Lett.* **55**, 408 (1989)
3. A. Mallikarjunan, S. P. Murarka, and T. M. Lu, *J. Electrochem. Soc.* **149**, F155 (2002)
4. A. Mallikarjunan, S.P. Murarka, and T.-M. Lu, *Appl. Phys. Lett.* **79**, 1855 (2001)
5. T.-M. Lu, Y. Ou, and P.-I. Wang, *MRS Proceedings* **990**, 0990-B09-05 (2007)
6. K.-L. Fang and B.-Y. Tsui, *J. Appl. Phys.* **93**, 5546 (2003)

Patterned wafers backside thinning for 3-D Integration and multilayer stack achievement by direct wafer bonding.

Barbara Charlet, Antoine Chiteboun, Marc Zussy, Laurent Bally, Patrick Leduc, and Myriam Assous
MINATEC, CEA - DRT/LETI, 17, rue des Martyrs, Grenoble, F 38054, France

ABSTRACT

Scaling down the devices to keep increasing the integrated circuits (ICs) performance at the rate defined by Moore's [1] law becomes more and more difficult and so costly that new circuits architectures and new integration technologies are investigated. One of the most promising ways in integration technology is the vertical stacking of circuits, also called "3D Integration". One of the challenges in this technology is the patterned substrate backside thinning. Compatibility with the whole 3D Integration process has to be guaranteed, the existing circuit has to be kept intact and the bonding interface mustn't be damaged.
In this study we discuss some experimental results of wafer thinning by grinding and polishing of molecular bonded silicon wafers applied to 3D Integration [2-4]. The wafer with patterned copper interconnections are stacked by direct SiO_2 bonding and thinned down on one backside. These stacks are then bonded again to one or two circuits via a deposited oxide on the thinned surface. The top bulk Si surface was thinned down again on one backside, giving a multi layers stack. This wafer level vertical assembly demonstrates the possibility to adjust the remaining Silicon thickness to small values ($<15\mu m$) and then bond the thinned surface to achieve multiple layer 3D structure.

INTRODUCTION

3-D Integration is motivated by device and system improvement requirements: ICs density increase, interconnections length reduction, new above IC functions development, multiple functions integrated in smart hetero-structures and systems, etc [5]. Recent advances in materials and process engineering give the opportunity to enhance the 3D approaches by the heterogeneous substrates with patterned circuits or structures stacking and their post-processing for the final functionality of 3D integrated system [6]. The key processes are developed for efficient and precise vertical stacking, like: direct wafer bonding, eutectic layer adhesion, polymer layer gluing etc, and also wafer or die accurate alignment, stacked layer backside bulk substrate thinning or removing and finally stacked chips interconnect by inter-strata connections, for example by implementing the deep via technique [7].
ICs and MEMS integration is usually done on standard several hundred micrometer thick substrates which are relatively thick in comparison with integrated devices thickness which have sub micrometer thickness. The backside wafer thinning is commonly applied on the finished circuits, before the dies separation by a saw for its packaging. One other application of backside substrate thinning emerged with vertical integration of patterned wafers or dies is 3D integration which requires the small distance between the circuit layers in order to decrease the interconnect path between the elements of stacked structure. For this aim, the backside thinning of patterned wafer or die is generally done after its temporary transfer on an adapted for this handler or after its definitive vertical stacking on an other one circuit. The stacked circuits' inter-chips connections improvement requires important backside surface thickness reduction, and an adapted thinned surface finishing. Depending on circuit type an optimum thickness value is required for effective devices functionality and reliability.

Most commonly used enabling processes for wafers thinning and thinned layer stress relief can be resumed in following process categories which can be applied in suitable order and as much as the thinned structure need:
- Grinding – mechanical coarse and/or fine grinding for most important parts of bulk substrate thinning;
- Etching – chemical wet and/or plasma etch;
- Polishing – chemical mechanical (CMP) and/or dry polishing;
- Edge finishing - mechanical or chemical outlining;
- Surface scrubbing and cleaning.
Figure 1 shows schematically a process flow applied for bonded wafers backside thinning on one side of the stack. The most fragile parts of the thinned structure are the wafer's edges which curvature is excessively represented on the wafers drawing.

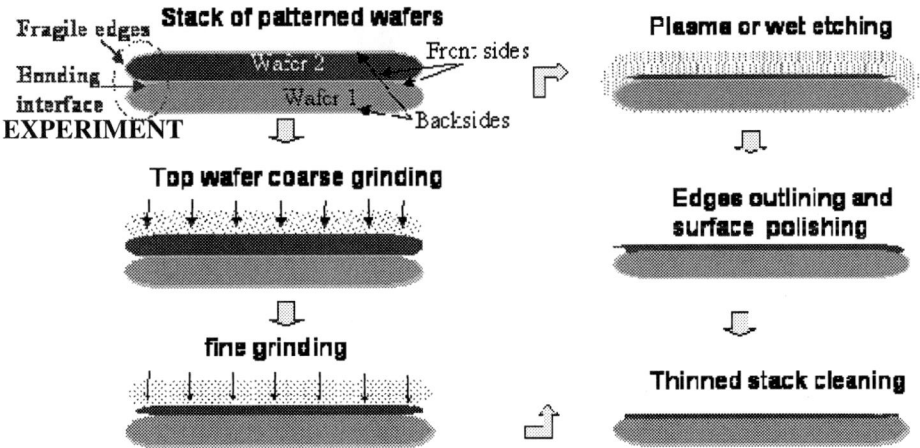

Figure 1. Schematically represented steps of bonded wafers thinning; applied on one backside of the stack.

In all cases of ICs device backside thinning, the bulk substrate thickness removing processes must be adapted to preserve the remaining layers integrity and quality. Furthermore the thinned wafers or die remaining bulk material and surface must have the possibility of the appropriated post-processing applying to this remaining layer, including the stacking of next strata integration. For this, it is necessary to define the thinned layer requirements, which must have a strong relationship with the type of stacked circuits. Generally it will be summarised as:
- Controlled remaining layer thickness and its uniformity (TTV);
- Regular and smooth surface;
- Remaining layer and stacked circuits without induced stress and deformations;
- No crack, failing or crystalline defects in the thinned layer;
- No edge crack and no chipping on the stacked wafers;
- Post processing compatibility of the thinned layer (no contamination, no interfacial or embedded layer deformation and defects);
- Thinned wafers stack compatible with further possibility to add some other layers by vertical stacking and post processing.

113

EXPERIMENT

The processing and characterization tools used in this study are standard machines usually dedicated to single silicon wafer thinning and very often having an adapted methodology for bonded wafers processing. For several ones, a comparative evaluation was done for their compatibility with stacked wafers.

Two types of stacked substrates involved in backside thinning and multi-circuit stack achievement are used in this study. The first type consists in direct bonded (via SiO_2 layer) not patterned bulk silicon wafers, and the second type also bonded by same method, but having two level Cu damascene patterned wafers (diameter 200mm, <100>orientation Si). Figure 2 shows the drawing of the backside thinning process which was divided into several steps: coarse grinding, fine grinding, and chemical mechanical polishing (CMP). The machine used for coarse and fine grinding was a tool from Okamoto. The ceramic wheel used in this study has a grain size around 50μm for the coarse grinding and few micrometers for the fine grinding. The CMP tools used were from Alpsitec and/or from Applied Material.

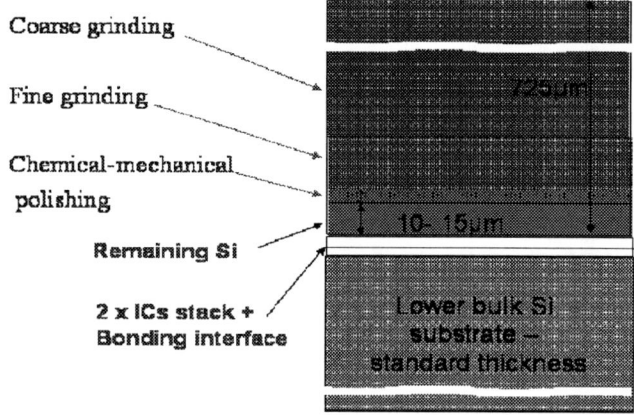

Figure 2. Schematic drawing of the thickness removed by different steps from the backside of two stacked wafers.

Several characterisations of the remaining bulk substrate performed at different steps of thinning and also for the bonded interface and stacked circuit integrity evaluation. The bonding interface quality characterization is performed with two tools: the IR camera and the acoustic wave scanning microscope. Both are non-destructive techniques but they are different in terms of resolution and simplicity of use. The main surface and stack characteristics after the successive thinning process steps are measured by different set-up:
- average thickness value - by capacitive measurement probe, FTIR interferometer and mechanical comparator;
- thickness uniformity of the thinned layer - by capacitive measurements, FTIR interferometer;
- wafer deformation in bow and warp - by capacitive measurements, laser interferometer;
- surface roughness -by atomic force microscopy and SEM;
- subsurface defectiveness by SEM after the wafer cleaving and chemical defects decoration.

RESULTS AND DISCUSSION

The bonded wafer stack deformation [8] during the thinning steps was measured for stress evaluation on the unpatterned and backend patterned wafers. Both stacks are elaborated using the same process of direct wafer bonding [9] and the both batches of wafers had the same 400nm thick PECVD SiO_2 as a bonding layer. Furthermore all the stacks had exactly the same pre-bonding surface conditioning and thermal treatment (@T<400°C) for stack stabilisation. On figure 3 a & b is shown the evolution of warp and bow as a function of the successive thinning step: coarse grinding, fine grinding, polishing. The first value – oxide deposition - concerns the wafers before the bonding and represents their deformation after the oxide layer deposition. On figure 3a, the curves represent bow and warp measured on the unpatterned wafer stacks and, on the figure 3b, the same characteristics are represented for bonded patterned wafers. In the both cases of thinned stacks the top layer thickness ranges from 10µm to 20µm.

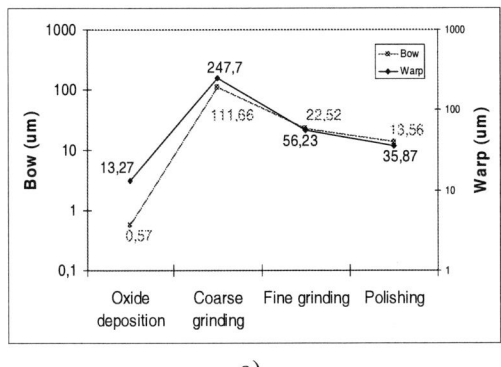
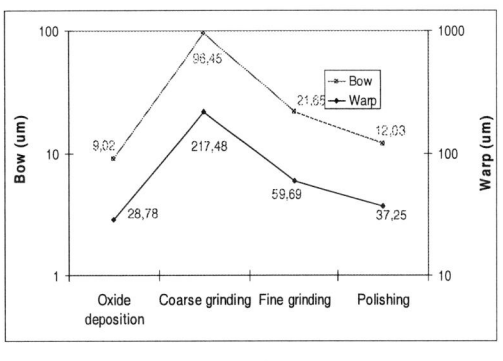

a)   b)

Figure 3. Bow and warp values measured at different steps of stacked wafers backside thinning -a: unpatterned stack, -b: patterned stack.

In order to know the effect of the stress relief on the remaining top silicon layer by CMP process, bulk silicon test wafers are grinded by coarse and fine grinding and then polished during the appropriate time to remove 1 to 4µm of bulk Silicon from the grinded surface. Figure 4-a, shows the bow and warp versus polished thickness measured after the fine grinding and polishing steps. It can be seen that after 1µm of Si removed from the surface by polishing, the bow and warp are very close to their initial values of bulk wafers and have the same range of whatever the supplementary thinning time of the stack. It can be concluded that after 1µm polishing liberates almost of stress accumulated during coarse and fine grinding. In order to evaluate the subsurface crystalline integrity of these thinned substrates it has been characterised by RBS method. The results are reported on the figure 4-b, which shows the single crystal Si reference spectrum and those measured on post-CMP thinned surface. The shapes of all measured surfaces are similar which suggests the crystalline defect free subsurface from the first micrometers of Silicon removed by CMP.

During the grinding process applied on the surface forces are considerable [10; 11] and their actions induce a significant stacked wafers deformation during the top wafer thinning as shown on figure 3. For that it was crucial to verify that the thinning step doesn't create nor amplified the bonding interface defects. On the pictures given by figure 5 is shown the SiO_2 interface of stacked wafers, checked by acoustic scanning microscope, before and after the thinning process (fig.5 a & b respectively). In order to have better scanning conditions the bulk silicon wafer stack was evaluated in first. Then the interconnections patterned wafers stack was observed after coarse grinding (fig. 5. picture –c).

It appears that the few interface defects, checked by this method on the wafer interfaces are not changed in dimension and location after the top wafer thinning down. Furthermore no new defect is observed on the interface for the both type of wafers. It can be concluded that the grinding process doesn't affect the SiO$_2$ direct bonding interface for bulk silicon top layer thickness down to 15µm. In this way elaborated stack was then processed for through silicon via (TSV) and inter-chip connection patterning as reported in [12; 13].

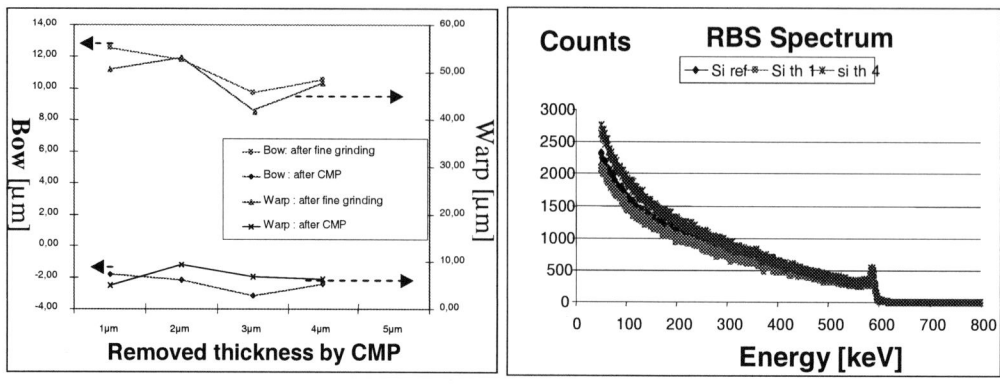

a   b

Figure 4. a) -Bulk wafers bow and warp variation versus polishing depth of grinded bulk Silicon surface, b) – RBS spectrum of crystalline Si surface corresponding at a reference Si wafer and two thinned and CMP finished surface (Si channel effect is <4% for all cases of spectrum).

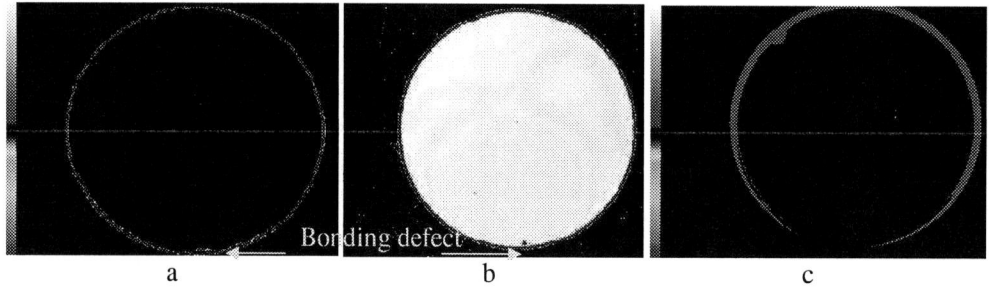

a   b   c

Figure 5. Bonded wafers interface checked by acoustic microscope. Picture – a and b represent the same unpatterned wafer, respectively - before and after top wafer thinning down, picture -c is an interconnection patterned wafer after the coarse grinding process. The contrast colour on different pictures is changed due to its top wafer thickness.

Multi-strata staking – feasibility demonstration at wafer level.

The feasibility of three and four very thin strata (multi-strata) stack carried out by direct wafer bonding and backside bulk silicon thinning was evaluated using the Cupper double metal level (with via) patterned wafers, elaborated by damascene integration and having the via and line dimensions from few micron to submicron (scheme figure 6-a). The patterned circuits having the resistances structure, Kelvin structures, leakage current structure etc and thin part of bulk wafer are successively stacked without accurate pattern alignment but simple mechanical notch alignment. Implemented for this process steps were the same as the ones developed for double strata patterned wafers achievement. Each additional strata stacking on the thinned top surface needed its covering by 400nm of low temperature PECVD oxide, SiO2 surface conditioning, bonding, thermal stabilization and backside thinning. It is very important in successive strata stacking process application to preserve the quality of the previously stacked layers and assure the efficient bonding and thinning process of the latest one.

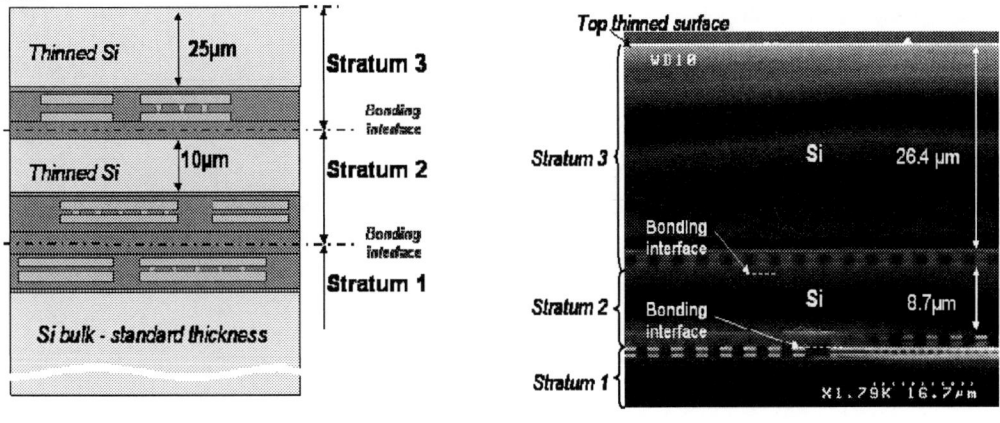

a   b

Figure 6. Cross sectional view of three-strata vertically integrated patterned wafers: a-schematically represented stack, b- SEM cross section.

After the double wafer stack achievement by face to face at wafer level bonding and thinning, the thinned backside surface was covered by same as before, low temperature oxide and conditioned for direct bonding. Then the third wafer was bonded, stabilized and backside thinned down. By this approach, we have built a three strata stack integrated at the wafer level. As shown on the SEM picture given by figure 6-b, the three-strata stack has the patterned layer without damages despite the very thin second stratum bulk silicon, thinned at < 10μm and the third one slightly thicker than 25μm. This picture shows also the good patterned copper behavior no visible deformation and shifting of Cu patterned layers. Furthermore the strata adhesion has been preserved giving good interface stability at the second and third wafer interface.

CONCLUSIONS

Backside wafer thinning process was investigated as a key step in the 3D Integration allowing the bulk patterned wafer stacking using direct SiO_2 bonding. Very thin silicon layers have been achieved (< 10μm) without damaging the bonding interfaces and circuits. The stresses in the thinned stacks were efficiently released after 1μm SI removal by CMP from fine grinded surface. The crystalline quality of the thinned sub-surface has been preserved. These results obtained applying the standard technological steps and relatively simple stacking processes can qualify described 3D Integration process as compatible with low cost technology applications.
Thereby stacked ICs and thinned bulk Si substrates decrease the path length in vertical direction between the integrated strata and allow their interconnection by high density TSV. In this work we also demonstrated the feasibility of multi-strata stacking at wafer level using the direct bonding process and bulk silicon thinning process. This multi-strata stacked at wafer level structure, has also shown the required potentialities for further integration processes such as high density TSV.

ACKNOWLEDGMENTS

This work was supported by ALLIANCE - STMicroelectronics, Freescale and NXP. CEA-Leti technology and characterization teams are kindly acknowledged for sample elaboration.

REFERENCES

1. C.H. Yu, The third dimension-More Life for Moore's Law, Advanced Module Technology Division, R&D, Taiwan Semiconductor Manufacturing Company (2006).
2. B. Charlet, 3-D Integration Latest Developments at LETI, Mater. Res. Soc. Symp. Proc. Vol. 970 (2007).
3. S.H. Christiansen, R. Singh, U. Gösele, Wafer direct bonding: From advanced substrate engineering to future applications in micro/nanoelectronics, Proceedings of the IEEE, Vol. 94, No. 12, December 2006.

4. R.S. Patti, Three-dimensional integrated circuits and the future of system-on-chip designs, Proceedings of the IEEE, Vol. 94, No. 6, June 2006.

5. A. W. Topol et al., Three-dimensional integrated circuits, IBM J. Res. & Dev., Vol. 50 no. 4/5 July/September 2006.
6. Ph. Garroud, - 3D Integration: A status Report" - 3D architectures for Semiconductor Integration and Packaging, Tempe, Arisona June 2005.
7. M. Koyanagi and al – Three-Dimensional Integration Technology Based on Wafer Bonding with Vertical Buried Interconnections; IEEE Transactions on Electron Devices, vol. 53,No.11, Nov. 2006.
8. P.H. Townsend and al. Elastic relationships in layered composite media with approximation for the case of thin on a thick substrate. J. Appl. Phys. 62(11), December 1987.
9. H. Moriceau et al., 7th Int.Symp.on Semiconductor Wafer Bonding. ECS Proceedings PV2003-19 p.49.
10. Z.J. Pei, S.R. Billingsley, S. Miura, Grinding-induced subsurface cracks in silicon wafers, International Journal of Machine Tools Manufacture 39 (7) (1999) 1103–1116.
11. Z.J. Pei, A study on surface grinding of 300 mm silicon wafers, International Journal of Machine Tools & Manufacture 42 (2002) 385–393.
12. B. Charlet et al. 3-D IC Integration: Technology and Integration, Willey VCH book edited by P. Garrou, P. Ramm & C. Bower - in press.
13. P. Leduc et al. Enabling technologies for 3D chip stacking, VLSI-TSA 2008, - to be published.

Mater. Res. Soc. Symp. Proc. Vol. 1079 © 2008 Materials Research Society 1079-N08-09

Dependence of thermal stability of NiSi and Ni(Pt)Si /Si on crystal orientation

Kazuya Okubo[1], Kazuo Kawamura[1], Shinich Akiyama[1], Yasutoshi Kotaka[2], Tsukasa Itani[2], Hirofumi Watatani[1], Kenichi Yanai[1], Masafumi Nakaishi[1], and Masataka Kase[1]

[1]Fujitsu Limited, Tokyo, 197-0833, Japan

[2]Fujitsu Laboratories, Tokyo, 197-0833, Japan

ABSTRACT

We report NiSi and Ni(Pt)Si films with excellent thermal stability showing a particular crystal orientation on Si(001). The NiSi and Ni(Pt)Si films with a particular crystal orientation formed through Ni or Ni(Pt) deposition at a temperature range from 200 to 240°C consists of a conformal domain structure. We examined detailed crystallographic analysis of silicide and clarified that the psudo-epitaxial growth of NiSi(202)//Si(220) [or NiSi(211)//Si(220)] was the key scheme of superior thermal stability. NiSi films with a particular crystal orientation is thermally stable up to 650 °C and shows low fluctuation in sheet resistance in narrow lines and low junction leakage current in electrical measurements by using optimized NiSi formation process. This process is a promising candidate for future silicidation technology.

INTRODUCTION

For sub-90 nm technology node, NiSi is widely used as a contact material because of several advantages such as small distribution of sheet resistance in narrow lines [1]. However, there are several problems in realizing NiSi-based MOSFETs, because of inferior thermal stability such as agglomeration of NiSi and the formation of $NiSi_2$ spikes [2]. In order to suppress the NiSi agglomeration, Ni-alloys silicide have been used[3]. The addition of Pt is one of the candidates of Ni-alloy silicide because Ni and Pt are homologous elements, and both NiSi and PtSi have an orthorhombic structure. However, the thermal stability of thinner Ni-alloy silicide films used for future CMOS devices is not sufficient. So, improvement of the thermal stability of Ni(-alloy) silicide films must be the key to realize the future CMOS devices. In this work, we report NiSi and Ni(Pt)Si with superior thermal stability showing a particular crystal orientation on Si(001). We examined detailed crystallographic analysis of silicide and investigated initial Ni/Si reaction.

EXPERIMENT

10-nm-thick Ni or $Ni_{0.95}Pt_{0.05}$ film was deposited with deposition temperatures T_{Ni} of 50, 200, 220 and 240 °C. To achieve the formation of Ni-silicide, the samples were annealed using a rapid thermal processor in the temperature range from 400 to 650 °C for 30 s in nitrogen ambient. The phases and crystallographic structures of the films were investigated with an X-ray diffractometer (XRD) in θ-2θ geometry. The sheet resistance (Rs) was monitored with a four-point probe system, and transmission electron microscopy (TEM) was used to examine the surface morphology and microstructure of the films. For the electrical property measurements of Ni-silicide films, (001) Si substrates patterned using a 65-nm process (L_G= 30nm) were used. 10-nm-thick Ni films were deposited and the first annealing process for silicidation was performed. After stripping any unreacted metal, NiSi formation was completed with a second silicidation .

119

RESULTS and DISCUSSION

<u>Thermal stability and crystal structures</u>

Figure 1 shows values of R_s of the (a)NiSi and (b)Ni(Pt)Si films for various deposition temperatures of Ni or Ni(Pt) as a function of annealing temperature. In the case of NiSi sample (Fig. 1(a)), the value of R_s with $T_{Ni}= 50$ °C increased above 600 °C due to agglomeration. However, there was no degradation of R_s even after annealing at 650 °C in the case of deposition at 200-240 °C, indicating that these films had excellent thermal stability. In the case of Ni(Pt)Si sample(Fig. 1(b)), though themal stability improved a little by Pt addition, the Rs value with $T_{Ni}= 50$ °C increased above 650 °C. However, there was no degradation of R_s in the case of deposition at 220-240 °C. As well as NiSi, thermal stability of Ni(Pt)Si improves in the case of deposition at high temperature.

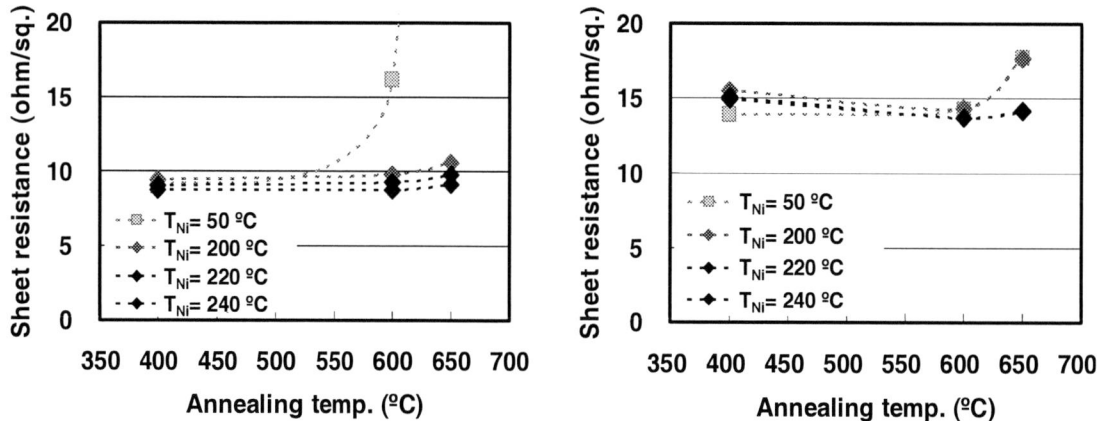

Figure 1. Sheet resistance of (a) NiSi and (b) Ni(Pt)Si with the nickel deposition temperature of 50, 200, 220 and 240 °C in the temperature range from 400 to 650 °C for 30 s. A 10-nm Ni or $Ni_{0.95}Pt_{0.05}$ film was deposited on p-type (001) single-crystalline Si substrates.

To clarify the differences in thermal stability, TEM observation of the NiSi films was performed. Plan-view TEM images of the NiSi films with $T_{Ni}= 50$ °C and 240 °C are shown in Figs. 2(a) and 2(b), respectively. The NiSi sample with $T_{Ni}= 50$ °C had two types of regions. One was a domain structure with grains partly satisfying the Bragg condition [4]. The other showed a structure that did not satisfy the Bragg condition, with localized small grains. On the other hand, the NiSi sample with $T_{Ni}= 240$ °C showed only the domain structure satisfying the Bragg condition preferentially parallel to Si<110>.

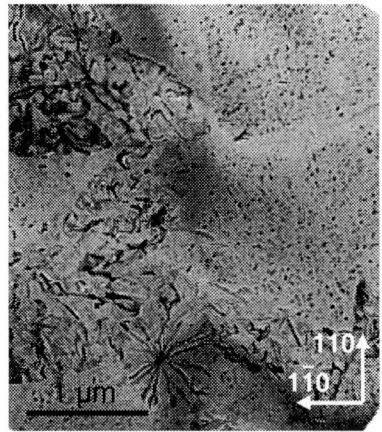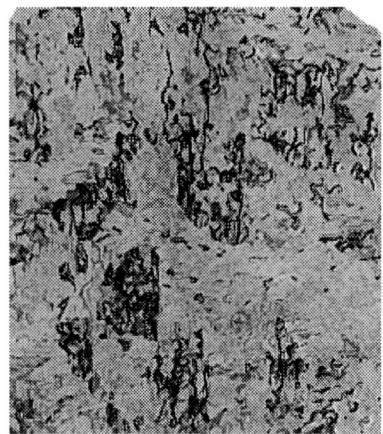

Figure 2. Plan-view TEM images of the NiSi films formed at different nickel deposition temperatures: (a) 50 °C and (b) 240 °C.

Figure 3 shows XRD spectra of (a) nickel silicide and (b) nickel platinum silicide films after annealing at 400 °C, formed at different deposition temperatures. Preferential orientation was observed in the samples with excellent thermal stability. Random orientation was observed in NiSi at 50 °C while a strong (111) [or (102)] peak was observed in NiSi at 200-240 °C. Because NiSi(111) is nearly perpendicular to NiSi(20-2) [or NiSi(2-11)] with the lattice space of 0.192 nm, we speculate that a psudo-epitaxial structure of NiSi(202)//Si(220) [or NiSi(211)//Si(220)] is formed, enhancing the thermal stability of the films. This is confirmed by TEM showing NiSi domains running along Si<110> [or Si<1-10>].

On the other hand, in the case of Ni(Pt)Si, NiSi(111) [or (102)] and additional NiSi(202) [or (211)] peaks are observed in Ni(Pt)Si at 200-240 °C. Since NiSi(202) is almost perpendicular to NiSi(20-2) [or Ni(211) is almost perpendicular to NiSi(2-1-1)], it is deduced that Ni(Pt)Si has a similar psudo-epitaxial structure of Ni(Pt)Si(202)//Si(220) [or Ni(Pt)Si(211)//Si(220)]. This is confirmed by HR-TEM images of Ni(Pt)Si sample with T_{Ni} =240°C in Fig. 4. The mechanism of the improved thermal stability of Ni(Pt)Si is due to the psudo-epitaxial structure similar to that of NiSi. However, the detected peaks in XRD is different between NiSi and Ni(Pt)Si samples. There is little change in a theoretical lattice constant in the case that Pt is put by several %. The reason why preferential orientation is different between NiSi and Ni(Pt)Si is not so clear, but the strain effects such as the difference of thermal expansion of NiSi and PtSi may affect the increase of lattice space of NiSi by Pt addition[5].

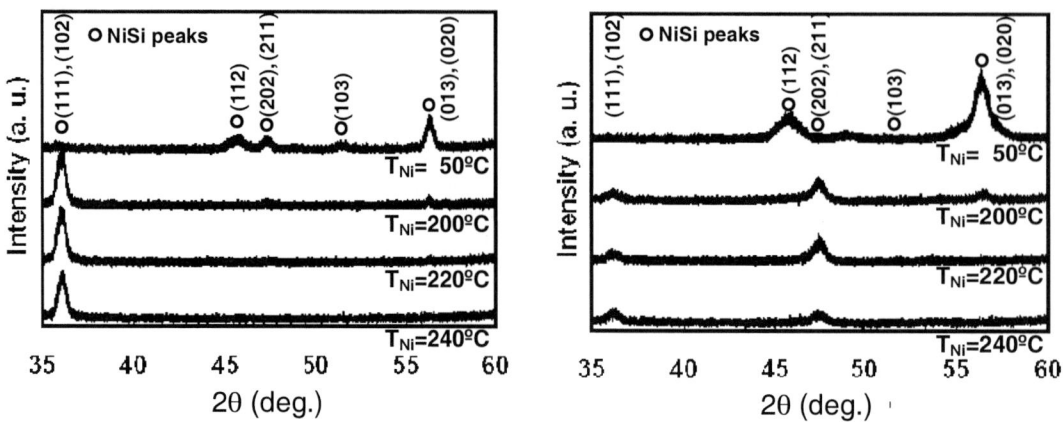

Figure 3. XRD spectra of (a) nickel silicide and (b) nickel platinum silicide films after annealing at 400 °C, formed at different deposition temperatures.

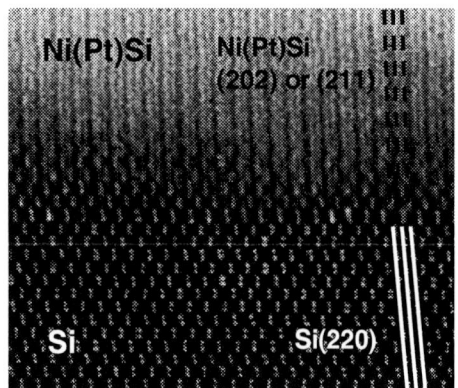

Figure 4. Dark field HR-TEM image of the Ni(Pt)Si sample with nickel deposition temperature of 240 °C. Ni(Pt)Si(202)[or (211)] grew epitaxially on Si(220).

Initial reactions

We further investigated the formation of thermally stable NiSi structure. We focused on the initial reaction when Ni atoms were deposited on the Si substrate. Figure 5 shows Rs value of Ni(10nm)/Si samples as a function of annealing time with T_{Ni}= 50 °C and 240 °C.
In the case of Ni/Si sample with T_{Ni} =50°C, the Rs value of as-deposited sample is low of 23 ohm/sq, which is almost the same as that of Ni films deposed on SiO_2. So, there is no reaction between Ni and Si in the case of deposition at 50°C. Then, the values of Rs become higher with the annealing time and saturated for 120s. XRD spectra of Ni/Si films for various annealing time are shown in Fig. 6. In the case of Ni deposition at 50°C, Ni(111) peak is detected only for no-anneal sample and Ni_2Si peaks are detected for annealed samples. Thus, Ni_2Si layer was formed due to annealing, so Rs value was increased. On the other hand, in the case of Ni/Si samples with T_{Ni}= 240 °C, all of the samples exhibited high Rs value of less than 50 ohm/sq. (see Fig. 5), and no crystal structure could be observed regardless of annealing time in Fig. 6(b). From these

results, we deduced that the Ni atoms diffused into the Si substrate and formed an amorphous Ni-Si layer [6][7]. As in the case for Ni deposition at high temperature, each Ni atom diffused easily into Si before the next Ni atom arrived. At substrate temperatures of over 400°C, Ni atoms could react at once and formed $NiSi_2$ because they had sufficient thermal energy to form the most stable structure in the Ni/Si system. On the other hand, at a deposition temperature of about 200°C, Ni atoms diffused easily but did not react easily with the Si substrate. This amorphous layer may lead to the formation of NiSi films having a particular orientation through control of the Ni reaction. NiSi with this particular crystal orientation, probably with low bulk and interfacial energy, is thermally stable and resistant to agglomeration.

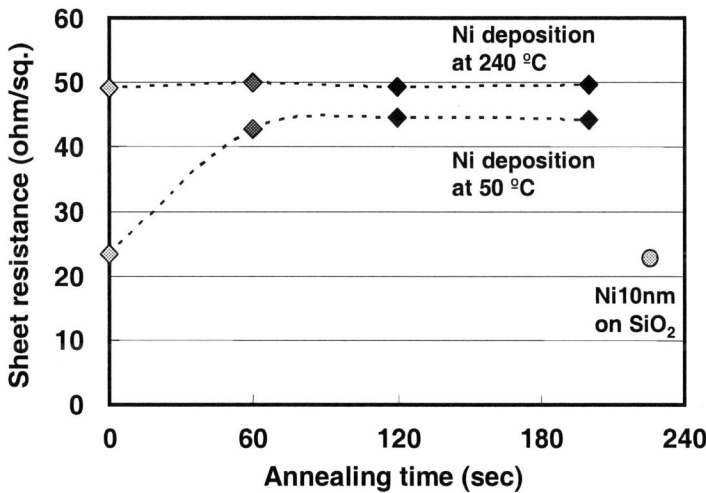

Figure 5. Sheet resistance of Ni/Si samples as a function of annealing time with the nickel deposition temperature of 50 and 240 °C. The annealing temperature is 240°C

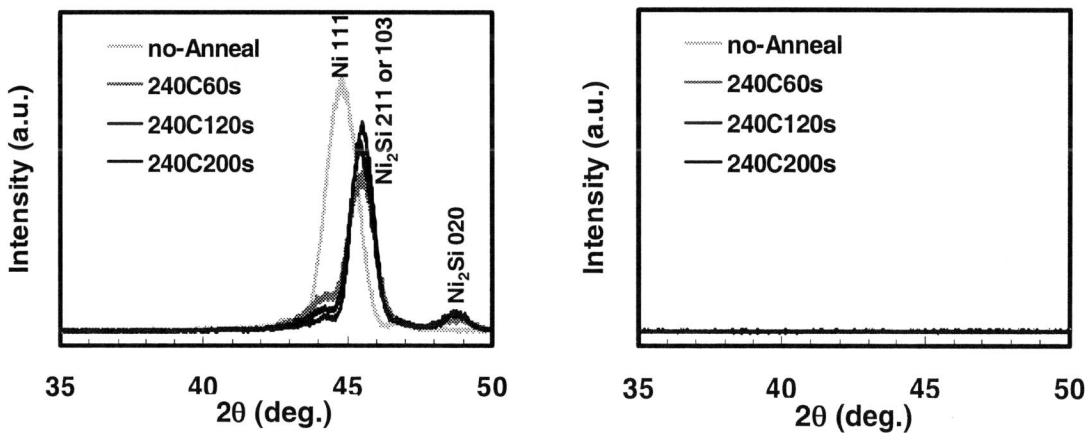

Figure 6. XRD spectra of Ni(10nm)/Si samples with the nickel deposition temperature of (a)50 and (b)240 °C, after annealing at 240°C for various annealing time.

Electrical properties measurements

We tried various conditions to improve the thermal stability of NiSi on a patterned wafer. Rs value of the NiSi gate is shown in Fig. 7. The R_s value of the improved NiSi was low and stable compared with the conventional process. The leakage current of the SRAM monitor is shown in Fig. 8. The leakage current for the improved condition decreased to 10% of the conventional one. The improvement in thermal stability of NiSi films may have retarded the diffusion of Ni atoms during annealing, so the fluctuation in sheet resistance and leakage current was successfully suppressed.

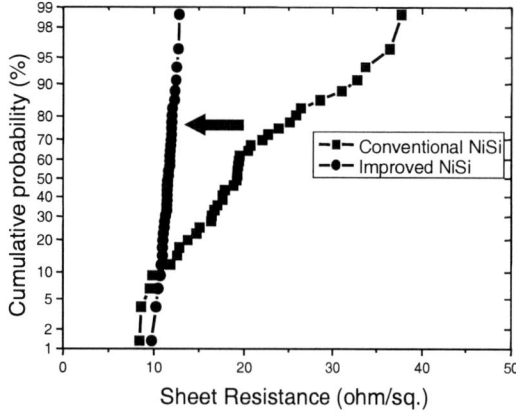 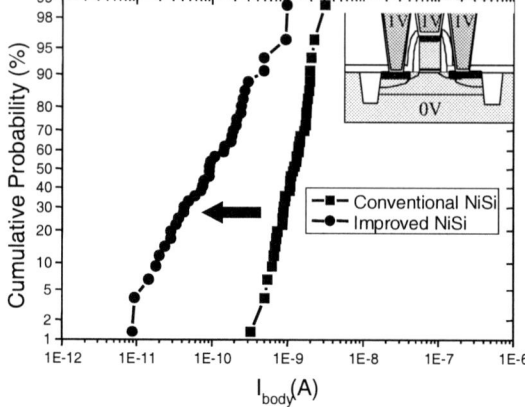

Figure 7. Sheet resistance of NiSi films on p^+ poly-gate region: (a) conventional and (b) improved NiSi formation processes.

Figure 8. Leakage current of 1k SRAM monitor with (a) conventional and (b) improved NiSi formation processes.

CONCLUSIONS

We have successfully formed NiSi and Ni(Pt)Si films with excellent thermal stability by high-temperature sputtering. Although the detected peaks in XRD is different in NiSi and Ni(Pt)Si samples, psudo-epitaxial structure of NiSi(202)//Si(220) [or NiSi(211)//Si(220)] is formed because of the initial reaction of an amorphous Ni-Si layer, enhancing the thermal stability of the films. By using this optimized Ni-silicide formation process, low fluctuation in sheet resistance and low leakage current in electrical measurements is realized. We suggest that this method is a promising candidate for future logic devices to realize nickel silicide contacts with low resistance and leakage current.

REFERENCES

1. J. P. Lu et al., IEDM Tech. Dig. 2002, (2002) 371
2. V. Teodorescu et al., J. Appl. Phys., 90 (2001) 167
3. K. Park, et al., JES 2007, 154, H557 (2007).
4. A. Alberti, et al., Acta Crystallographica, B 61, 486 (2005).
5. C. Detavernier et al., Appl. Phys. Lett., 84, 18 (2004).
6. G. B. Kim et al., J. Vac. Sci. Technol. B 21, 319 (2003).
7. L. J. Chen et al., Mater. Sci. Eng., R 29, 115 (2000).

Mater. Res. Soc. Symp. Proc. Vol. 1079 © 2008 Materials Research Society

Modelling and Characterization of Ultrasonic Consolidation Process of Aluminium Alloys

Amir Mohammed Siddiq, and Elaheh Ghassemieh
Mechanical Engineering, University of Sheffield, Mappin street, Sheffield, S1 3JD, United Kingdom

ABSTRACT

Ultrasonic consolidation process is a rapid manufacturing process used to join thin layers of metal at low temperatures and low energy consumption. In this work, finite element method has been used to simulate the ultrasonic consolidation of Aluminium alloys 6061 (AA-6061) and 3003 (AA-3003). A thermomechanical material model has been developed in the framework of continuum cyclic plasticity theory which takes into account both volume (acoustic softening) and surface (thermal softening due to friction) effects. A friction model based on experimental studies has been developed, which takes into account the dependence of coefficient of friction upon contact pressure, amount of slip, temperature and number of cycles. Using the developed material and friction model ultrasonic consolidation (UC) process has been simulated for various combinations of process parameters involved. Experimental observations are explained on the basis of the results obtained in the present study. The current research provides the opportunity to explain the differences of the behaviour of AA-6061 and AA-3003 during the ultrasonic consolidation process. Finally, trends of the experimentally measured fracture energies of the bonded specimen are compared to the predicted friction work at the weld interface resulted from the simulation at similar process condition. Similarity of the trends indicates the validity of the developed model in its predictive capability of the process.

INTRODUCTION

Ultrasonic consolidation is a rapid manufacturing process in which ultrasonic energy is used to create a solid state bond among different layers of composites, metals and alloys. It is a solid state joining process with low temperature and requires low process energy. Main advantages of ultrasonic consolidation include, absence of liquid-solid transformations, no atmosphere control required, low energy consumption, and low temperature allows embedding of electronics, such as sensors and actuators. Ultrasonic power required to join two components increase as the size (thickness) of the specimens being joined increase. Therefore, this limitation of thickness has restricted its use to microelectronics industry. Typical applications include, metal encased sensors, metal composite shields, fibre reinforced metal/matrix composites, satellite panels, electrical and electronic joints.

A number of researchers [1-6] have performed experimental studies on ultrasonic consolidation process. However, very little effort has been made to develop theoretical models to simulate ultrasonic consolidation process [7]. In all the theoretical and simulated works, the effect of ultrasonic vibration is attributed in the friction coefficient rather than taking into account both surface and volume effects. In the presented work, a material model based on cyclic plasticity theory has been proposed to take into account the volume effects as well as surface effects. Also, a kinematic friction model has been proposed to include the contribution of surface effects during ultrasonic consolidation process. The process of UC is simulated using developed finite element (FE) model with the introduced material and friction models that are especially formulated for UC condition. Simulations are performed for both UC of monolithic welding of aluminum foils and UC of embedment of Sic fiber in between the aluminum foils. The effect of UC process parameters on the quality of the weld is investigated and the results for aluminum alloy 3003 and 6061 are compared.

METHODOLOGY

Finite element Model

Finite element analyses of the ultrasonic consolidation process have been performed using coupled temperature-displacement analysis. Typical ultrasonic consolidation setup of monolithic structure is shown in Figure 1. The welding setup consists of three main components, a foil, a substrate and a sonotrode (attached to the ultrasonic welding unit). The substrate is fixed to an anvil at the bottom surface and it consists of ten layers of previously welded foils. A foil which acts as next layer of the structure is placed at the top surface of the substrate.

The sonotrode vibrates at a frequency of 20 kHz in the direction perpendicular to the rolling (welding) direction. A simultaneous vertical load is applied to the sonotrode to keep the surfaces of the sonotrode and foils in close contact.

The process parameters during ultrasonic welding are applied load ($P_{applied}$), velocity of sonotrode (V) and amplitude of vibration. These are varied in order to investigate the effect of each of these parameters on weld quality. The frequency of the ultrasonic vibration (f) is kept constant at 20 KHz in this study. The applied load is varied between 25-175 MPa. The amplitude of vibration is chosen in the range of 8.4-14.4 μm while velocity of sonotrode is between 27.8-38.8 mm/sec.

The geometrical dimensions are as follows: the width of the specimen (w) 20 mm, the thickness of the substrate (t_s) 1 mm, the thickness of the foil (t_f) 100 μm. The radius of the sonotrode is 25 mm [2].

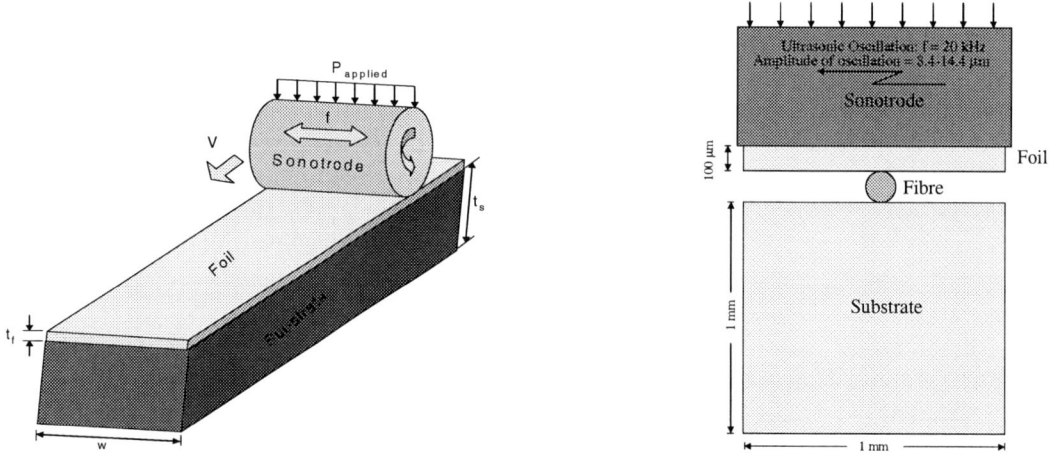

Figure 1: Ultrasonic metal welding setup   **Figure 2: Two dimensional model of the UC fiber embedding**

In the case of simulation of fiber embedding, a two dimensional model is used which is shown in Figure 2. The specimen consists of a foil, fibre and substrate. Foil and substrate has been modelled using the material model described below while the sonotrode and Sic fiber are considered as rigid materials [10].

Materials and Friction models

Material properties of the aluminium alloys (AA-6061 and AA-3003) have been assigned using a developed material model. The material model is based on the combined nonlinear isotropic/kinematic hardening model for time independent cyclic plasticity presented by Chaboche and coworkers [8, 9]. The modified isotropic and kinematic hardening equations and the parameters derived using inverse modeling are defined in another article under review and are not reported for brevity. Comparison of experimental [14-16] and simulated response is plotted Figure 3.

Thermomechanical interaction properties (normal and tangential interaction) at various interfaces are as follows. The normal interaction at various interfaces (i.e. foil/sonotrode, foil/fibre, fibre/substrate, and foil/substrate interfaces) is defined using hard contact formulation available in ABAQUS [10]. The friction (Tangential behavior) properties between foil (aluminium alloy) and sonotrode (steel) are defined using pressure dependent isothermal coefficient of friction in coulombs friction model [10]. While, the friction between foil and substrate is defined using thermomechanical friction model introduced in this work (table 1). This model is based on the dependence on the coefficient of friction μ, number of cycles N, temperature T, and parameters a and b which are affected by the magnitude of slip and contact pressure. Summary of the friction model is reported in table 1. Comparisons between the results of friction coefficient versus temperature and number of cycles measured by experiment [17-18] and simulated using the friction model in table 1, are plotted in Figure 4. Friction model (table 1) has been implemented in user friction subroutine (FRIC) available in ABAQUS [10].

Table 1: Friction model

Isothermal COF:	$\mu_{iso} = \mu_s + \mu_s \cdot \left(a \cdot \log(N) + b\right)$

where a and b are friction parameters depend upon the magnitude of the slip amplitude and contact pressure, while μ_s is the initial static coefficient of friction, and N is number of cycles.

Thermomechanical COF:	$\mu = \mu_{iso} \cdot \left(p \cdot T^4 + q \cdot T^3 + r \cdot T^2 + s \cdot T + t\right)$

The additional friction parameters p, q, r, s, t are identified using the experimental results [11 for aluminum alloy.

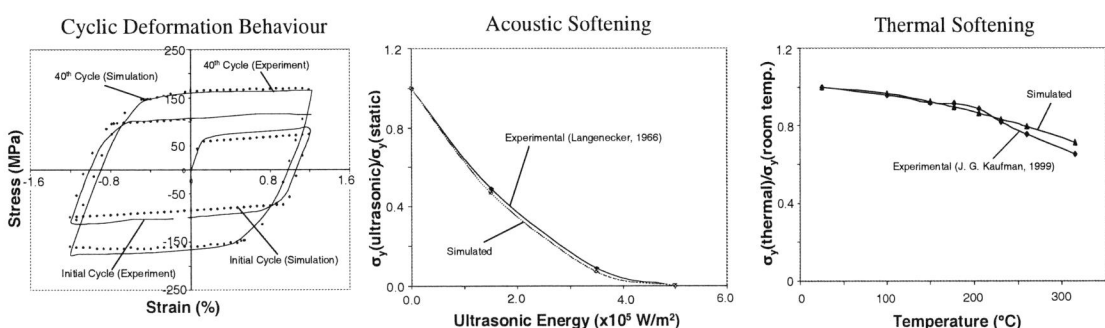

Figure 3: Comparison of experimental and simulated response during parameter identification

Friction properties at foil/fiber and substrate/fiber interfaces are defined using coulomb's friction model with coefficient of friction of 0.2. Throughout this work it is assumed that all the friction energy generated between different contacting surfaces is converted into thermal energy. Results of the thermomechanical analyses of the ultrasonic consolidation process for monolithic and fiber embedding specimen have been discussed in the following.

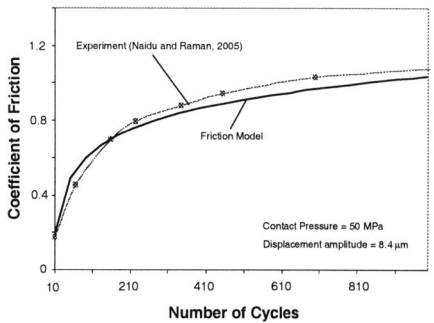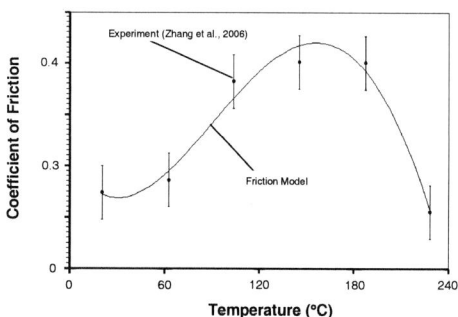

Figure 4: Comparison of experimental and simulated response during parameter identification

RESULTS and DISCUSSION

Monolithic Specimen

Temperature profile during ultrasonic consolidation of monolithic AA-6061 and AA-3003 is plotted in Figure 5(a). The plot in Figure 5(a) is only for on set of process parameters; other process parameters were also simulated and found to show similar behaviour. It has been found that maximum temperature at foil/sonotrode interface is highest due to the highest amount of friction dissipation at this interface. It is also found that maximum temperature attained during ultrasonic consolidation process is well below melting temperatures (30-60% of melting temperature). Results also show that temperature in AA-3003 is found to be higher than AA-6061, which is due to the fact that AA-3003 is much harder than AA-6061. It must also be noted that initial yield strength of AA-3003 is approximately 4 times larger than that of AA-6061. Temperature results are also in agreement with the experimental results of Cheng and Li [11], where temperature for the case of copper and nickel were found to be in the range of 100-250 °C at a distance of 200 µm (at Point A in Figure 5(a)) from the weld interface. Similar values of the temperature, with slight variations, are found for other sets of process parameters.

The contour plots of equivalent plastic strain are shown in Figure 5b. It is found that regions of the foil near foil/sonotrode interface undergo severe plastic deformation. This high plastic deformation is due to high friction dissipation and ultrasonic energy transferred to the foil near foil/sonotrode interface. It can also be inferred from Figure 5b that amount of plastic deformation in foil near the foil/substrate interface is higher than the plastic deformation in the substrate. This is due to the dual effect, i.e. surface (friction dissipation at the foil/sonotrode and foil/substrate interface) and volume (ultrasonic softening) effects. On the other hand the substrate has the dominating surface effects, i.e. friction dissipation at the foil/substrate interface, and very small amount of volume (acoustic softening) effects to cause plastic deformation. It is also found that plastic strains in AA-6061 at foil/sonotrode interface are almost twice as high as AA-3003, as AA-6061 is 4 time softer than AA-3003. Also, plastic strains at the foil/substrate interface for AA-6061 is lower than AA-3003.

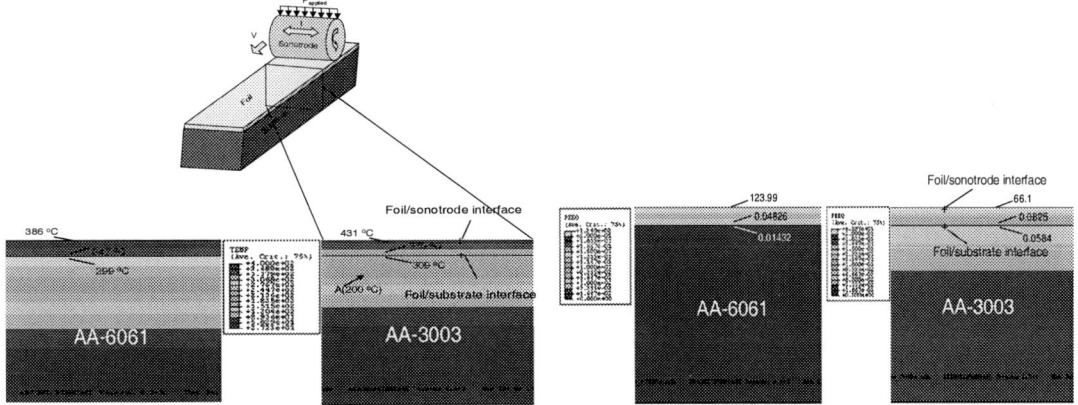

Figure 5: (a)Temperature profile in monolithic specimen (velocity=27.8 mm/sec; Load = 175 MPa; Amplitude = 8.4 μm), (b) Equivalent plastic strain in substrate and foil at foil/substrate interface (velocity=27.8 mm/sec; Load = 175 MPa; Amplitude = 8.4 μm)

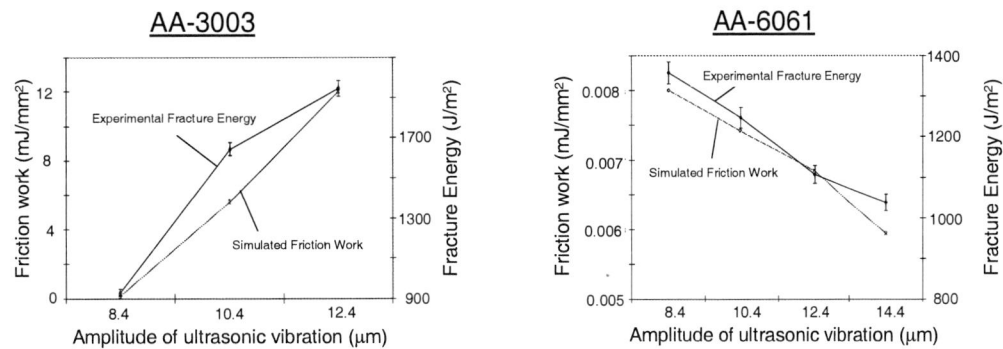

Figure 6: Friction work and experimental fracture energies as a function of amplitude of ultrasonic vibration

This is due to the higher friction work for the case of AA-3003 which is found to be 10-53 time higher than that of AA-6061 for various combinations of process parameters (see Figure 6). A comparison among the friction work at the weld interface and fracture energies computed from peel test curves for both alloys (AA-3003 and AA-6061) has been made in. For the case of AA-3003, experimental fracture energies are found to be increasing with increasing amplitude of ultrasonic vibration, similar trends can be seen in Figure 6 (left figure) for friction work at the weld interface. Simulation results conclude that increase of fracture energy is mainly due to the increasing friction work at the weld interface. The reason for the increase could be understood by considering the fact that as the amount of applied loading is always lower than the static yield stress of AA-3003, the friction dissipation starts well before yielding can occur as a result of applied load and ultrasonic oscillation. Therefore the friction work for the case of AA-3003 increases with the increasing applied load and amplitude of ultrasonic vibration. On the other hand, experimental fracture energy and friction work show an opposite trend for AA-6061 (Figure 6). It can be inferred from Figure 6 (right figure) that experimental fracture energies as well as friction work from simulation decrease with increasing amplitude of ultrasonic vibration. Simulation results conclude that the decrease in friction work causes less breaking off and dispersion of oxide layers, ultimately resulting into smaller number of bonds between the opposing surfaces (weaker joints). The decrease in friction work for the case of AA-6061 is due to the plastic dissipation which starts in less number of ultrasonic cycles. Another important conclusion that can be inferred from Figure 6 that friction work at the weld interface for the case of AA-3003 is higher than that of the AA-6061. This is why AA-6061 required extra cleaning process [3] before welding while AA-3003 was successfully welded without any prior treatment.

Fiber Embedding Specimen

Finite element analyses of Silicon Carbide (SiC) fibre embedding in AA-6061 have been performed using the material and friction models developed in this work. A qualitative comparison of plastic deformation with experimental results is shown in Figure 7a. Both simulation and experimental results indicate severe deformation at foil/sonotrode interface during ultrasonic consolidation experiments. It is also found that material flows completely around the fibre as is observed during ultrasonic consolidation experiments.

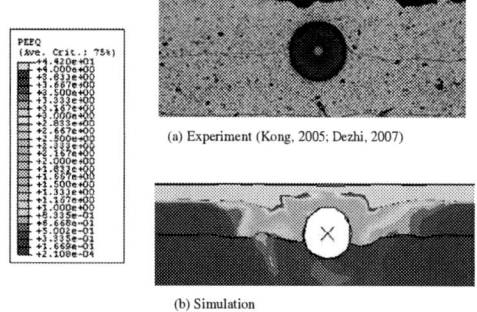

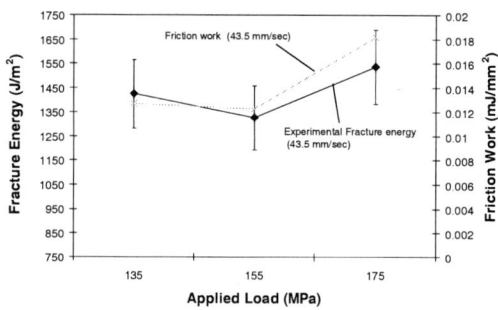

Figure 7: (a) Comparison between experimental and simulated material deformation in a UC specimen, (b) Comparison between experimental fracture energies and simulated friction work for various applied loadings and velocities of sonotrode (displacement amplitude of ultrasonic vibration = 10.4 μm)

In other words, during experiments [2, 3, 12, 13] material showed full closure of voids around the fiber (Figure 7a), results of the thermomechanical analyses of ultrasonic consolidation process show the similar effects. Maximum plastic deformation in fiber embedding specimen is found to be more than 50 times higher as compared to the monolithic specimens. This is due to the large plastic deformation in the aluminum alloy as a consequence of material flow around the fiber.

A comparison between experimental fracture energies obtained from peel tests and simulated friction work (obtained in the present work) is shown in Figure 7b. Results for the case when velocity of sonotrode is 43.5 mm/sec, have been plotted for the cases when displacement amplitude of the ultrasonic vibration was 10.4 μm. Simulated friction works show similar trends as experimentally measured fracture energies [2, 12].

CONCLUSIONS

The presented work shows the capability of the developed material and friction model in simulating the ultrasonic consolidation process for monolithic as well as fibre embedding specimen. The results of thermomechanical analyses showed that ultrasonic consolidation of monolithic AA-6061 behaves differently than AA-3003. Maximum temperature is found to be 30-60% of the melting temperature. A comparison of friction work values of AA-6061 and AA-3003 was also made and it is found that friction work of AA-3003 is always higher than the friction work of AA-6061. Comparison of experimental fracture energies and simulated friction works show similar trend for different process parameters.
Thermomechanical analyses results of fibre embedding using ultrasonic consolidation process showed that material flows completely around the fibre as is observed during ultrasonic consolidation experiments. Plastic flow is found to be maximum at foil/sonotrode interface. Comparison of experimental fracture energies and simulated friction work showed similar trends

ACKNOWLEDGMENTS

The authors thankfully acknowledge the financial support of EPSRC (Engineering and Physical Sciences Research Council) and MOD (Ministry of Defence) through the grant (GR/T19988) and the collaborative support of the Solidica Ltd.

REFERENCES

1. Joshi, K.C., *The formation of ultrasonic bonds between metals.* Welding Journal, 1971. **50**: p. 840-848.
2. Kong, C.Y., *Investigation of Ultrasonic consolidation for embedding active/passive fibres in aluminium matrices*, in *Rapid Manufacturing Center*. 2005, Loughborough University: Loughborough, UK. p. 1-207.
3. Kong, C.Y., R.C. Soar, and P.M. Dickens, *Characterisation of aluminium alloy 6061 for the ultrasonic consolidation process.* Materials Science and Engineering A, 2003. **360**: p. 99-106.
4. Kong, C.Y., R.C. Soar, and P.M. Dickens, *Optimum process parameters for ultrasonic consolidation of 3003 aluminium.* Journal of Materials Processing Technology, 2004. **146**: p. 181-187.
5. Krzanowski, J.E., *A Transmission Electron Microscopy Study of Ultrasonic Wire Bonding.* IEEE Trans. CHMT, 1990. **13**: p. 176-181.
6. Tucker, J.C., *Ultrasonic welding of copper to laminate circuit board*, in *Materials Science & Engineering*. 2002, Worcester Polytechnic Institute. p. 125.
7. Doumanidis, C. and Y. Gao, *Mechanical Modelling of Ultrasonic Welding.* Welding Journal, 2004. **4**: p. 140-146.
8. Chaboche, J.L., *Time independent constitutive theories for cyclic plasticity.* International Journal of Plasticity, 1986. **5**: p. 247-302.
9. Chaboche, J.L., *Constitutive equations for cyclic plasticity and cyclic viscoplasticity.* International Journal of Plasticity, 1989. **5**: p. 247-302.
10. ABAQUS, *ABAQUS version 6.5, Online Documentation.* 2006, Hibbit & Karlsson.

11. Cheng, X. and X. Li, *Investigation of heat generation in ultrasonic metal welding using micro sensor arrays.* J. Micromechanics and microengineering, 2007. **17**(2): p. 273-282.

12. Li, D., *Defining Optimum Parameters for Embedding SiC fibres and influence on bond strength in Al 6061 matrix.* 2007, Rapid Manufacturing Research Group, Loughborough University: Loughborough, UK. p. 1-19.

13. Li, D., *Report for machine performance validation through comparison of peel strength and linear weld density of samples made before and after machine maintenance.* 2007, Rapid Manufacturing Research Group, Loughborough University: Loughborough. p. 1-14.

14. Davis, J. R., *ASM specialty handbook: Aluminium and aluminium aloys.* 1993 ASM international Ohio.

15. Kaufman, J. G., *Properties of Aluminium Alloys: Tensile, Creep, and Fatigue Data at High and Low Temperatures,* 1999 ASM, Metals Park, OH 44073-0002, USA.

16. Hopperstad, O. S., Langseth, M., Remseth, S., *Cyclic stress-strain behaviour of Alloy AA6060, Parth I: Uniaxial experiments and modeling.* Int. J. Plasticity, 1995. 11: p. 725-739.

17. Naidu, N. K. R, and Raman, S. G. S., *Effect of contact pressure on fretting fatigue behaviour of Al-Mg-Si alloy.* International Journal of Fatigue, 2005. 27: p. 283-291.

18. Zhang, C. B., Zhu, X. J., and Li, L. J., *A study of friction behaviour in ultrasonic welding (consolidation) of aluminium.* AWS Conference: Session 7: Friction and resistance welding/materials bonding process, 2006.

Growth and Integration of High-Density CNT for BEOL Interconnects

Ainhoa Romo Negreira[1,2], Daire J. Cott[1,2], Anne S. Verhulst[1,2], Santiago Esconjauregui[1,2], Nicolo' Chiodarelli[1,2], Johan Ek Weis[1], Caroline M. Whelan[1], Guido Groeseneken[1,2], Marc Heyns[1,3], Stefan De Gendt[1,4], and Philippe M. Vereecken[1]

[1]IMEC, Kapeldreef 75, Leuven, B-3001, Belgium

[2]Department of Electrical Engineering, Katholieke Universiteit Leuven, Kasteelpark Arenberg 1, Leuven, B-3001, Belgium

[3]Metallurgy and Materials Engineering Department, Katholieke Universiteit Leuven, Kasteelpark Arenberg 44 bus 2450, Leuven, B-3001, Belgium

[4]Chemistry Department, Katholieke Universiteit Leuven, Celestijnenlaan 200f- bus 2404, Leuven, B-3001, Belgium

ABSTRACT

The integration of high-density CNT bundles as via interconnects in a CNT/Cu-hybrid BEOL stack is evaluated. CNT via-conduits may greatly improve heat dissipation and as such lower interconnect resistance and improve electromigration resistance. Each carbon shell of the nanotube contributes to electrical and thermal conduction and densities as high as 5×10^{13} shells per cm^2 are estimated necessary. CNT growth processes on BEOL compatible metals are presented with tube densities up to $10^{12} cm^{-2}$ and shell densities approaching $10^{13} cm^{-2}$ on blanket substrates. Selective growth of CNT bundles with carbon shell densities around $10^{12} cm^{-2}$ is demonstrated with high yield. Ohmic behavior of TiN/CNT/Ti contacts is shown with a CNT via resistivity of 1.2 mΩ cm.

INTRODUCTION

Carbon nanotubes (CNT) are one-dimensional hollow nanostructures with extreme aspect ratios and exceptional mechanical, thermal and electrical properties [1]. One could picture a CNT by rolled-up sheets of graphene: a single graphene layer forms a single-walled (SW-), a double-layer forms a double-walled (DW), and multiple layers form a multi-walled (MW) tube. A single CNT can be as narrow as 0.4nm for the smallest SW-CNT and several hundred nanometer in diameter for MW-CNT [2]. Electrical conductance of ideal metallic CNT is ballistic and thus independent of length in contrast to metal conductors such as copper where conductance is determined by electron scattering. In addition, CNT can carry extreme current densities up to 10^9 A/cm^2 compared to about 10^6-10^7 A/cm^2 for (capped) interconnect lines. Maybe even more important for interconnect applications is the high thermal conductivity which is expected to be 5 to 10 times higher than copper. In this paper, the possibility to replace copper via interconnects in the BEOL stack by bundles of densely-packed CNT will be presented.

A CNT bundle can be considered a series of conducting shells in parallel and its resistance is determined by the total number of shells in the bundle. For example, a bundle of two SW-CNT and two DW-CNT has a total of four CNT (N=4) and six shells ($\sum_{N}^{0} n_{sCNT} = 6$), with n_{sCNT} the number of shells in a particular CNT. For CNT with not too large diameter (d<20nm), the resistance of a single CNT shell (R_{sCNT}) can be taken equal to the resistance of a single-walled CNT (diameter independent) and the total resistance is then given by:

$$R_{CNT,bundle} = \frac{R_{sCNT}}{\sum_{N}^{0} n_{sCNT}} \qquad . \qquad (1)$$

The resistance of a CNT shell is given by:

$$R_{sCNT} = R_{quantum,metallicSWCNT} + \rho_{ac.phon}l + \rho_{opt.phon}l \qquad (2)$$

with a first term for the fixed quantum resistance (6.5kΩ), and two terms for acoustic and optical phonon scattering, respectively, which are both linear with length, l, of the CNT [3]. Similar results were obtained by N. Srivastava et al [4, 5]. From best know literature values [6-8], a quantitative relation for equation (2) was derived:

$$R_{sCNT} = 6500\Omega + 5500\tfrac{\Omega}{\mu m} \times l + \frac{6500\Omega \times l}{0.2/_V \times l + 0.01\mu m} \qquad . \qquad (3)$$

Note that the term for optical phonon scattering is dependent on the voltage over the CNT interconnect bundle. Based on equations (1) and (3) comparison can be made with copper and the example of 9nm lines will be discussed.

From equation (1) it follows that the lowest resistance is obtained for the highest CNT shell density. Higher shell densities can be obtained by stacking many SWCNT closely together in one space than by having one MW-CNT in the same space. For example, in a box of 2.5nm by 2.5nm one can fit fourteen SW-CNT of 0.4nm diameter with a minimum spacing of 0.34nm in a close-packed arrangement compared with one 2.5nm MW-CNT with a maximum of four walls. The former case represents the upper limit of stacking density with 2×10^{14} shells per cm^2. Figure 1 shows the variations in resistance with interconnect length for cylindrical 9nm Cu line and CNT bundles with two packing densities. The resistance of the CNT bundles is much less

134

dependent on wire length due to the small scattering rate in CNT compared to Cu. Hence, CNT bundles could lower the resistance of longer interconnect structures especially for highest shell densities. Unfortunately, CNT bundles as horizontal interconnect lines are currently technologically not possible and CNT bundles are only considered at the via or contact levels. For very short lengths (the 9nm via would have a height of 20nm), however, the CNT resistance is worse than copper even for maximum packing density. Since both the delay and dynamic power, and thus the performance, are dominated by the interconnect capacitance and not by the interconnect resistance (for typical resistance values), a reasonable increase in resistance is tolerable. Even so, what would be the benefit of CNT bundles when they can not match the resistance of the copper vias? Well, there is of course the much higher current-carrying capacity (2-3 orders of magnitude better than Cu) which will improve electromigration. More importantly, there is the high thermal conductivity of the CNT vias that will significantly improve heat dissipation and as such have a beneficial impact on the resistance as well as on the electromigration of the total Cu-CNT hybrid stack. N. Srivastava et al have calculated that the integration of high- density CNT bundles at the via level would keep the temperature of the Cu-CNT hybrid stack below 300C for the 22nm NODE whereas a Cu-only BEOL stack would run at temperatures above 600C [4, 5]. To keep the resistance of a few micron long CNT-via/Cu-line/CNT-via structure comparable to that of a Cu-only via-chain structure, an estimated CNT shell density of 5×10^{13} cm^{-2} or higher will be necessary. Such shell densities can be obtained by a closed packed stacking of either SW-CNT with diameters between 0.4 and 1.2 nm, DW-CNT with diameters between 1.08 and 1.8nm, triple-walled CNT with diameters between 1.76 and 2.4nm, quadruple-walled CNT with diameter between 2.44 and 2.7nm and even quintuple-walled CNT with a diameter of 3.12nm.

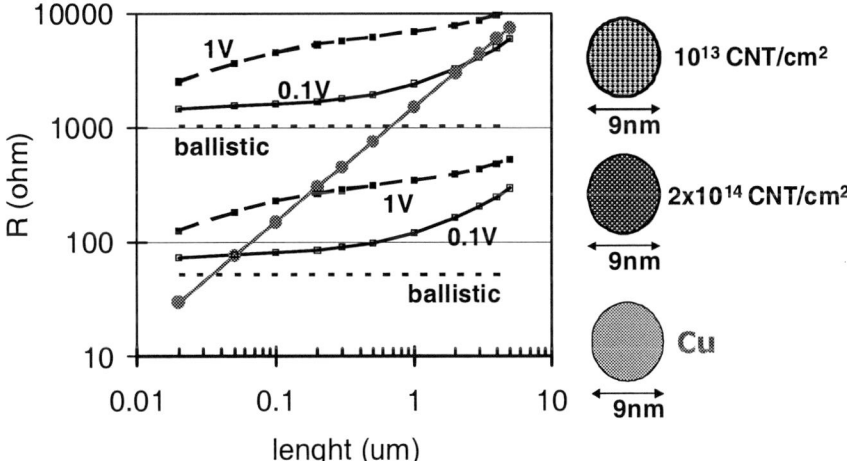

Figure 1. Resistance for 9nm diameter cylindrical Cu and CNT conduits as a function of length. CNT resistance calculated according to eqn.(1) for maximum carbon shell density of 2×10^{14} cm^{-2} and for a lower density of 10^{13} cm^{-2} for 0.1V and 1V. Also the ballistic limit for both densities is shown.

Growth of densely-packed CNT carpets can be achieved by catalytic CVD directly on a substrate surface after formation of Fe, Co or Ni based catalyst nanoparticles [9]. Most growth studies are done on SiO$_2$ surfaces where SW and DW CNT with densities of 10^{12} cm^{-2} and higher are easily obtained after proper catalyst activation step [10-12]. By using sandwiched layers with aluminum (which oxidize into alumina) very high densities were obtained [13]. For interconnect applications the CNT bundles need to be grown on metals and preferably liner materials such as Ta, TaN and TiN. Unfortunately, the obtained CNT densities are typically lower on metal substrates [14]. At IMEC a process was developed for high-density growth on Ti-based substrates [15]. Growth of CNT bundles in via holes down to 150nm in diameter with shell densities of 10^{12} cm^{-2} was demonstrated. Ohmic current-voltage characteristics were obtained for the TiN/CNT/Ti contacts with a minimum equivalent resistivity of 1.2mΩ cm.

RESULTS and DISCUSSION

CNT growth on blanket substrates: effect of catalyst activation

Figure 2a shows a cross-sectional SEM image of a densely packed CNT carpet grown from a 0.5nm Fe film deposited with ion-beam sputtering on top of Ti(5nm)/SiO$_2$/Si wafers. Since the wafers had been exposed to air for considerable time before Fe deposition, the actual stack was Fe/TiO$_2$/Ti on SiO$_2$. Catalyst activation was done right before growth by treatment in hydrogen plasma (1 Torr H$_2$, 200W high-frequency plasma) at 650C for 5 minutes. Growth was achieved by thermal CVD (same chamber) in 1 Torr acetylene (C$_2$H$_2$) at 650C. Growth of densely aligned and straight CNT was observed with a length of 28um for 60min of deposition. Figure 2b shows a high-resolution TEM image of a single CNT plucked from the CNT mat. The CNT diameter was typically between 6 and 8 nm with 7 to 10 shells per CNT. The carbon density of the CNT film was estimated at 0.43g/cm^3 by simulation of the X-ray reflectivity (XRR) spectra. From the measured carbon density, an average CNT diameter of 7nm, and an average of 8 shells per CNT, the density of tubes was estimated 9×10^{11} cm^{-2} and thus approaching 10^{13} shells per cm^2.

Plasma activation of the catalyst was found to be a crucial factor for growth and determined both the CNT density and growth rate, under similar growth conditions. In the absence of plasma (thermal activation in H$_2$ only) sparse growth was observed. Conversely, extended plasma activation times also degraded growth as the catalyst material was eroded away. When the catalyst activation conditions were changed to 4 Torr H$_2$, 300W low-frequency plasma for the same time (5 minutes), the CNT density dropped by a factor of 10 while the growth rate increased by a factor of 10 as compared with the CNT in figure 2. Hence, for each plasma-enhanced catalyst activation condition, an optimum time needs to be found for the plasma-assisted formation of nanoparticles to achieve high densities while avoiding undesired erosion of the catalyst material. Note that achieving fast growth rates does not necessarily imply optimum conditions since it typically indicates lower density (under same growth conditions). When changing the growth conditions while keeping the catalyst activation the same, the CNT density was found the same while the growth rate was changing. Hence, the catalyst activation step determines the effective CNT density (and diameter) whereas the growth rate (for same density) is determined by the CNT growth conditions.

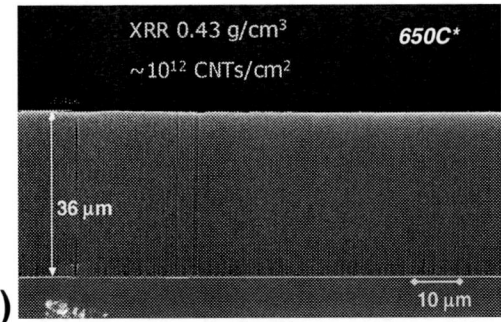 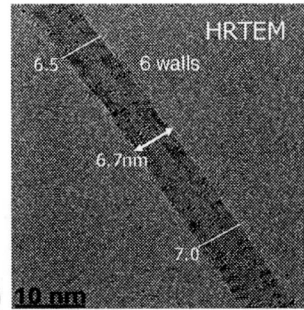

Figure 2. Cross-sectional SEM image of densely aligned CNT carpet (a) and high-resolution TEM image of single MW-CNT (b). CNT were grown from 0.5nm Fe catalyst deposited with ion-beam sputtering on 5nm PVD Ti layer on SiO_2/Si substrate: 5 minutes plasma-enhanced catalyst activation at 650C in 1Torr H_2 with 200W high-frequency plasma and 60 minutes growth at 650C in 1 Torr acetylene (C_2H_2).

CNT growth on blanket substrates: effect of under layer

CNT growth with 0.5nm Fe catalyst was compared for Al, Ti, TiN and Ta/TaN layers on SiO_2/Si wafers. The first two combinations are prone to oxidation and the metal surfaces will be completely transformed in Al_2O_3 and TiO_2 surfaces, respectively (Al and Ti will also react with the SiO_2 underlayer at higher temperatures). The nitride compounds are more stable to oxidation and will be primarily metallic. In all cases straight aligned growth of multi-walled CNT was obtained, with higher tube density for the oxide surfaces (~10^{12} cm^{-2}) compared to the metallic surfaces (~10^{11} cm^{-2}). The average diameter of the grown tubes was 6-9 nm on the oxide surfaces compared to 8-12 nm on the metallic surfaces. Since surface diffusion is inhibited on oxide surfaces, higher densities of smaller catalyst particles can be obtained during the plasma-enhanced thermal treatment during catalyst activation step. Interestingly, the CNT growth mechanism was found to be base growth (i.e. Fe remains on surfaces) for the Al_2O_3 and TiO_2 surfaces, while tip growth (i.e. Fe particles at the tip of the CNT) was observed for the Ta and TiN surfaces. Hence, the iron may also form compounds with the oxide thus being anchored on the surface.

CNT growth on blanket substrates: ECD versus PVD catalyst

Thus far we have shown that CNT density is largely affected by the conditions of catalyst activation and by the underlaying substrate. Also the nature of the catalyst itself will determine the density as well as growth kinetics. In general, for the same catalyst activation and growth conditions, the CNT growth was the fastest for Fe and the slowest for Ni, with Co in-between the two. The density (as well as growth rate) depended strongly on the amount of deposited catalyst, and typically less catalyst material was necessary for Fe and Co than for Ni to achieve similar densities. Two methods for deposition of the catalyst have been evaluated: physical vapor deposition (PVD) and electrochemical deposition (ECD). In the first case, (incomplete) thin films are deposited which are then transformed into particles by combination of thermal and plasma treatment. In the latter case, particles are immediately formed during the electrochemical nucleation and growth process: the particle density is determined by the applied potential and the particle size is determined by the amount of deposited charge (corrected for the current efficiency) [16].

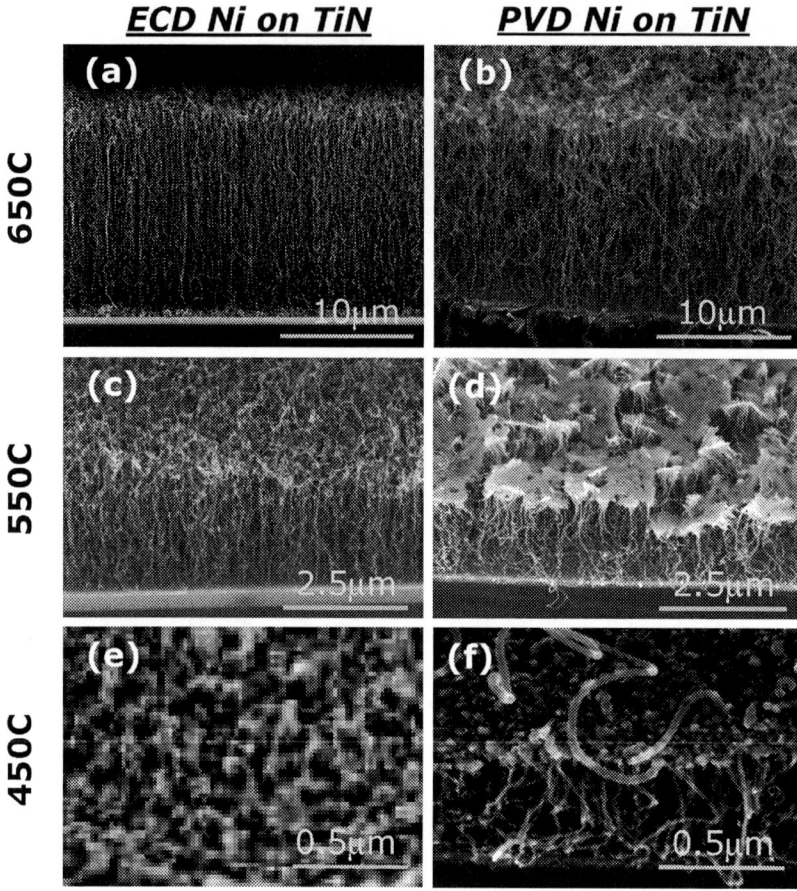

Figure 3. Cross-sectional SEM images of CNT carpet layers deposited from Ni catalyst with 5nm equivalent film thickness deposited with ECD ((a), (c) and (e)) and PVD ((b), (d) and (f)): plasma-enhanced catalyst activation in 1 Torr H_2 at 200W high-frequency for 5 min + thermal growth in 1 Torr acetylene for 10 minutes at 650C ((a) and (b)), 550C ((c) and (d)) and 450C ((e) and (f)); the temperature for catalyst activation and growth were the same.

To evaluate both deposition methods CNT growth was compared for samples with same amount of deposited catalyst material on TiN (70nm) and Ti(5nm)/TiN(70nm) substrates. Figure 3 shows SEM images for PVD and ECD Ni catalyst samples for catalyst activation and growth at different temperatures for the TiN substrates. In contrast to Fe catalyst, similar results were obtained for Ni on TiN and Ti/TiN substrates. At 650C, no significant difference between catalyst deposition methods was seen: aligned growth of carpet layer of densely aligned CNT about 13 to 14 um in length. At 550C, the CNT growth rate is lower resulting into 2-3 um long CNT carpets after 10 minutes. The ECD samples still produced aligned and dense CNT carpets. For the PVD samples, the CNT density was somewhat lower, their morphology more bended and, moreover, large clusters were present on top of the CNT. At 450C, growth has not really started yet after 10min for the 5nm ECD sample; however a thin CNT layer was grown for 10nm

ECD Ni (not shown). For the PVD samples, a crusted layer is now clearly visible on top of the about half a micron thick CNT carpet. Also for 0.8nm PVD Ni a thin crust is still distinguished and HRTEM revealed that in this case carbon nanofibers (CNF) were grown instead of tubes. These results indicate that the PVD layer was not completely broken up into particles at temperatures ≤ 550C. The agglomeration of PVD films is next to time, initial film thickness and ambient pressure, strongly dependent on temperature [17]. SEM inspection of the samples with the catalyst activation step only indeed showed clustered and connected Ni islands for the PVD films plasma treated at 550C, whereas the ECD Ni particles were similar to the as-deposited samples (see also Figure 5). Since the BEOL temperature budget is limited, the room temperature ECD technique has a significant industrial advantage.

CNT integration: selective catalyst placement and growth in via holes

The test structures used for integration experiments had large arrays of contact holes in 330nm oxide dielectric landing on a 70nm TiN layer serving as a common electrode. Arrays with hole diameters ranging between 300nm and 150nm were available with relative pitch of 2, 3 and 8. Figure 4 shows schematically the process flow followed for integration experiments.

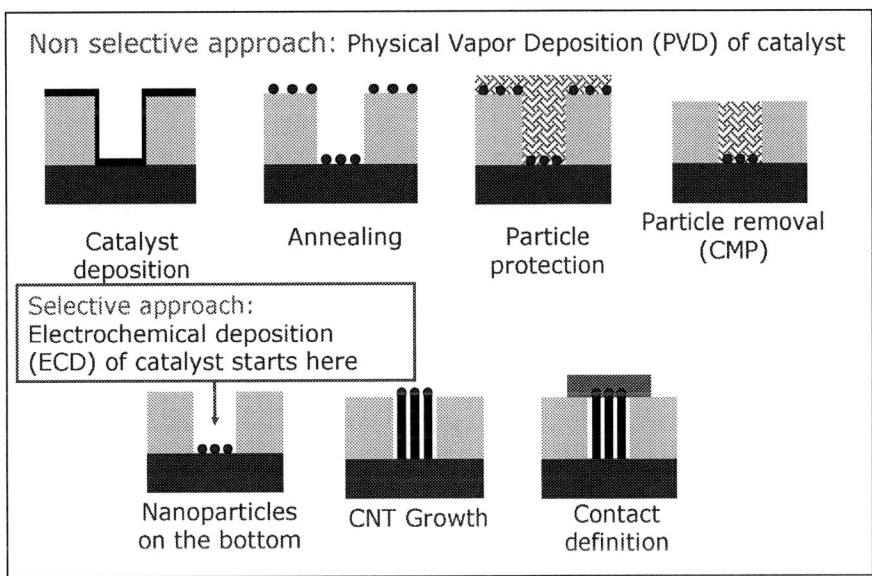

Figure 4. Process flow for selective catalyst placement, CNT growth and top-contact formation for test structures with via holes landing on a TiN layer which serves as common bottom electrode. For PVD deposition the catalyst particles are removed by CMP after a resist protection layer is applied. For ECD the nanoparticles is deposited directly on the exposed bottom electrode without additional process steps needed.

Since the PVD catalyst film covers the whole wafer, the catalyst nanoparticles, formed by combined thermal and plasma treatment are removed from the oxide field area by CMP. To protect the catalyst nanoparticles inside the via during CMP, a protective resist layer is spin-coated on top, which is removed again after the CMP step. After an additional cleaning step the nanoparticles are ready for CNT growth. Details of the process steps are given in [18]. Note that in this selective CMP approach, catalyst particles will be also present on the side walls of the via.

The CNT which will grow from the side-walls will not contribute the electrical conduction in this case. Since ECD will only occur on the exposed TiN electrode, particles will be deposited on the bottom of the via only and one can proceed immediately to CNT growth. Note however that the selective placement of the ECD catalyst is only valid in the proposed "through-mask" approach. In case of damascene liner coverage the particles will be also on the sidewalls and in the field of the wafer. The advantage of ECD as low-temperature deposition technique remains however.

Figure 5 shows Ni nanoparticles as-deposited inside 150nm and 300nm vias with dense and isolated pitch. For comparison also Ni particles on blanket TiN are shown. Similar particle density and size were obtained independent of hole diameter and pitch and deposition conditions obtained for blanket substrates are directly transferable to patterned substrates. Due to the very short deposition times (only nucleation), diffusion fields for Ni^{2+} ions are still small and not affected by its neighbors yet, which explains the absence of pattern density effects.

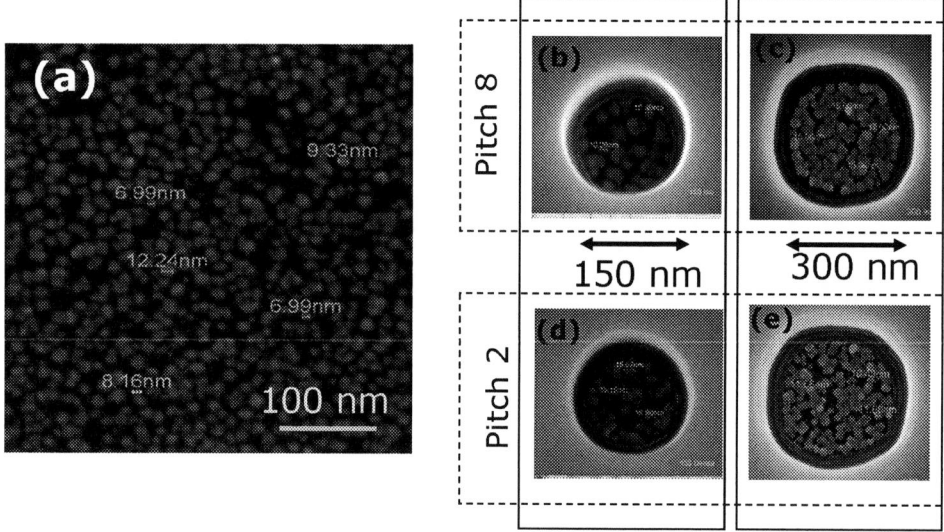

Figure 5. Top-down SEM images of electrodeposited Ni nanoparticles from an aqueous solution of 0.01M $Ni(NO_3)_2$ + 1M NaCl +1M NH4Cl with pH adjusted to 8 with NH4OH at a constant potential of -1.6V vs. Ag/AgCl reference for 0.6s on blanket TiN (a) and for 0.1s in patterned TiN wafer ((b)-(d)). The average particle diameter, d, and particle density, N_p, were similar for blankets and via holes with diameter between 150nm and 300nm , and relative pitches between 2 and 8: (a) d=12±3 nm, $N_p=3x10^{11}$ cm^{-2}, (b) d=15±7 nm, $N_p=2x10^{11}$ cm^{-2}, (c) d=17±5 nm, $N_p=2x10^{11}$ cm^{-2}, (d) d=15±3 nm, $N_p=2x10^{11}$ cm^{-2}, (e) d=14±4 nm, $N_p=2x10^{11}$ cm^{-2}.

Figure 6 shows the selective growth of CNT with the electrodeposited Ni in the vias. CNT growth is selective for the via only and with 100% yield. The density of CNT inside the via is estimated $\sim10^{11}$cm^{-2} or similar to the measured Ni nanoparticle density. For multi-wall tubes with 8-10 shells this means a CNT shell density of $\sim10^{12}$ cm^{-2} or still more than one order of magnitude below target.

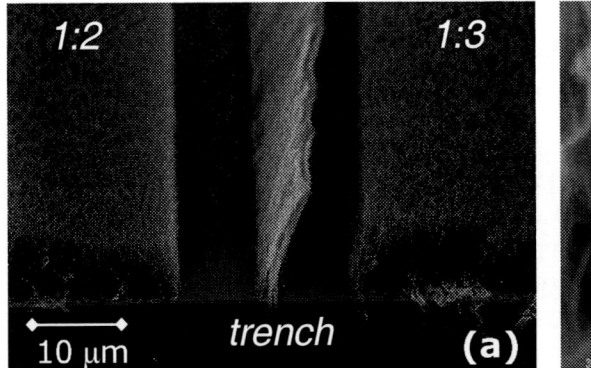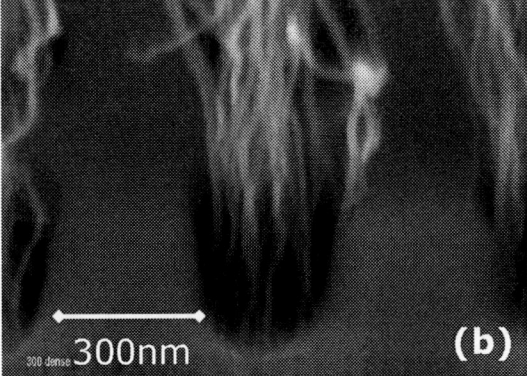

Figure 6. Cross-sectional SEM images of CNT bundles grown selectively grown inside 300nm via holes from ECD Ni catalyst particles (plasma-enhanced catalyst activation in 1 Torr H_2 at 200W high-frequency for 5 min + thermal growth in 1 Torr acetylene for 10 minutes at 650C). (a) overview of dense array of filled holes (relative pitch 2) and semi-dense array (relative pitch 3) separated by a trench which is also filled with CNT, and (b) detail of cleaved 300nm via in dense array, the density of CNT is estimated $\sim 10^{11} cm^{-2}$.

CNT integration: electrical characterization

For quick assessment of the electrical characteristics, Ti pads were formed on top of the CNT by the following steps: PVD of blanket Ti film, photolithography of photoresist pads, chemical etch of the exposed Ti and removal of the photoresist. The resulting Ti pads were 50um or 100um square and contacted several hundred vias in parallel. In all cases, Ohmic current-voltage characteristics were obtained for the TiN/CNT/Ti contacts. For a quantitative measurement of the CNT via resistance, an electrical test structure with Kelvin probe was used. 300nm vias in 650nm thick oxide were made on top of M1 Cu level. Ta/TiN/Ni was deposited with PVD and the top surface was removed by CMP using the selective protection procedure shown in Fig. 4 [18]. After CNT growth (thermal catalyst activation + thermal growth) Ti pads were formed by PVD and chemical etch for 4-point probe measurements. Contact with the bottom electrode (M1-Cu) was made through about 10,000 CNT via landing on the contact pad. Two other Ti pads were connected to a Ti line crossing a single via (Kelvin structure). The best measurement was obtained for 4nm Ni catalyst film with a via resistance of 0.1kΩ, which corresponds to about 1.2mΩ.cm when converted into an equivalent via resisitivity. Y. Awano et al measured a resistance of 0.6Ω for a via with 2um in diameter and 350nm high, which corresponds to an equivalent via resistivity of 0.6mΩ.cm [19, 20].

CONCLUSIONS

CNT interconnects at the via level could improve heat dissipation of the BEOL stack and as such reduce line resistance and electromigration issues, provided CNT-shell densities higher than $5x10^{13} cm^{-2}$ can be achieved. Growth of densely aligned CNT carpets with tube density $\sim 10^{12} cm^{-2}$ and shell density approaching $\sim 10^{13} cm^{-2}$ was demonstrated on Ti layers. Process steps for selective growth of CNT into via holes with high yield and shell densities $\sim 10^{12} cm^{-2}$ were developed. Electrical characteristics for TiN/CNT/Ti contacts were Ohmic and via resistivities down to 1.2 mΩ cm were achieved. Additional work is needed to increase carbon shell density in vias and to implement the integration scheme to a (dual) damascene approach. Low resistance metal contacts are needed to reduce contact resistance.

ACKNOWLEDGMENTS

This work was partly supported by the EU projects CARBonCHIP (NMP4-CT-2006-016475) and SEA-NET Pro-Nano (IST-027982) and by METACEL project (IWT-SBO project 060031) funded by IWT-Vlaanderen. The authors would like to thank Hefin Griffiths of Oxford Instruments Plasma Technologies U.K. for experiments in their PECVD reactor, Jean Dijon of Liten, Grenoble, France for IBS deposition of Fe, Christophe Detavernier of University of Ghent for PVD of various catalysts, Youssef Travaly of IMEC for XRR measurements, Hugo Bender and Olivier Richard of IMEC for HRTEM. Part of this work was done in collaboration with Nanocyl SA, Namur, Belgium.

REFERENCES

1. R. Saito, G. Dresselhaus, M. S. Dresselhaus. *Physical Properties of Carbon Nanotubes* Imperial College Press, London (1998)
2. Lu-Chang Qin, Xinluo Zhao, Kaori Hirahara, Yoshiyuki Miyamoto, Yoshinori Ando, Sumio Iijima. Nature, **408**, 50 (2000).
3. A. S. Verhulst, M. Bamal, G. Groeseneken, IMEC internal report (2005).
4. N. Srivastava, R. V. Joshi and K. Banerjee, Carbon Nanotube Interconnects: Implications for Performance, Power Dissipation and Thermal Management, IEDM (2005)
5. Hong Li, Navin Srivastava, Jun-Fa Mao, Wen-Yan Yin and Kaustav Banerjee, Carbon Nanotube Vias: A Reality Check, IEEE (2007)
6. Raychowdhury and K. Roy, Carbon Nanotubes as Interconnects of the Future: A Circuit Perspective, Proc. of the Advanced Metallization Conference, San Diego, October 2004
7. S. Li, Z. Yu, C. Rutherglen, and P.J. Burk, Nano Lett. **4**, 2003 (2004).
8. J-Y. Park, S. Rosenblatt, Y. Yaish, V. Sazonova, H. ¨Ust¨unel, S. Braig, T.A. Arias, P.W. Brouwer, and P.L. McEuen, Nano Lett. **4**, 517 (2004).
9. Z.P. Huang, D.Z. Wang, J.G. Wen, M. Sennett, H. Gibson, Z.F. Ren, *Appl Phys A* **74**, 387 (2002).
10. M. Cantoro, S. Hofmann, S. Pisana, C. Ducati, A. Parvez, A.C. Ferrari, J. Robertson, Diamond and Rel. Mat. **15**, 1029 (2006)
11. G. Zhang et al. PNAS **102**,16141 (2005)
12. M. Cantoro, S. Hofmann, S. Pisana, V. Scardaci, A. Parvez, C. Ducati, A.C. Ferrari, A.M. Blackburn, K.Y. Wang, J. Robertson, Nano Lett **6**, 1107 (2006)
13. G. F. Zhong, T. Iwasaki, K. Honda, Y. Furukawa, I. Ohdomari, H. Kawarada, Jpn. J. Appl. Phys. 1 **44**, 1558 (2005).
14. T. de los Arcos, M. G. Garnier, P. Oelhafen, D. Mathys, J. W. Seo, C. Domingo, J. V. Garcıa-Ramos, S. Sanchez-Cortes. Carbon **42**, 187 (2004).
15. D.J. Cott, P.M. Vereecken, A.R. Negeira, H. Griffiths, S. DeGendt, (*in preparation*)
16. A. Romo Negreira, P.M. Vereecken, C. M. Whelan, K. Maex, ECS transactions, **2**, 409 (2007).
17. S. Esconjauregui, C.M. Whelan and K. Maex, *Nanotechnol.* **18**, 015602 (2007).

18. S. Esconjauregui, C.M. Whelan and K. Maex, *Nanotechnol.* **19**, 135306 (2008).
19. Y. Awano, S. Sato, D. Kondo, M. Ohfuti, A. Kawabata, M. Nihei, N. Yokoyama, Phys. Stat. Sol. A 203, 14 (2006).
20. D. Yokoyama, T. Iwasaki, T. Yoshida, H. Kawarada, S. Sato, T. Hyakushima, M. Nihei, Y. Awano, APL **91**, 263101 (2007).

Mechanical Integrity Study of Air Gap Structures Assisted by FE Simulations

Stephane Moreau[1], Frédéric Gaillard[1], Jean-Charles Barbé[2], Raphaël Gras[3,4], Gérard Passemard[3], and Joaquin Torres[3]

[1]DRT/LETI/LBE, CEA-LETI MINATEC, 17 rue des Martyrs, Grenoble, 38054, France

[2]DRT/LETI/LSCDP, CEA-LETI MINATEC, 17 rue des Martyrs, Grenoble, 38054, France

[3]STMicroelectronics, 850 rue Jean Monnet, Crolles, 38926, France

[4]LTM/CNRS, 17 rue des Martyrs, Grenoble, 38054, France

ABSTRACT

In this paper, mechanical reliability of "air gap" structures has been evaluated when a copper line is completely surrounded with air. Different Finite Element (FE) simulation models have been used on a 2-metal level structure to study the M2 copper line bow evolution as a function of its dimensions if complete air cavities are generated underneath (*i.e.* at via level). Design rules information may therefore be obtained to optimize "air gap" integration considering the 65 nm and 22 nm technology nodes. Thus, we not only highlight that M2 copper line can not collapse considering our failure criterion but that M2 bow variation may also be improved when a tensile SiCN capping layer is deposited on top of the structure. The influence of the interline spacing vs. M2 bow has also been studied and we show that its increase is a beneficial parameter for the air gap structure. In opposite, we demonstrate that buckling can occur when a compressive SiCN layer is used. Finally, we accurately predict the M2 bow variation for air gap structures of the 65 nm and 22 nm technology nodes, but stress and strain distribution can complementary be provided. Those results highlight interesting criteria for designers to build reliable air gap structures.

INTRODUCTION

In order to meet the ITRS requirements for advanced interconnects (22 nm technology node and below) [1], new materials with ultra low dielectric constant (ULK) or new integration architectures are necessary to improve RC delay, crosstalk and power consumption. Nowadays, the introduction of air cavities as a dielectric material has been considered to achieve those performances. To create "air gaps", we currently investigate a "full oxide" approach where a sacrificial SiO_2 layer is used in the whole stack and further removed by an HF chemical etching agent. This HF chemistry diffuses through out patterned apertures localized in an upper SiCN capping layer deposited at the end of the integration. In such approach, one concern deals with the potential collapse of a M2 copper line on the M1 structure on long M2 line patterns when air cavities are made under the copper line. To prevent this phenomenon, a SiO_2 pillar is experimentally left between two adjacent SiCN apertures during the chemical etching process. This idea was successfully processed on a dual damascene integration scheme and exhibits promising results [2].

The intent of this paper is to define design rules to optimize mechanical reliability and, by extension, electrical performances of this previous discussed air gap approach considering the 65 nm and 22 nm technology nodes. To achieve this goal, Finite Element (FE) simulation is used.

STRUCTURE, MATERIAL PROPERTIES AND BOUNDARY CONDITIONS

The reference structure is a basic structure composed of 2 copper metal lines interconnected with vias located on each M2 extremity and surrounded with air. Except the presence of the SiCN capping layer at the upper metal level, this configuration represented the one we expected to obtain on our dual-damascene integration scheme after air gap formation. It will be useful to validate our FE simulations model. Only one half of the structure was modeled due to its symmetry. Structure and boundary conditions are depicted on the figure 1.

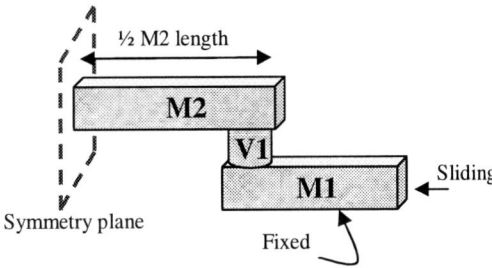

Figure 1. Structure and boundary conditions.

The copper is supposed to follow a linear elastic law *i.e.* the stress is proportional to the applied deformation (Hooke's law). The copper residual stress was beforehand evaluated on thin films by FleXus stress measurements and fixed to +250 MPa (at 25 °C) in the simulations.

FINITE ELEMENT MODELS

Because no analytical model was found, different FE models were developed in order to minimize the errors and estimate the accuracy of our simulations. Thus, 3-D, 2-D and mixed mode models were developed (see figure 2). 3-D model is meshed with either only 3-D 20-node solid elements or these elements in conjunction with *contact* elements. 2-D model uses beam elements. Beam elements are used to create a mathematical one-dimensional idealization of a 3-D structure. They offer computationally efficient solutions when compared to solid elements (3-D model). The mixed mode model use both mesh element types: solid elements for M1, V1 and M2 closed to V1 and beam elements for M2 far from V1.
ANSYS v.11.0, a Finite Element (FE) software [3], was used to perform all simulations.

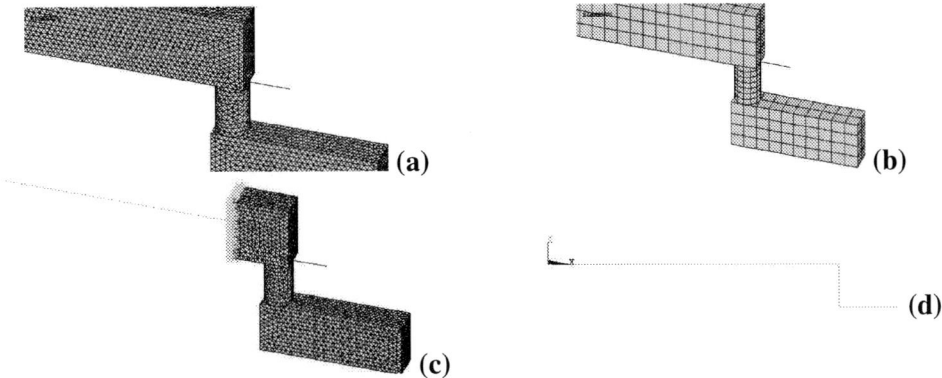

Figure 2. FE models: (a) 3-D model with its continuous mesh, (b) 3-D model with its discontinuous mesh using *contact* elements, (c) 3-D/2-D model with its mesh and (d) 2-D model.

FINITE ELEMENT SIMULATIONS' RESULTS

Failure criterion

Because, the M2 level collapse may be a reliable issue in our "air gap" approach, the effect of M2 bow as a function of M2 length has been studied when the structure is surrounded by air. So, the mechanical integrity is arbitrarily discussed regarding to the M2 bow. In this context, the failure criterion is defined as a critical bow value equal or larger than one via height.

65 nm technology node

M1-V1-M2 without the SiCN layer

Figure 3 shows the evolution of the M2 bow versus the M2 length for the four FE models. The results for 3-D model are limited to M2 lengths smaller than 7 μm because of the important CPU time consumption. For convergence reasons, the results for 2-D model are restricted to M2 lengths smaller than 130 μm.

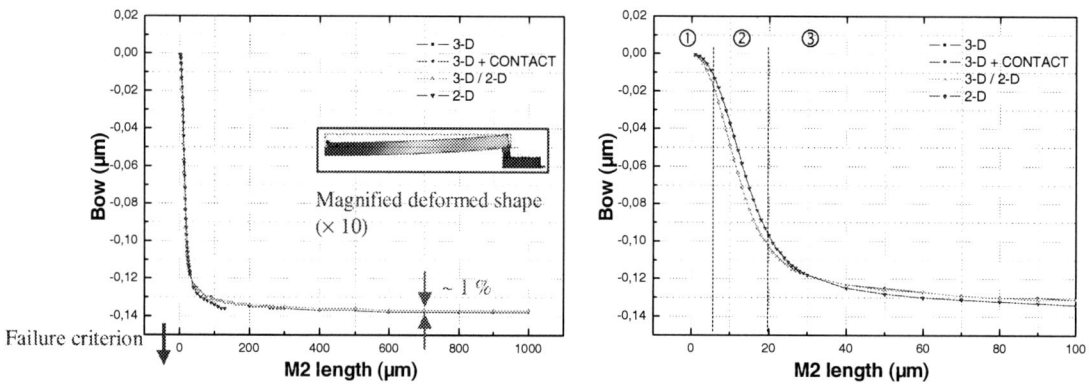

Figure 3. Evolution of the M2 bow versus the M2 length for the 4 FE models for the 65 nm technology node (global view and zoom).

It can be noticed that the same behavior is obtained for the four different models. The M2 bow response follows an exponential decay. A smooth S-shape is first observed and the curve can be divided into 3 parts:

1. M2 length ∈ [1; 5 μm]: the via stiffness controls the behavior of the M2 level;
2. M2 length ∈]5; 20 μm]: elasticity phase: proportionality between M2 length and its bow;
3. M2 length ∈]20; 1000 μm]: the M2 stiffness controls the behavior of the M2 level,

and, at the end, the M2 bow tends to a horizontal asymptote (~140 nm).

The difference between each model is lower than 1 % on the M2 length range. Only the 2-D model exhibits an error which can reach 50 %. It is observed for M2 lengths smaller than 17 μm (difference > 10 %) and can be explained by the non-validity of the beam theory for this lengths' range. Otherwise the difference is near 2 %. As the difference between each model is very small, we can be confident in our models validity.

With the defined failure criterion, the structure is not failing. Nevertheless, it could be interesting, in the future, to verify the stress intensity reached in the structure. Indeed, the copper ultimate strength could be exceeded.

M1-V1-M2 with the SiCN layer

In our approach, the last metal level is encapsulated with a silicon carbide nitride (SiCN) layer (see figure 4-a): an excellent barrier against HF-based chemistry diffusion. Different apertures are patterned in this layer to generate HF diffusion pathways and localized air cavities in specific areas *e.g.* where high signal propagation performance is required [2]. Until now, the SiCN capping layer was not considered in our FE models and its influence on the M2 bow must be studied.

The SiCN layer is 40 nm thick and its mechanical behavior is considered to be elastic. Its residual stress can be compressive or tensile depending on the process conditions [4]. It has been shown that SiCN residual stress can vary between -200 MPa [4, 5] and +200 MPa at 25 °C [4].

The impact of the interline spacing parameter at M2 level on the M2 bow response has been also investigated (see figure 4-b).

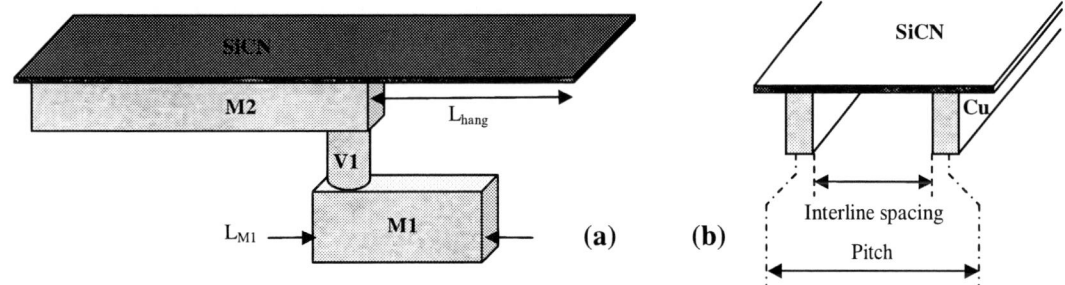

Figure 4. (a) 2-metal level structure with the upper SiCN capping layer and (b) the interline spacing parameter definition.

<u>Tensile SiCN layer</u>

Figure 5 shows the evolution of the M2 bow versus the M2 length for the 3-D FE model with *contact* elements considering a tensile SiCN layer (+200 MPa) and a 100 nm interline spacing.

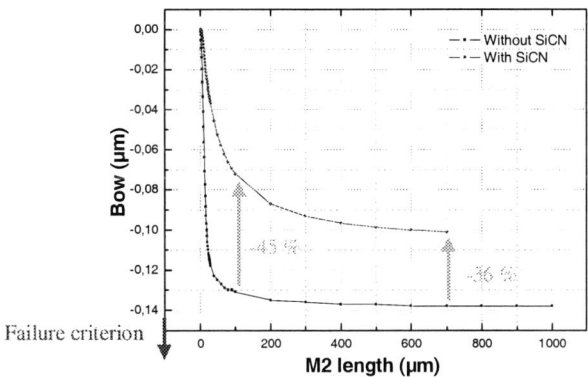

Figure 5. Effect of a tensile SiCN layer (+200 MPa) on the M2 bow for the 65 nm technology node and a 100 nm interline spacing.

It is shown that the mechanical integrity of air gap structure is considerably improved using a tensile SiCN capping layer. Thus, a decrease of 36 % on the bow evolution is obtained comparing a structure with or without a tensile SiCN layer.

147

Figure 6 presents the impact of the interline spacing on the M2 bow with a tensile SiCN capping layer (+200 MPa) for the 3-D FE model with *contact* elements. The larger the interline spacing is, the lower the M2 level deflection is. For example, multiplying the interline spacing by a 2× factor leads to a M2 bow reduction of 22 % for a M2 length of 100 µm.

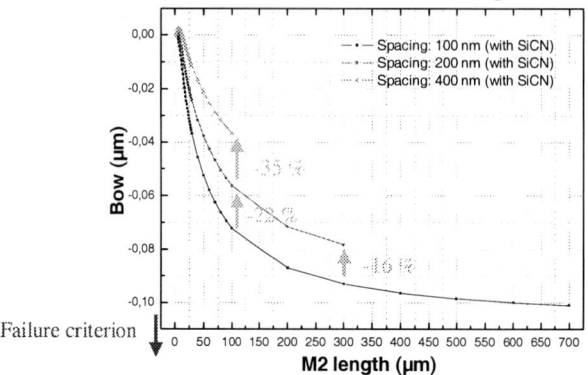

Figure 6. Effect of the interline spacing on the M2 bow with a tensile SiCN capping layer (65 nm technology node).

Compressive SiCN layer

A value of -200 MPa is used for the compressive SiCN residual stress. When $L_{hang} > L_{M1}$ (see figure 4-a) and a compressive SiCN layer is used, the structure can buckle (with a small deformations hypothesis) whatever the M2 length *i.e.* structure becomes unstable. From this specific buckling analysis [3], we can extract the coefficient by which we have to multiply the stress stiffness matrix [S] to reach the first buckling eigenmode. This *buckling coefficient* is critical when it lies under unity [3].

Figure 7 shows the evolution of the buckling coefficient for air gap structures with different M2 length. This graph illustrates the fact that buckling may occur for air gap structures when a compressive SiCN capping layer is used on top of the integration scheme.

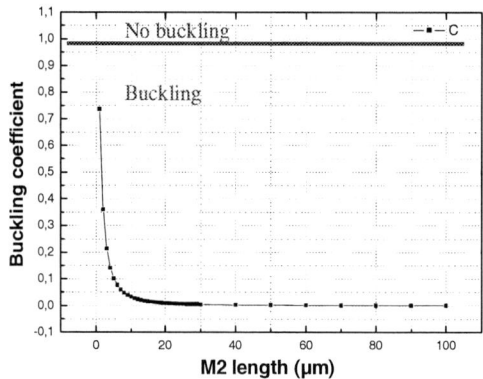

Figure 7. Evolution of the buckling coefficient for different M2 length of air gap structures with a compressive SiCN capping layer.

22 nm technology node

The same study has been performed for the 22 nm technology node. Identical conclusions as for the 65 nm technology node can be drawn: an exponential decay of the M2 bow, a model difference near 1 % and a safe structure for the whole M2 length range. A decrease of 66 % of the bow is noticed between the structure without the SiCN layer and the structure with a tensile one (see figure 8).

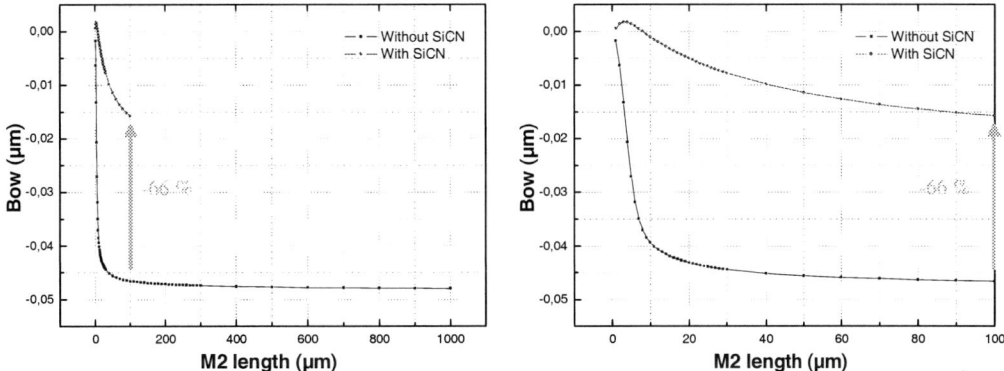

Figure 8. Effect of a tensile SiCN layer (+200 MPa) on the M2 bow for the 22 nm technology node and a 35 nm interline spacing (global view and zoom).

CONCLUSIONS

FE models were developed in order to confirm/predict the mechanical integrity of the "air gap" approach in the future interconnections. Structures with 2 metal levels were studied. M2 bows were investigated according to its length when the metal line is completely surrounded with air. From this study, we can not only predict the M2 bow accurately but also provide stress and strain data. If the via height is used as a failure criterion, simulations show the "air gap" structure can not collapse even more with a tensile SiCN capping layer. Furthermore, the increase of the interline spacing is beneficial for the "air gap" structure. In case of a compressive SiCN capping layer, buckling can occur and then damage the integration. Finally, SiO_2 pillars can be adequately placed to improve the mechanical reliability of "air gap" structures.

ACKNOWLEDGMENTS

This work has been carried out in the frame of the CEA-LETI / Crolles2 Alliance collaboration.

REFERENCES

[1]   http://www.itrs.net/Links/2005ITRS/Home2005.htm: International Technology Roadmap for Semiconductors (ITRS), 2005.

[2]   F. Gaillard, D. Bouchu, R. Gras, S. Moreau, G. Sale, P. Lyan, B. Icard, J. C. Le-Denmat, L. Pain, J. Bustos, P. Brun, M. Rivoire, C. Euvrard, S. Olivier, G. Imbert, G. Passemard, and J. Torres, "Air Gap Integration for Edge Interconnection Technologies," in *VLSI/ULSI Multilevel Interconnection Conference*, Fremont, CA, 2007, pp. 299-304.

[3]   "ANSYS 11.0 Documentation," ANSYS, Inc., 2007.

[4]   K. Goto, D. Kodama, N. Suzumura, S. Hashii, M. Matsumoto, N. Miura, T. Furusawa, M. Matsuura, and K. Asai, "Stress Engineering in Cu/Low-k Interconnects by using UV-Cure of Cu Diffusion Barrier Dielectrics," in *International Interconnect Technology Conference*, 2006, pp. 95-97.

[5]   N. Chérault, "Caractérisation et modélisation thermomécanique des couches d'interconnexions dans les circuits sub-microélectroniques," Ecole des Mines de Paris, 2006, p. 286.

Mater. Res. Soc. Symp. Proc. Vol. 1079 © 2008 Materials Research Society 1079-N02-10

Dielectric Recovery of Plasma Damaged Organosilicate Low-k Films

Hualiang Shi[1], Junjing Bao[1], Huai Huang[1], Junjun Liu[1], Ryan Scott Smith[1], Yangming Sun[1], Paul S. Ho[1], Michael L. McSwiney[2], Mansour Moinpour[2], and Grant M Kloster[2]

[1]Laboratory for Interconnect and Packaging, Microelectronics Research Center, Pickle Research Campus, The University of Texas at Austin, Austin, TX, 78758

[2]Logic Technology Development, Intel Corporation, Hillsboro, OR, 97124

ABSTRACT

Methyl depletion and subsequent moisture uptake have been found to be the primary plasma damages leading to dielectric loss in porous organosilicate (OSG) low-k dielectrics. A vacuum vapor silylation process was developed for dielectric recovery of plasma damaged OSG low-k dielectrics. The methyl or phenyl containing silylation agents were used to convert the hydrophilic -OH groups to hydrophobic groups. Compared with Trimethylchlorosilane (TMCS) and Phenyltrimethoxysilane (PTMOS), Dimethyldichlorosilane (DMDCS) was found to be more effective in recovering surface carbon concentration and surface hydrophobicity. But the carbon recovery effect was limited to the surface region.

Alternatively, UV radiation with thermal activation was applied for dielectric recovery of plasma damaged OSG low-k dielectrics. The combined UV/thermal process was found to be efficient in reducing –OH, physisorbed water, and C=O bonds. The dielectric constant was recovered within 5% of the pristine sample and the leakage current was also much reduced. Aging test in air showed that no moisture retake was observed, indicating the repaired film was stable.

INTRODUCTION

Ultra low-k dielectrics with porosity are being incorporated to reduce the capacitance coupling in copper interconnects beyond 45 nm technology node [1, 2]. During plasma etching and ashing processes, carbon depletion, surface hydrophilisation, and surface densification have been observed, increasing the effective dielectric constant and leakage current [3-11]. These plasma damages limit the further scaling of low-k structures and have stimulated considerable interests in the study of dielectric recovery by silyation methods, using vapor-phase [12], liquid-phase [13-14], and supercritical CO_2 processes [15-17].

Since silylation agents are susceptible to water molecules, moisture should be removed before the silylation process. Otherwise, even a small amount of moisture will react with the silylation agents and form byproducts which can be adsorbed on the wafer surface to reduce the dielectric recovery and complicate the reaction mechanism. In the first part of this paper, the vapor-phase silylation process was first demonstrated under high vacuum with in-situ heating. This improved the control of moisture uptake, making the process easy to be integrated. To investigate the silylation mechanism, several key parameters were studied including substrate temperature, functionality of silylation agent, and the effect of chemical desorption. Three silylation agents were compared: Trimethylchlorosilane (TMCS), Dimethyldichlorosilane (DMDCS), and Phenyltrimethoxysilane (PTMOS).

Because the recovery of the vapor-phase silylation process was limited to the surface, other methods were developed to improve the silylation process. UV radiation and thermal

150

activation are commonly used to remove porogen to improve the mechanical properties of as-deposited low-k dielectrics [18]. But in this study, they were used for bulk recovery of plasma damaged OSG low-k dielectrics. Since the recovery process may involve chemical bonds different with as-deposited OSG low-k dielectrics, the mechanism of UV radiation with thermal activation was investigated.

EXPERIMENT

The pristine OSG film was deposited with about 25% porosity to yield a dielectric constant of 2.5, refractive index of 1.33 at 633 nm, and thickness of 100 nm. Before vapor-phase silylation, Plasma-Therm 790 RIE was used to prepare O_2 plasma damaged OSG samples under the condition of 150mtorr, 15sccm, 150W, and 5minutes. The samples were baked in a UHV chamber around 120°C for about 1 hour to remove the physisorbed water molecules. The silylation agents were delivered with CO_2 carrier gas into the vacuum chamber where the chamber wall and gas delivery system were kept around 100°C to minimize moisture uptake and the bubblers were kept around the boiling point of silylation agents, such as 70°C for DMDCS.

The OSG films were first subjected to O_2 plasma in an Oxford RIE chamber under the condition of 30mtorr, 30sccm O_2, 150W, and 10minutes. This followed with UV radiation and thermal activation at 350°C for about 300seconds. The UV lamp wavelength was selected to be above 200 nm to minimize the influence on Si-CH$_3$ bonds.

The changes in molecular bonds were observed using a Nicolet Magna 560 Fourier transform infrared (FTIR) spectrometer. An angle-resolved X-ray photoelectron spectroscopy (ARXPS) was used to study the surface chemical composition. A CA100 Ramé-Hart Goniometer was used to evaluate the water contact angle and surface hydrophobicity. XPS depth profiling was performed in a PHI 5700 XPS system to examine the composition change in the OSG film. Standard MIS capacitors were used for C-V and I-V measurements.

DISCUSSION

Damages induced by O_2 plasma

To evaluate the bonding changes induced by O_2 plasma in Oxford RIE, differential FTIR was obtained from the FTIR spectra of the OSG before and after O_2 plasma treatment, as shown in Fig. 1. After O_2 plasma treatment, the peak intensities of -OH/H_2O (3100cm^{-1}~3800cm^{-1}, 935cm^{-1}), C=O (1732cm^{-1}), and Si-O-Si network (1074cm^{-1}) bonds increased, while those of CH$_x$ (~2974cm^{-1}), Si-H$_x$ (2251cm^{-1}, 2185cm^{-1}), Si-(CH$_3$)$_x$ (1280~1250, 870~760cm^{-1}), and Si-O-Si suboxide (1021cm^{-1}) bonds were reduced [19-22]. Moreover, the dielectric constant at 1 MHz increased from 2.50 to 3.26.

In a previous study [23], the origin of dielectric loss induced by O_2 plasma on OSG films was investigated by combining FTIR and SE with the Kramers-Kronig dispersion relation. The dielectric loss was found to be dominated by the dipolar contribution compared with the electronic and the ionic polarizations. Quantum chemistry calculations showed that the physisorbed H_2O molecule was mainly responsible for the dipolar contribution to the dielectric loss. Si-CH=O and Si-O- also contributed to the dielectric loss, but much less than that due to moisture uptake.

Dielectric recovery by vacuum vapor silylation process

The methyl or phenyl containing silylation agents were expected to convert the hydrophilic -OH groups to hydrophobic groups by the following reactions [13,24-25]:

$$R_1R_2R_3Si - X + H - OSi \equiv \rightarrow R_1R_2R_3Si - OSi \equiv +HX$$

where X is the desorbed group, Cl for TMCS and DMDCS, and OCH_3 for PTMOS, and R_i is CH_3 for TMCS, CH_3 or Cl for DMDCS, and C_6H_5 or OCH_3 for PTMOS. When PTMOS was used, a small amount of TMCS was added as catalyst. The HCl from the interaction between TMCS and Si-OH can enhance the interaction between PTMOS and Si-OH.

In order to improve the recovery effect, vacuum vapor silylation process should be optimized. As shown in Fig. 2 and Fig. 3, with increasing silylation treatment time or substrate temperature, the XPS C/Si ratios and water contact angle increased, indicating further recovery of surface carbon concentration and surface hydrophobicity. The temperature dependence can be explained by a thermally activated diffusion process of silylation agents or chemical reaction between silylation agents and silanol group. In ARXPS, a higher C/Si at 30^0 scan indicated more recovery on the surface than in the bulk. One possible reason is that the small pore size or densified surface induced by plasma treatment or the interaction between silylation agents and surface Si-OH bonds can block the penetration of silylation agents.

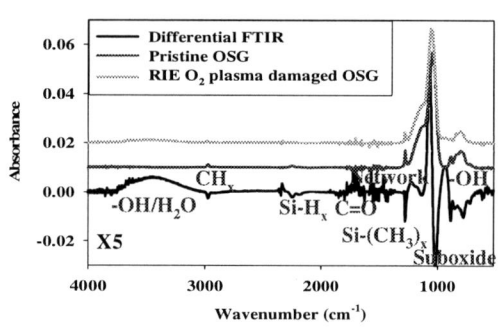

Figure 1. FTIR spectra for pristine and O_2 plasma damaged OSG

Figure 2. Time dependence of vacuum vapor TMCS silylation process

Under similar conditions, DMDCS had the best recovery effect compared with TMCS and PTMOS, as indicated by the largest XPS C/Si ratio and water contact angle in Fig. 4. There could be three possible reasons [26-27]. First, O_2 plasma treatment induced different types of silanols into OSG film including isolated silanol and disilanol (also named as geminal silonal) and TMCS can react with the isolated silanol easily. But once a TMCS molecule reacts with the first silanol of a disilanol, the extra CH_3 group can cause steric hindrance to block further interaction between other TMCS molecule and the second silanol. Second, DMDCS has two chlorine desorbed groups and a smaller molecule size, so that it has a bigger chance to interact with silanol groups. Third, PTMOS has the largest molecule size and the least effective leaving group. The results indicated that the recovery by vapor silylation process depends on the functionality, size, and effectiveness of leaving group.

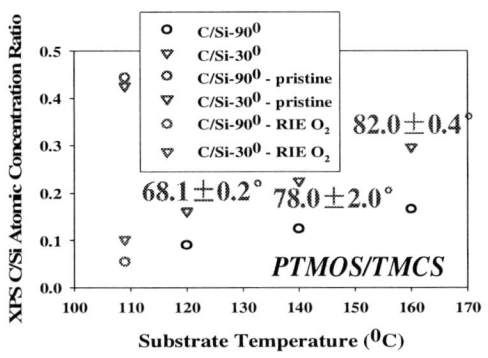
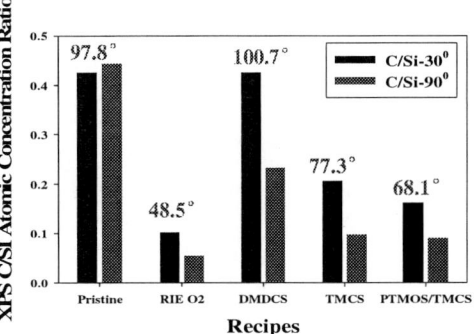

Figure 3. Substrate temperature dependence of vapor PTMOS/TMCS silylation process

Figure 4. Comparison of vapor TMCS, DMDCS, and PTMOS/TMCS silylation processes

The recovery depth of carbon concentration was evaluated by combining XPS depth profiling and film thickness from SE measurement. As indicated by Fig. 5, the recovery effect was limited to the surface within 5nm. A simple model with serial capacitors was used to evaluate the dielectric recovery of low-k structures:

$$\frac{s}{k_{eff}^*} = \frac{2r}{k_r} + \frac{2(d-r)}{k_d} + \frac{s-2d}{k_p}$$

where k_{eff} is the effective dielectric constant, k_r and r are the dielectric constant and thickness for the repaired layer respectively, k_d and $d-r$ for remained damaged layer, k_p for undamaged layer, and s the line spacing. As indicated by Fig. 6, if the plasma damage was limited to the surface, a full dielectric recovery was possible by the vapor-phase silylation process. But, if the plasma damage was deep into the bulk, this process was not sufficient for the full dielectric recovery. This problem will become more serious with the scaling of low-k structures. At the 32nm technology node, the spacing is about 45nm. If the sidewall damage layer thickness is 10nm, the vapor-phase silylation process in this study can only recover the effective k to about 2.7. Other methods will be needed to assist silylation process. In the following, UV radiation with thermal activation was applied to O_2 plasma damaged OSG film without silylation process.

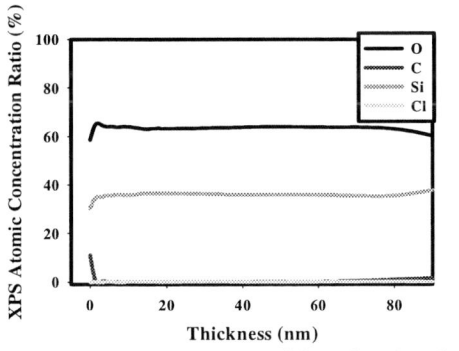
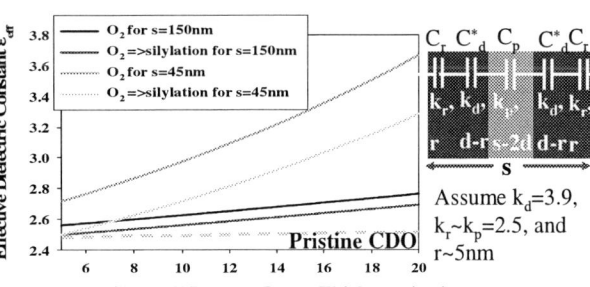

Figure 5. XPS depth profiling for O_2 plasma damaged OSG with DMDCS silylation process

Figure 6. Evaluation of recovery of effective dielectric constants of low-k structures

153

Dielectric recovery by UV radiation and thermal activation

After UV treatment, the film thickness, the refractive index, and the dielectric constant of the O_2 plasma damaged OSG film were reduced, as indicated in Tab. I . In particular, the dielectric constant was recovered to within 5% of the pristine film. In Fig. 7, differential FTIR was used to deduce the mechanism for dielectric recovery. After UV treatment, -OH, physisorbed H_2O molecules, and C=O bonds were almost completely removed, which contributed to dielectric recovery. At the same time, there was an increase of Si-O-Si network and suboxide bonds. The detailed mechanisms will be further investigated.

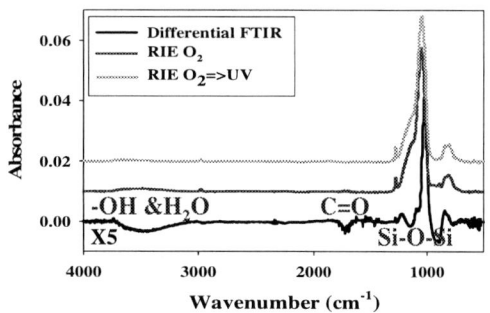

Figure 7. FTIR spectra for O_2 plasma damaged and UV repaired OSG

Table I. Dielectric recoveries induced by UV treatment

Sample	RIE O_2	O_2=>UV
Thickness (nm)	83.1	78.5
R.I. @633nm	1.333	1.323
ε@1MHz	3.26	2.55

CONCLUSIONS

The dielectric recovery by vapor-phase silylation process has been investigated as a function of the functionality, size, and effectiveness of leaving group. Under similar conditions, DMDCS was most effective for surface carbon and hydrophobicity recovery among the silylation agents of TMCS, DMDCS and PTMOS. The recovery, however, was limited to the surface within 5nm. UV radiation with thermal activation can remove most of the –OH, H_2O, and C=O bonds in the film to improve the electrical properties. Since the surface hydrophobicity was only partially recovered, a combination of vacuum vapor silylation and UV processes seems to be promising for the dielectric recovery.

ACKNOWLEDGMENTS

The authors would like to thank Dr. M. D. Goodner, Dr. J. Bielefeld, Dr. B. Boyanov, and Dr. M. Baklanov for their discussions. This work was supported in part by Intel Corp. and performed at the Microelectronics Research Center at UT Austin of National Nanofabrication Infrastructure Network supported by National Science Foundation under award # 0335765.

REFERENCES

1. M. Morgen, E. T. Ryan, J. Zhao, C. Hu, T. Cho, and P. S. Ho, Annu. Rev. Mater. Sci., Vol. 30, 645-680 (2000).
2. K. Maex, M. R. Baklanov, D. Shamiryan, F. Lacopi, S. H. Brongersma, and Z. S. Yanovitskaya, J. Appl. Phys., 93, 8793-8840 (2003).
3. J. Sun, D. W. Gidley, Y. Hu, W. E. Frieze, and E. T. Ryan, Appl. Phys. Lett., 81, 1447 (2002).

4. D. L. Moore, R. J. Carter, H. Cui, P. Burke, S. Q. Gu, H. Peng, R. S. Valley, D. W. Gidley, C. Waldfried, and O. Escorcia, Journal of the Electrochemical Society, 152 (7) G528-G533 (2005).

5. X. Hua, M. Kuo, G. S. Oehrlein, P. Lazzeri, E. Iacob, M. Anderle, C. K. Inoki, T. S. Kuan, P. Jiang, and W. Wu, J. Vac. Sci. Technol. B, 24, 1238 (2006).

6. H. Lee, C. L. Soles, E. K. Lin, W. Wu, and Y. Liu, Appl. Phys. Lett., 91, 172908 (2007).

7. M. A. Worsley, S. F. Bent, N. C. M. Fuller, T. L. Tai, J. Doyle, M. Rothwell, and T. Dalton, J. Appl. Phys., 101, 013305 (2007).

8. A. M. Urbanowicz, M. R. Baklanov, J. Heijlen, Y. Travaly, and A. Cockburn, Electrochemical and Solid-State Letters, 10(10) G76-G79 (2007).

9. S. Uchida, S. Takashima, M. Hori, M. Fukasawa, K. Ohshima, K. Nagahata, and T. Tatsumi, J. Appl. Phys., 103, 073303 (2008).

10. Y. Yin and H. H. Sawin, J. Vac. Sci. Technol. A 26, 151 (2008).

11. J. Bao, H. Shi, J. Liu, H. Huang, P. S. Ho, M. D. Goodner, M. Moinpour, and G. M. Kloster, J. Vac. Sci. Technol. B, 26, 219 (2008).

12. T. Rajagopalan, B. Lahlouh, J. A. Lubguban, N. Biswas, S. Gangopadhyay, J. Sun, D. H. Huang, S. L. Simon, D. Toma, and R. Butler, Applied Surface Science, 252 (18), 6323-6331 (2006).

13. J. C. Hu, C. W. Wu, W. C. Gau, C. P. Chen, L. J. Chen, C. H. Li, T. C. Chang, and C. J. Chu, Journal of the Electrochemical Society, Vol. 150 (4) F61-F66 (2003).

14. P. G. Clark, B. D. Schwab, J. W. Butterbaugh, H. J. Martinez, and P. J. Wolf, Semiconductor International August 2003.

15. R. A. Orozco-Teran, B. P. Gorman, Z. Zhang, D. W. Mueller, and R. F. Reidy, MRS Proceedings Vol. 766 (2003).

16. B. P. Gorman, R. A. Orozco-Teran, Z. Zhang, P. D. Matz, D. W. Mueller, and R. F. Reidy, J. Vac. Sci. Technol. B, 22, 3, 1210-1212 (2004).

17. B. Xie, A. J. Muscat, Microelectronic Engineering, 80, 349 (2005).

18. F. Iacopi, Y. Travaly, B. Eyckens, C. Waldfried, T. Abell, E. P. Guyer, D. M. Gage, R. H. Dauskardt, T. Sajavaara, K. Houthoofd, P. Grobet, P. Jacobs, K. Maex, J. of Appl. Phys., 99, 053511 (2006).

19. G. Socrates, Infrared Characteristic Group Frequencies, 2nd edition, John Wiley & Sons, (1994).

20. A. Grill and D. A. Neumayer, J. Appl. Phys., 94, 6697 (2003).

21. Y. Lin, T. Y. Tsui, and J. J. Vlassak, Journal of the Electrochemical Society, 153(7), F144-F152 (2006).

22. N. Posseme, T. Chevolleau, T. David, M. Darnon, O. Louveau, and O. Joubert, J. Vac. Sci. Technol. B, 25 (6), 1928-1940 (2007).

23. H. Shi, J. Bao, J. Liu, H. Huang, R. S. Smith, Q. Zhao, P. S. Ho, M. D. Goodner, M. Moinpour, and G. M. Kloster, the Advanced Metallization Conference 2007, VB.3.

24. J. Liu, W. Kim, J. Bao, H. Shi, W. Baek, and P. S. Ho, J. Vac. Sci. Technol. B, 25 (3), 906 (2007).

25. T. C. Chang, P. T. Liu, Y. S. Mor, T. M. Tsai, C. W. Chen, Y. J. Mei, F. M. Pan, W. F. Wu, S. M. Sze, J. Vac. Sci. Technol. B, 20 (4), 1561 (2002).

26. S. V. Nitta, S. Purushothaman, N. Chakrapani, O. Rodriguez, N. Klymko, E. T. Ryan, G. Bonilla, S. Cohen, S. Molis, K. McCullough, Advanced Metallization Conference 2005.

27. V. M. Gun'ko, M. S. Vedamuthu, G. L. Henderson, and J. P. Blitz, Journal of Colloid and Interface Science 228, 157-170 (2000).

Mater. Res. Soc. Symp. Proc. Vol. 1079 © 2008 Materials Research Society 1079-N07-04

Changes of UV Optical Properties of Plasma Damaged Low-k Dielectrics for Sidewall Damage Scatterometry

Premysl Marsik[1,2], Adam Urbanowicz[1], Klara Vinokur[3], Yoel Cohen[3], and Mikhail R Baklanov[1]

[1]AMPS, IMEC, Kapeldreef 75, Leuven, 3001, Belgium

[2]UFKL, Masaryk University, Kotlarska 2, Brno, 61137, Czech Republic

[3]NOVA, Weizmann Science Park, Bldg 22, Rehovot, 76100, Israel

ABSTRACT

Porous low-k dielectrics were studied to determine the changes of optical properties after various plasma treatments for development of scatterometry technique for evaluation of the trench/via sidewall plasma damage. The SiCOH porogen based low-k films were prepared by PE-CVD. The deposited and UV-cured low-k films have been damaged by O_2Cl_2, O_2, NH_3 and H_2N_2 based strip-plasmas and $CF_4/CH_2F_2/Ar$ etching plasma. In this work, blanket wafers were studied for the simplicity of thin film optical model. The optical properties of the damaged low-k dielectrics were evaluated the using variable-angle spectroscopic ellipsometry in range from 2 to 9 eV. Multilayer optical model was applied to fit the measured quantities and the validity was supported by other techniques: The atomic concentration profiles of Si, C, O and H were stated by TOF-SIMS and changes in the overall chemical composition were derived from FTIR spectra. Toluene and water based ellipsometric porosimetry was involved to examine the porosity, pore interconnectivity and internal hydrophilicity.

INTRODUCTION

Porous low-k dielectric films are introduced as interconnect-dielectric in integrated circuits below 45nm technology node. For the successful implementation of porous films, many technology challenges have to be faced. The desired material must keep low dielectric constant, while being mechanically tough and chemically stable at the same time, it must be hydrophobic and resistant against damage during plasma treatments. The effect of various plasma chemistries on low-k dielectrics is intensively studied [1].

To control the level of damage in production conditions, nondestructive and fast optical methods must be used. For correct implementation of optical models for high precision in-situ scatterometry and ellipsometry, optical properties of the low-k materials and their changes during plasma treatments have to be known and understood.

We focused our study to damage that low-k suffers during the photoresist strip and trench/via etching. In most of the cases, the exposition to plasma causes hydrophilisation of the material and subsequent water adsorption from the clean room air humidity. This process is highly undesired, because it leads to dramatic increase of k-value, as water has k~80.

The optical properties of SiCOH porogen based low-k dielectrics [2,3] in our range of interest are mostly determined by the properties inherited from silica and by the presence of organic compound of the SiCOH matrix itself as well as the residuals of decomposed porogen. Photoresist strip-plasmas act typically on the organic part of the low-k and cause carbon depletion from the low-k volume due to the accessibility through open pores. In some cases, the plasma exposure can cause pore sealing and reduce the volume damage of the dielectric.

156

EXPERIMENT

We studied two different porogen based low-k dielectrics (denoted as low-k A and low-k B) with target k value 2.5 in the form of non-patterned thin films (with thickness around 200 nm) deposited by CVD on silicon substrates. For both dielectrics, the plasma processing recipes are under development and we applied a selection of actual recipes on our samples.

For low-k A, one etching plasma and two different strip-plasma chemistries were used. We prepared set of samples based on the low-k dielectric A with varying deposition and curing conditions and damaged the samples by 10 sec flash of NH_3 plasma in ICP (inductively coupled plasma) reactor. Limited number of samples was damaged by O_2 and O_2/Cl_2 based strip-plasmas in ICP chamber. To study effect of $CF_4/CH_2F_2/Ar$ ICP etching recipes on the optical properties, the CF_4:CH_2F_2 ratio was modified to obtain different etching rates and to control the by-production of polymer residues [4].

In the case of the low-k dielectric B, two different stripping recipes were applied on the samples with varying time of treatment to observe the evolution of the changes. First recipe was based on O_2, and performed in CCP (capacitively coupled plasma) reactor during 20 and 30 seconds. Second recipe was based on H_2N_2 chemistry and we treated the samples for 10, 20 and 30 seconds.

The set of samples was completed with referential non-damaged samples of the dielectrics and also initialization layers, based on more dense and non-porous low-k material.

To evaluate the optical characteristics of the samples, we performed spectroscopic ellipsometry measurement in the range from 2 eV to 9 eV, using various angles of incidence between 55 degrees and 85 degrees on nitrogen-purged Sopra GES5 PUV-SE in rotating analyzer configuration.

The measured ellipsometric angles Ψ, Δ were fitted by layered optical models using following algorithm: 1) The optical properties and thickness of initialization layers were obtained independently from single layer samples and were fixed for all next fitting steps. 2) Non-damaged layers were measured and the data were fitted by proper harmonic oscillator optical model. 3) The thickness of the treated film in total was estimated, using the optical model of undamaged material. 4) The model layer was sliced to two (or three) sub-layers, keeping the total thickness as a starting point and then all the thicknesses and the properties of the top layer (resp. top two) were optimized in iterative steps, while the properties of the bottom layer were kept fixed.

Additional knowledge was gathered from Fourier-transformed infra-red (FTIR) absorption measurements on a Biorad QS2200 ME FTIR system, from water-based and toluene-based ellipsometric porosimetry (EP) using our EP10 tool [5] and the atomic concentration (Si, C, O, H) profiles from were obtained TOF-SIMS (time-of-flight secondary ion mass spectroscopy) measurements.

RESULTS

Effect of NH_3 strip-plasma on low-k A

Set of samples of low-k dielectric A was prepared, varying the deposition and curing conditions and reaching various porosities and compositions [6]. The samples were exposed for

10 seconds to NH_3 plasma in ICP reactor. All the samples were measured by spectroscopic ellipsometry and their properties were evaluated using a three layer model, because a model using two layers (top damaged and bottom non-damaged) was not representing the data well. The authors are aware of approximate character of such model. The typical example of the optical properties of the material is plotted in fig. 1a. After strip, the absorption band between 3 eV and 7 eV is substantially reduced resulting in porous SiO_2-like top hydrophilic layer. The very top layer exhibit increased refractive index and this effect can be attributed to densification. The FTIR absorbance spectra (not shown) show clear reduction of Si-CH_3 bonds and increase of -OH groups in damaged material [7].

Water based ellipsometric porosimetry experiment was performed to track changes of the layers upon change of the ambient humidity. In this case, the most damaged sample from the set was chosen for the experiment and two-layer model was used for the evaluation (fig. 1b).

Although the set of samples resulted in heterogeneous set of results, a relation between porosity and depth of damage (thickness of two top layers) can be stated. The samples prepared with higher porogen load for target porosity 35% (±3%) exhibit depth of damage 73 ± 5 nm and the samples with target porosity 27% (±2%) exhibit depth of damage 58 ± 6 nm.

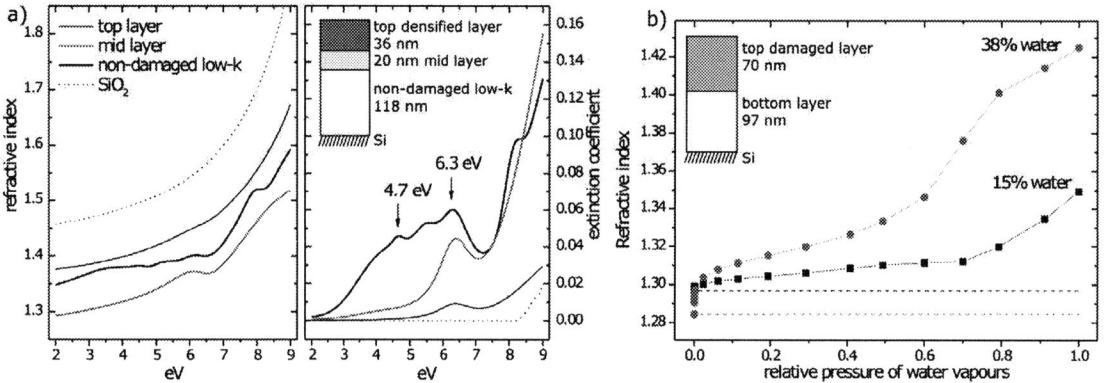

Figure 1 a) Optical properties of NH_3 plasma damaged low-k dielectric (left). The three-layer model reveals reduction of absorption band between 3 eV and 7 eV (attributed to organic compound of the material) in the top layers and densification of the very top layer. Optical properties of SiO_2 are plotted for comparison. b) The optical properties of low-k dielectric change with air humidity (right) as the damaged film becomes hydrophilic. The water based ellipsometric porosimetry (WEP) can detect higher hydrophilicity in the top layer. (Note: the sample and optical model used for WEP experiment differs from the one plotted in graph a)

Effect of O_2/Cl_2 strip-plasma on low-k A

The effect of 10 sec exposition of O_2 and O_2/Cl_2 strip-plasmas in CCP reactor on dielectric samples is similar to the one of NH_3 plasma. Top carbon depleted layer has been detected by the spectroscopic ellipsometry, but no further densification (fig. 2a). The difference between damaged and non-damaged layer can be observed by TOF-SIMS (fig. 2b) as decrease in C and H concentration and small increase of Si and O concentration. Ellipsometric porosimetry measurements detected increased pore size and porosity as the organic material is removed from the pore interior.

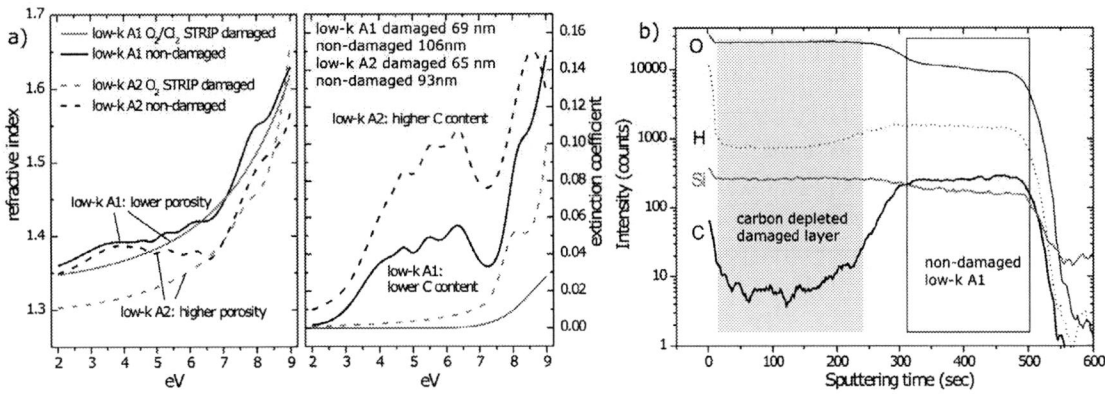

Figure 2 a) Optical properties of two samples of low-k with higher (A1) and lower (A2) porosity damaged by O_2 and O_2/Cl_2 strip-plasma (left). Removal of organics from top layer is observed in the ellipsometric spectra as well as in b) the TOF-SIMS profiles of atomic concentrations of Si, C, O and H (right).

Effect of $CF_4/CH_2F_2/Ar$ etch plasma on low-k A

$CF_4/CH_2F_2/Ar$ based CCP chemistries are used for patterning the low-k materials by anisotropic etching. The process is known for leaving fluorocarbon byproducts on the sidewalls and bottoms of the trenches/vias in non-favorable cases. On the blanket wafers, it was observed, that the etching rate can vary, when the CF_4 and CH_2F_2 ratio is changed and also that the process can result as pure deposition of the fluorocarbon for ratio 4:4 [5]. We studied a sample of dielectric A prepared by 16 sec treatment by such plasma and observed 31nm thick layer on the non-damaged material by spectroscopic ellipsometry (fig. 3a). This layer is also detected in the TOF-SIMS profile; however, the thickness can not be directly estimated from the profile due to unknown sputtering rates. Fluorocarbon layer of this thickness seals the pores, as can be proven by means of ellipsometric porosimetry.

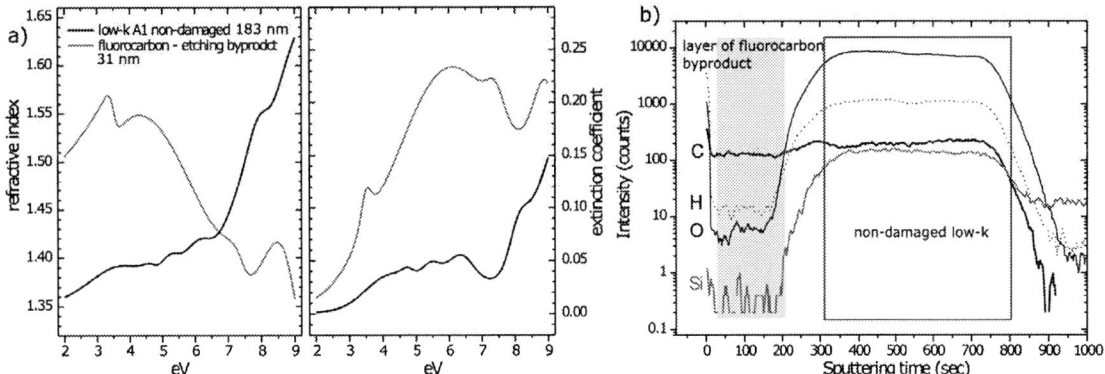

Figure 3 a) Optical properties of fluorocarbon by-product from (4:4:2) $CF_4/CH_2F_2/Ar$ etch plasma. The polymer forms 31nm thick layer on top of the non-damaged film. b) The TOF-SIMS depth profiles shows carbon-rich layer on the sample.

For the standard etching chemistry with ratio (7:1:2 $CF_4/CH_2F_2/Ar$), removal of material is observed without causing significant changes to the optical properties of the low-k. However, the EP experiments sings pore sealing, therefore thin layer of fluorocarbon byproduct is expected on the top of the etched blanket low-k. This layer is not clearly visible in the TOF-SIMS profile and direct trial of ellipsometric modeling was not successful, but if the optical properties of the top layer are fixed as obtained from the thicker layer, the model can be improved, resulting in thickness of the fluorocarbon layer equal to 4 nm. It has to be mentioned that the quality of this layer can be different from the deposited fluorocarbon, containing the etching byproducts and diffusing into the near-surface pores.

Effect of O_2 strip-plasma on low-k B

We treated the low-k dielectric B by the O_2 plasma in CCP reactor for 20 and 30 seconds. The toluene EP reveals pore sealing and the water EP shows 5% of absorbed water in saturation pressure for both samples (while the non-damaged sample absorbs less than 1%). The determination of top damaged layer from ellipsometric measurements was not clear for this set of samples because of low contrast between refractive index (RI) of damaged and non-damaged sub-layers, but a reduction of UV absorption in the top layer can be reported (fig. 4a).

Effect of H_2N_2 strip-plasma on low-k B

Alternative strip-plasma for dielectric B based on H_2N_2 CCP chemistry was applied for 10, 20 and 30 seconds on the samples. The EP implicates sealing of pores and low hydrophilisation, resulting in 3% of absorbed water in saturation pressure. The spectroscopic ellipsometry can clearly detect densified top layer with increasing RI with longer treatment time and formation of absorption band around 7.6 eV (fig 4b). The origin of this absorption is not understood and will require further study. The absorption band at 4.1 eV is no longer observed in plasma treated samples.

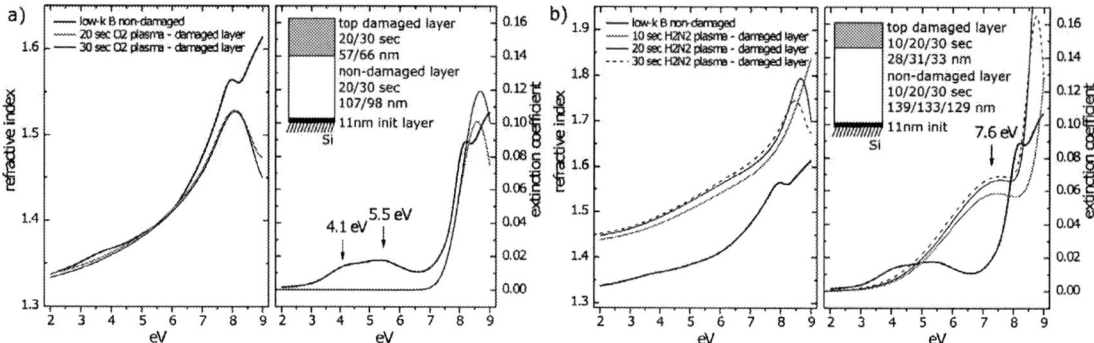

Figure 4 a) Optical properties of non-damaged low-k B and of top layer treated by O_2 plasma in CCP reactor for 20 and 30 seconds. The reduction of absorption bands between 3 and 7 eV is observed, but no significant evolution of RI. b) The optical properties of non-damaged low-k B and of top layer treated by H_2N_2 CCP for 10, 20 and 30 seconds. Increasing RI is detected with increasing treatment time as well as formation of 7.6 eV band.

DISCUSSION

The ellipsometric experiments and modeling efforts are limited by the nature of low-k materials. The inhomogeneities created by non-uniform curing, gradual effect of plasma damage, and also the presence of water in the hydrophilic samples lead to limited accuracy of exact optical constants measured by ellipsometry. All these fact have to be taken into account and studied in more details for exact model construction for scatterometry.

On this stage, first simulations on scatterometry sensitivity were performed using measured properties of low-k A damaged by O_2 plasma in ICP chamber. The feasibility study involved gradient damage on the sidewall and predicted good sensitivity to gradient step parameter, although the gradient profile had to be fixed and linear function was chosen.

Attention has to be given to difference between sidewall damage and modification observed on the blanket wafers. The uniformity of the damaged layer will no longer be present on the sidewall and additional study using electron microscopic techniques has to be performed.

CONCLUSIONS

Collection of low-k films was treated by various plasmas and studied by spectroscopic ellipsometry to evaluate the changes of optical properties of plasma damaged layer. Three main effects of strip-plasmas responsible for the changes were found: 1) Removal of carbon content resulting in reduction of UV absorption and lowering the RI. 2) Hydrophilisation of the layer and following shifts in RI related to amount of absorbed water. 3) Densification of the layer resulting in increased RI. Properties of the fluorocarbon byproduct of $CF_4/CH_2F_2/Ar$ etch plasma has been estimated on special sample with fluorocarbon layer deposited by tuning the $CF_4:CH_2F_2$ ratio.

The observed changes of optical properties are, according to first feasibility simulations, sufficient for development of the sidewall plasma damage scatterometry.

ACKNOWLEDGMENTS

This work was supported by European Pull Nano project. The authors want to thank to Salvador Eslava Fernandez for useful discussions and Tomin Liu for his help with the measurements.

REFERENCES

1. D. Shamiryan et al., J. Vac. Sci. Technol. B **20** (5) (2002)
2. P. Marsik et al., Phys. Stat. Sol. (c), accepted (2008)
3. S. Eslava et al., J. Electrochem. Soc., accepted (2008)
4. A. Zaka, unpublished IMEC 2007
5. M. R. Baklanov, K. P. Mogilnikov, Microelectron. Eng., **64**, 335 (2002)
6. P. Marsik, in preparation 2008
7. A. Grill, D. A. Neumayer, J. Appl. Phys., **94**, 10, 6697 (2003)

Mater. Res. Soc. Symp. Proc. Vol. 1079 © 2008 Materials Research Society

Using a Barrier Layer to Inhibit Ti/Oxide Reaction to Reduce RC Delay and Improve Electromigration in Al-Cu/Ti/W Interconnect for High Power Analog and Mixed Signal Applications

William J Murphy, Tom C Lee, Jonathan Chapple-Sokol, Daniel A Delibac, Z. X. He, Stephen E Luce, Stephen A Mongeon, David C Thomas, Daniel S Vanslette, and Timothy D Sullivan
International Business Machines, 1000 River Road, Essex Junction, VT, 05452

ABSTRACT

In this paper, we report a novel method to improve aluminum interconnect electromigration performance, reduce metal line sheet resistance (Rs) and reduce via contact resistance (Rc) in a minimum pitch design. We report the effects of bottom redundancy layers on electrical line and via resistance and upstream electromigration performance. We studied 4 metallization stacks: (I) Ti/TiN/AlCu/Ti/TiN, (II) Ti/TiN/Ti/AlCu/Ti/TiN (annealed at 400°C for 20 minutes), (III) a flash process/stack I, (IV) stack III with aluminum deposition temperature of 250°C. Aluminum deposition temperature in stacks I, II, and III are 200°C. Bottom Ti/TiN, AlCu, and Top Ti/TiN layers have the same thickness for all these stacks. Metal line Rs and via contact resistances (Rc) of stack IV are 5% and 10% lower, respectively, than stack I. Stack II metal line sheet resistance is 15% higher than stack I, which is attributed to Al consumption in the $TiAl_3$ formation during 400°C/20minutes annealing. Electromigration performance is best with stack IV followed by III, II, then I.

INTRODUCTION

Aluminum interconnect technology is increasingly challenged at smaller technology nodes. While copper interconnect reliability has received much attention, most Analog/Mixed Signal applications are not being built in Cu due to cost [1]. Reducing thermal budgets, BEOL wiring/via requirements and defect density requirements have increased the manufacturing challenges of building reliable aluminum wires. In particular, technologies that have thermal budget constraints are limited in ways to improve electromigration resistance that do not increase metal line resistances or put constraints on designers by requiring via redundancies or metal wire extensions. Doping of aluminum wires can improve electromigration performance, but at the expense of metal line resistance. Annealing of wiring to allow for diffusion of dopant, grain growth or intermetallic formation, such as $TiAl_3$, have all been demonstrated to improve electromigration performance [2-4]. However such techniques are not practical in technologies that have thermal constraints or where BEOL R-C performance requires the lowest possible wire and via resistances. Additionally, redundant vias and line extensions [5] can be used to extend EM lifetimes but at the expense of silicon area and may impart layout constraints.

In this study, we report the effects of the bottom redundancy layer on the electromigration performance and wire/via resistances.

EXPERIMENT

Using a combined Cu and Al 180nm technology, electromigration and generic electrical test macros (kerf) structures were fabricated on 200mm wafers. Wafer processing was started at poly-silicon (PC) through CA (W) contact, then M1 was processed with Cu damascene. A V1 (W) via in SiO_2 connects to M1. M2 metallization experiments with the four metallization (I-IV) stacks were run. Table 1 provides the process descriptions for each stack. The aluminum composition is Al-0.5%wt Cu. All films were deposited by PVD in conventional DC Magnetron PVD equipment using planar sources. The aluminum deposition was temperature controlled using an electrostatic chuck. All other metal films have no temperature control. The wafers were initially degassed using lamp heating in a vacuum for water desorption. Stack III, IV used a 5 nm oxide equivalent preclean prior to W deposition. Figure 1 shows the schematic diagram of stacks I, II, and III, IV.

Stack	Process Description
I	150c degas, 15nm Ti, 20nm TiN, 370nm Al @200c, 5nm Ti/70nm TiN
II	150c degas, 15nm Ti, 20nm TiN, **10nm Ti**, 370nm Al @200c, 5nm Ti/70nm TiN
III	**250c degas, 5nm preclean, 10nm W, + air break +** **250c degas**, 15nm Ti, 20nm TiN, 370nm Al @200c, 5nm Ti/70nm TiN
IV	**250c degas, 5nm preclean, 10nm W, + air break +** **250c degas**, 15nm Ti, 20nm TiN, **370nm Al @250c**, 5nm Ti/70nm TiN

Table 1, Process Description

Fig. 1, Schematic of M2 stacks I-IV

Wafers were exposed to ambient atmosphere between the W PVD deposition in stacks III, IV and the balance of the stack (air break). The top Ti/TiN was deposited in the same PVD chamber with no target conditioning between wafers. The bottom Ti and TiN were deposited in separate PVD chambers. After M2 definition, a V2-M3 and final via/bond pad were built to replicate manufacturing process flow. For stack II, the M2 was annealed post M2 etch.

M2 Rs and V1 Rc were measured on kerf structures using M2 serpentines and M2-V1-M1 via chains, respectively. TEM analysis was conducted on structures to examine the wire to via and wire to insulator interfaces.

Electromigration was measured at M2 on minimum dimension, single via structures with no metal extensions over the vias [6]. The nominal M2 line is 0.28 µm wide and 200 µm long.

The V1 diameter is 0.24 µm. Electromigration evaluations were carried out on all four stacks at 250°C stress temperature and stress current densities between 10 to 15 mA/µm^2. Current flow is M1-V1-M2. Using 20% resistance increase (dR/R) as the failure criterion, the lifetime distributions of all four stacks exhibit bi-modal behavior. The normalized lifetimes of the early modes of three liner stacks are about 1X for stack I, 1.5X for stack II, 2X stack III and 3X for stack IV.

RESULTS AND DISCUSSION

Metal Wire and Via Resistances

Metal 2 Rs was measured on kerf structures. Figure 2 shows the effect of the stack composition on metal 2 wire resistance.

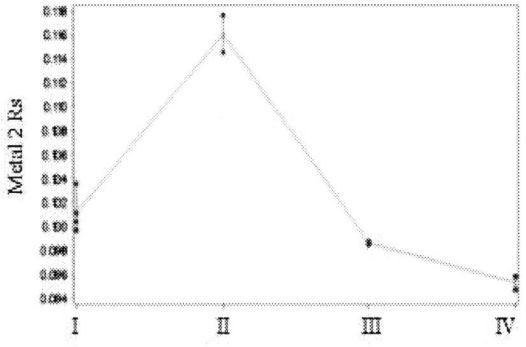

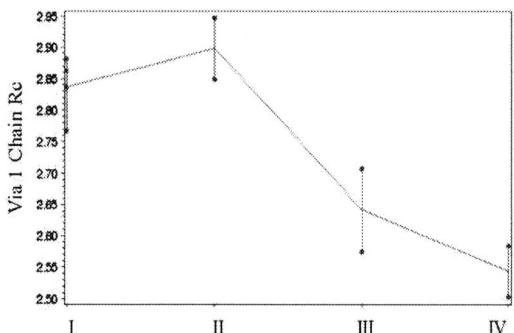

Fig. 2, Metal 2 Rs of Stacks I-IV   Fig. 3, Via 1 Chain Rc of Stacks I-IV

Clearly, adding the Ti in Stack II increases the Metal 2 Rs. This is due to TiAl$_3$ formation in the Al layer. The 5% improvement in Metal 2 Rs in Stack IV is due to the increased deposition temperature of the Al from 200°C to 250°C .

Fig. 3 shows the V1 chain Rc by stack. A significant reduction in via 1 chain Rc is observed when the top connection to the via is made with a W redundant layer (stacks III and IV). Usually, Ti is used to getter oxides from the top via surface to reduce via Rc, however our data shows that a via exposed to an Ar preclean then capped with W has 10% lower via Rc than when using Ti. The W layer also provides a low resistance shunt. Stack IV produces the lowest via Rc and lowest Al metal Rs.

Electromigration Performance

Electromigration lifetimes were evaluated on all four metal stacks at 250°C stress temperature and stress current densities between 10 and 15 mA/µm^2. For a 20% resistance increase as the failure criterion, the lifetime distributions of all four stacks exhibit bi-modal behavior (Fig. 4a) since the fail distribution does not lie on a straight line. Stack II allows for a single mode and is due to continued reaction of Ti with Al forming TiAl₃ driving resistance higher. However, the t50 time for stack II is below the 2.0mA/μm^2. Failure analysis (Fig. 4b) suggests that the two modes are differentiated by the void locations. For the early mode, the electromigration voids are located immediately at the cathode end of the test line. In this case, only the bottom redundant liner is left to carry current when the voids grow big enough to pass the edge of the bottom W via. In contrast, the voids of the late mode fails are located somewhat away from the cathode end of the line beyond the via such that both the bottom and the top liners are able to shunt the electrical current after voids of late mode fails have grown to extend over the entire AlCu cross sectional area. An inspection of Table 2 shows that the t50 times are improved by ~1.5x and the sigma reduced by 50% using stack IV compared to stack I. The resistance versus time curves, R(t), Fig. 5, of stack I have more abrupt resistance increases at shorter electromigration stress durations than those of either stack II or stack III/IV. The R(t) curve of stack II shows very good behavior, but, there is a resistance penalty of allowing titanium to react with aluminum. More accurately the stack II plot represents in part a resistance change due to Ti reacting with Al resulting in shorter t50 times. The anneal conditions during processing for stack II are not sufficient to convert all the Ti into TiAl₃. The anneal limitation is imposed by the presence of Cu wiring at M1. The R(t) plots of stack III/IV show very good stability, and perhaps some benefit to running at 250°C. Failure analysis of a fail with an abrupt resistance increase (Fig. 6) after the 20% fail criteria has been reached shows a thermal runaway signature at the bottom redundant layer of stack I. This type of failure will cause a catastrophic failure in the circuit.

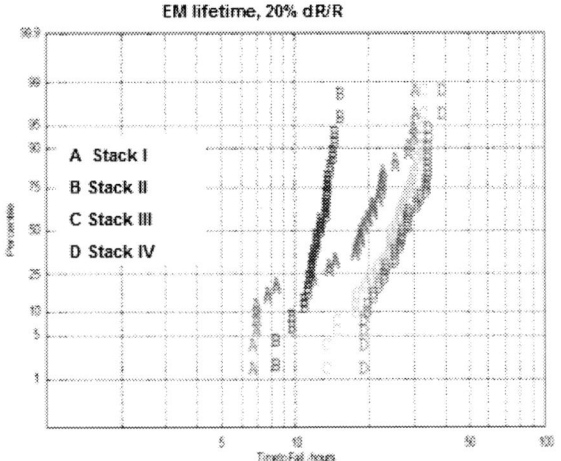

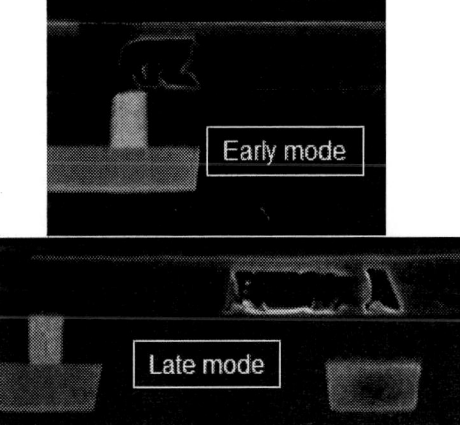

Fig. 4a EM lifetimes of Stacks I-IV   Fig. 4b, Early and Late mode fail

The void characteristics and resistance time behavior of electromigration imply that the bottom liner plays the most critical role.

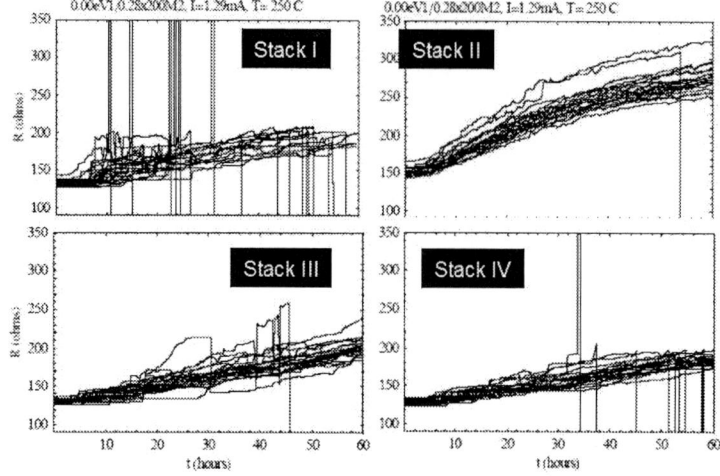

Stack	t50, hours	sigma
I	18.9	0.445
II	12.7	0.135
III	23.35	0.235
IV	30.35	0.235

Fig. 5 Resistance vs time of Stacks I-IV
(all y axis 100-350 ohms, all x-axis 0-60 hours)

Table 2, t50, sigma

Physical Analysis

There are no M2 extensions over V1 vias by design and the M2 wire is landed on a single via (Fig. 4b). Since the early mode fail is associated with the cathode end (void over the via) and the bottom redundant layer carries all the current, improving the bottom redundant layer will improve the early mode fail. Fig 4b shows the late mode fail with the void occurring away from the via. Fig. 6 shows melting of the redundant layer. It is noted that the metal wire is foreshortened on the via by observing the chamfering of the via. Since the design calls for a single via contacted by a minimum dimension wire, and no extensions, foreshortening over the via can occur. Clearly, the wire has failed. This suggests that the bottom Ti layer in Stack I is less conductive than designed for. Contamination of the bottom Ti with the V1 oxide is possible. Fig. 7 shows that the W redundant layer in an early mode fail is robust. Here too, the line is foreshortened on the via, meaning no connection is possible with the top redundant layer when the void exposes the via.

Fig. 6, Early Mode fail,
ruptured redundancy

Fig. 7, W redundant layer intact

EDX analysis of samples submitted for TEM show that the bottom titanium layer reacts with the SiO$_2$ insulator and in the early fail mode this results in earlier fail times. The reaction of the Ti with the SiO$_2$ results in a higher resistance redundant layer leaving less Ti cross section to carry current. Fig. 8, 9 shows elemental analysis for the bottom layers of stack I and III, respectively. It can be seen in Fig. 8 that oxygen is incorporated into the titanium, raising its resistance, while the tungsten, in Fig. 9 prevents reaction of titanium with the SiO$_2$ interface.

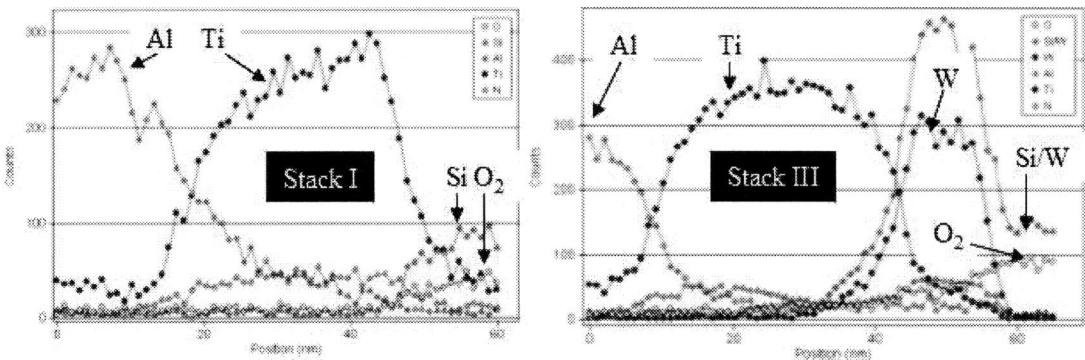

Fig. 8, EDX analysis of stack I   Fig. 9, EDX analysis of stack III

The Ti signal in Fig. 8 and Fig. 9 includes both Ti and TiN from the bottom layers. Limiting the Ti reaction with the oxide and providing a low resistance shunt to the via will improve the early mode fail time and improve the t50 time for EM resistance. Fig. 9 shows Ti and Al in the W and is due to beam interference caused by the W layer. Fig. 10 shows a TEM with EDX scan of a failed line over the stud using Stack I. The TEM shows TiAl mixing over the W stud. In addition, the TEM in Fig. 10 clearly shows the ruptured bottom redundancy adjacent to the stud. Joule heating results in a thermal runaway as void size increases. The resistance of the bottom redundancy is increasing as the void progresses away from the stud. As the void length increases, heat conduction is only through the bottom insulator resulting in thermal runaway [7]. This is seen as an electrical fail as in Fig. 5, Stack I, as shown by a near digital increase in wire resistance.

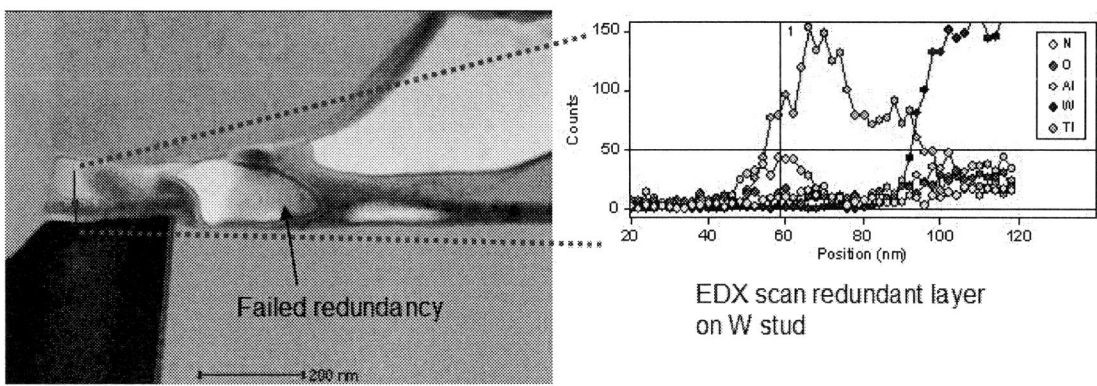

Fig 10, TEM/EDX of failure area, Stack I

CONCLUSIONS

We have shown that in high performance aluminum interconnect structures the bottom redundant layer is the most critical factor in improving early mode electromigration failure. Specifically, where thermal budget constraints exist, a 'No Anneal' metallurgy stack must be utilized, such as Ti/TiN/Al/TiN. This structure is common in the industry, but fails to meet EM requirements where minimum ground rules exist. We have shown the benefit of using W as a first redundant layer to improve the electromigration lifetimes and to eliminate the possibility of catastrophic circuit failure. The requirements, therefore, of a bottom redundant layer are to be chemically inert, and have a high melting temperature and low resistivity.

ACKNOWLEDGMENTS

The authors would like to thank Carl Farris for support in wafer fabrication, Chuck Cox for SEM and Phil Pokrinchak for TEM/EDX analysis and Chad M Burke, Kevin Lindstam and Randy Austin for EM testing support.

REFERENCES

1. http://www.eetimes.com/showArticle.jhtml;jsessionid=QH2WVHB4JSY2SQSNDLRSK H0CJUNN2JVN?articleID=205602370
2. R. Shohji, M. Uda, T. Nakamura, T. Yoda, Y. Itoh, IEEE Transactions, 1999, p. 302.
3. S. Matsumoto, R. Etoh, T. Ohtsuka, S. Ogawa, IITC, 1998, p. 113
4. C. Grass, H. Le, J. McPherson, R. Haverman, IEEE/IRPS, 1994, p.173
5. S. Skala, S. Bothra, IITC, 1998, p. 116.
6. M. Dion, IEEE IRW, 2000, p. 167.
7. D. Harmon, J. Gill, T. Sullivan, IRW, 1998, "Thermal Conductance of IC Interconnects Embedded in Dielectrics"

Mater. Res. Soc. Symp. Proc. Vol. 1079 © 2008 Materials Research Society 1079-N09-03

Semiconductor Film Bonding Technology and Application in Two-axis Hall Sensor Fabrication

Keishin Koh, Takashi Matushita, and Koji Hohkawa

Kanagawa Institute of Technology, Atsugi, 243-0292, Japan

ABSTRACT

We study basic problems with semiconductor film bonding technologies and propose a new release method for a large number of semiconductor films using a film photoresist to protect the semiconductor films. We investigated the basic process conditions, released many a large GaAs films, and bonded them to Si, LiNbO$_3$ substrates and a metal Au surface. We estimated the ctystallinity of semiconductor films by X-ray diffraction and Raman spectra. The results demonstrate the effectiveness of these processes. As an application of the semiconductor film bonding technology, we fabricated a two-axis Hall sensor with a planar structure. The two-axis Hall sensor can measure axial and radial magnetic filed components (*Bx and Bz*) with a sensitivity of about 9.2 Ω/G for *Bz* and 4.5 Ω/G for *Bx*.

INTRODUCTION

The epitaxial liftoff (ELO) process is an attractive material technology [1], as it allows the integration of different electronic or optical film materials on the same substrate and realization of novel functional electronic devices or sensors with new structure, small size, and low cost [2-4]. Recently, a new LSI manufacturing technology combined with bonding technology such as system on chip, and system in package has been developed. The semiconductor film bonding technology based on the ELO process is attracting attention, because it fuses the LSI manufacturing technology with bonding technology. In this paper, we report the results of a study on the basic problems of semiconductor film bonding technologies, such as release and bonding thicker semiconductor film, finding a protective material for semiconductor film, and its application in fabricating two-axis Hall sensors.

FILM BONDING TECHNOLOGY

The basic process of the semiconductor film bonding technology is based on the ELO process and consists of two process steps, the release of the semiconductor film and its bonding (Figure 1). In the release process, the GaAs compound semiconductor film, which is covered with a protective material, is released into an HF solution by selective etching of the AlAs sacrifice layer. In the bonding process, the GaAs film is bonded onto some other substrate or a metal with a smooth surface using van der Waals force or an adhesive material such as water glass (SiO$_2$: Na$_2$O) (Figure 1(b) or Figure 1(c)). Compared with the wafer bonding technology, the semiconductor film bonding technology has many features as follows:
1) It can release a large number of small semiconductor film devices from one compound semiconductor wafer simultaneously and the released film devices can be bonded onto a number of other substrate. There is no damage to the compound semiconductor substrate during release process, allowing it to be recycled.

2) The integration density of the functional device is high since the film semiconductor devices are small and thinner. After bonding, we can perform LSI processes such as planarization and wring, and complete the interconnection between devices. It is also possible to realize functional devices having three-dimensional structure.

3) Before the release process, we can nearly complete the fabrication of the compound semiconductor devices. Therefore, the fabrication process yields high quality devices.

4) Since semiconductor films are as thin as several micrometer and have weaker mechanical strength, a handling method and protection material are important for the semiconductor film during processing.

Enlarging the application area of the semiconductor film bonding technology requires overcoming some problems, which include the following: (1) finding an optimum protective material, (2) release and bonding of thicker semiconductor film, and (3) the bonding condition of the metal surface. During the entire process, the protective film not only controls stress on GaAs films and supports them, but also affect the feasibility of handling method for transferring the GaAs film from host substrate to another bonding substrate. In a previous study, the Apiezon wax, and a polyimide film have been are used as protective material [5 - 6]. Each material has various the merits and demerits in coating, mechanical and chemical properties and pattering possibilities. Therefore, to find the optimum protective material, we investigated film photoresist (SL-1129 Hitachikasei Co.). The film photoresist has many merits such as follows: (1) The coating is simple, and thickness of the film can be controlled over a wide range

(20μm to 100μm), (2) they can be removed easily by selecting appropriate solution, (3) low temperature

process (<120 °C), and (4) it offers good resistance to chemicals. Figure 2 shows the film bonding process using a film photoresist as protective material. The process steps were as follows. We covered a film photoresis on GaAs film that is growth on GaAs substrate epitaxially and patterned it. Next, we lightly bonded the sample on Si substrate for handling, which the film photoresis is covered and is used as adhesive material. We released the GaAs film by etching an AlAs layer from GaAs substrate and then, directly bonded the GaAs film on the Si, LiNbO$_3$ substrate or other substrate surface on which a metal pattern was formed. Finally, we removed the film photoresist using a solvent and held the sample at 100 °C for several hours. Table I shows the process parameters.

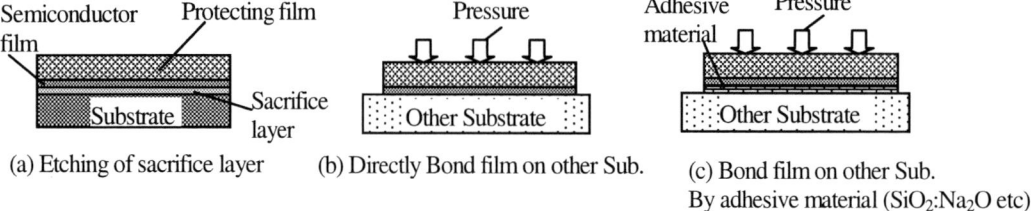

(a) Etching of sacrifice layer   (b) Directly Bond film on other Sub.   (c) Bond film on other Sub. By adhesive material (SiO$_2$:Na$_2$O etc)

Figure 1 Concept of semiconductor film bonding technology

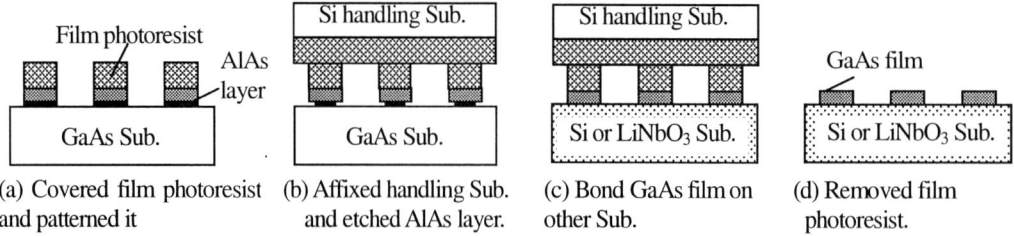

(a) Covered film photoresist and patterned it   (b) Affixed handling Sub. and etched AlAs layer.   (c) Bond GaAs film on other Sub.   (d) Removed film photoresist.

Figure 2 Film bonding process using film photoresist as protective material

Figure 3 shows experimental result of the GaAs film. The release rate of the GaAs film decreases with increasing thickness of the GaAs film. It is believed that stress in the thinner GaAs film caused by the film photoresist is larger than in thicker GaAs film; thus, thinner GaAs film is released more easily. For the thinner GaAs film, the release rate in the HF solution with an activator is about 1.2 times that in a solution without an activator. There is little dependence on the thickness of the film photoresist. Figure 4 shows a photograph of the GaAs film bonded on the LiNbO$_3$ substrate and striped Au film.

Table I. Process parameters for the film bonding process

GaAs film thickness	0.5 μm, 1.15 μm, 3.15 μm, 6.15 μm
AlAs film thickness	50 nm
Substrate size	GaAs 7 mm × 7 mm, LiNbO$_3$, Si 12 mm×12 mm
Film photoresist thickness	25 - 50 μm (SL-1129,negative type Hitachikasei Co.)
Concentration of HF solution	10%
Heating after bonding	100 °C, over 12hr
Removal of film photoresist	Acetone and remover (1165, Rohm and Haas Elec. Mater. Co.)

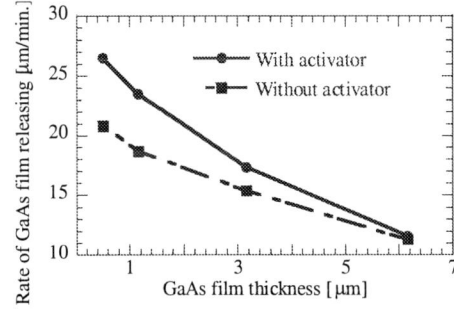

Figure 3 Relationship between the GaAs film release rate and GaAs film thickness (film photoresist thickness is 25 μm and GaAs size is 800 μm × 500 μm)

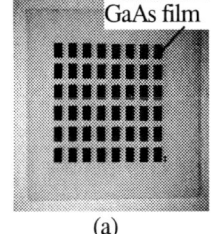

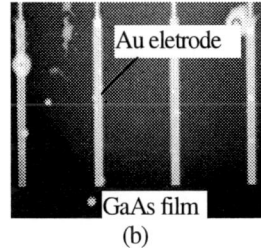

(a) (b)

Figure 4 Photograph of a bonded film (a) on the LiNbO$_3$ Sub., GaAs size is 800 μm × 500 μm, (b) on the striped Au surface 60 nm thick formed on the Si Sub.

We used an X-ray diffractometer and Raman spectra to estimate the stress in the GaAs film and deformation in the crystallinity of the GaAs film after bonding. By moving the sample in the y-direction, we measured the one-dimension X-ray diffraction pattern shown in figure 5(a). From the XRD pattern data, we calculated the average value, error and standard deviation of both the lattice constant, and full width at half maximum (FWHM). Table II shows the statistical results, which indicate that there is almost no variation of the lattice constant after bonding, and the variation of FWHM after bonding is obvious. With decreasing GaAs film thickness, the FWHM value after bonding tends to increase. On the other hand, the standard deviation of the lattice constant and the FWHM , these value are about $11 - 30 \times 10^{-5}$ and $1.4 - 5.6 \times 10^{-4}$ respectively. This indicates that the distribution of stress in the GaAs film due to bonding is highly uniform throughout entire GaAs film. We also measured Raman spectra of the bonded GaAs film as shown in figure 5(b). We observed two Raman shift peaks corresponding to the GaAs epitaxial crystal film.

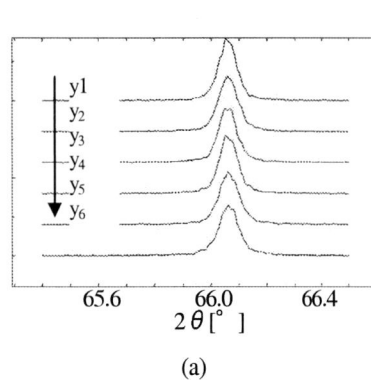 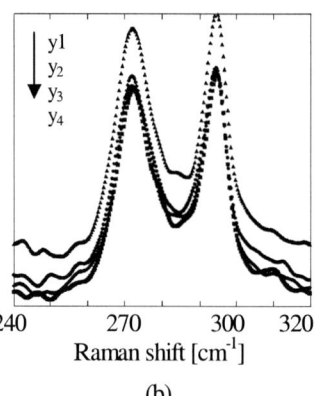

(a)   (b)

Figure 5 Experimental results of bonded film (a) X-ray diffraction pattern (b) Ramen spectra

Table II. Summary of the lattice constant and FWHM value for bonded GaAs film

Thickness of GaAs film (μm)	Pressure (kg/mm²)	Lattice constant			FWHM		
		Average (nm)	Error (×10⁻⁴)	Standard deviation (×10⁻⁵)	Average (deg)	Error (deg)	Standard deviation
0.5	1	0.5652	2.0	14	0.0770	0.0375	0.00588
	2.1	0.5653	0	16	0.0828	0.0433	0.00366
	3.1	0.5653	3.0	12	0.0737	0.0342	0.00501
1.15	3.1	0.5653	1.8	31	0.0443	0.0093	0.0014
3.15	3.1	0.5653	0.6	30	0.0463	0.0079	0.0043
6.15	3.1	0.5653	1.6	11	0.0389	0.0063	0.0025

FABRICATION OF TWO-AXIS HALL SENSOR ARRAY

In this section, we describe an application of the semiconductor film bonding technology. The Hall sensor is a very convenient, non- destructive, and contactless magnetic sensor [7]. It is widely used to estimate the current density distribution and the magnetic properties of superconducting materials [8-9]. To realize a highly functional real-time measuring system with a high spatial resolution, and estimate the properties of superconductor film with a complex structure or pattern, it is very important to develop a novel Hall sensor array with small size, a planar structure and two or three axes. We proposed a two-axis Hall sensor array fabricated on an Si substrate as shown in figure 6. The sensor for measuring B_z and B_x magnetic fields are integrated on the Si substrate with a planar structure. When the bias current flows in the y-direction, we can measure the Hall voltage V_z across the right and left electrodes corresponding to the B_z component, and the Hall voltage V_x across the top and bottom electrodes corresponding to the B_x component. LSI manufacture technology allows easy fabrication of a Hall sensor array consisting of numerous Hall sensors of micrometer size.

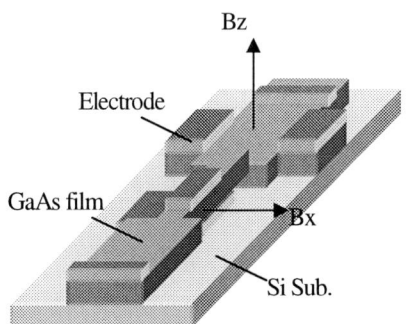

Figure 6 Two-axis Hall sensor with film bonding technology

We fabricated test devices using the semiconductor film bonded technology and GaAs devices fabrication processes. Table III shows the sensor parameters and key processes conditions. First, we formed a bottom ohmic electrode on the Si substrate, and then bonded the GaAs film (for the Hall sensor) on the Si substrate using semiconductor film bonding technology. Next, we patterned the GaAs film and formed the top ohmic electrode. Then we heated the ohmic contact. Finally, we coated polyimide on the GaAs film as insulation and planarization layer, and formed pad electrode for wiring. Due to the limitation of etching equipment, we used a wet etching process to form the GaAs pattern. Figure 7(a) shows the photograph of two-axis Hall sensor on the Si Substrate. We obtained good ohmic contact between top and bottom electrode for different contact area size, with a contact resistance of about 8.6×10^{-5} $\Omega \cdot cm^2$ which is the same as for bulk GaAs.

We estimated the basic characteristics of the test Hall sensor and the results are shown in figure 7(b) and (c). These results show that the Hall voltage is in proportion with the magnitude of the magnetic field Bz and Bx when the bias current is constant. When the magnitude of magnetic field is constant, the Hall voltage is almost in proportion with the magnitude of the bias current for both Bz and Bx sensor. The sensitivity is about 9.2 Ω/G for Bz and 4.5 Ω/G for Bx respectively. The difference in the sensitivity between Bz and Bx is because the thickness of the GaAs film in x-direction is about two times that in the z-direction. If we use dry etching process and improved fabricating precision of the sensor pattern, we can obtain the same sensitivity for both. We also measured two-axis magnetic component using test sensor and compared with marketed products. We kept the sample at 77 K for several hours and observed almost no change in the mechanical property of the bonded GaAs film, indicating that the device is stable at low temperatures.

Table III. Test Hall sensor parameters and process conditions

GaAs film (nm)	n+GaAs (50)/ n-GaAs (3000)/n+GaAs (100)
	The carrier concentration of n-GaAs is about 1×10^{17} cm^{-3}.
Sensor size (µm)	For Bz: $14 \times 12 \times 3$, for Bx: $3 \times 12 \times 6$
Ohmic electrode (nm)	Top: Au-Ge-Ni, thickness 60 , bottom: Au-Ge-Ni, thickness 100
Heating of Ohmic contact	370 -380 °C 1min. at N_2
Polyimide	Thickness 3 µm
Heating condition	Cure: 170 °C 30min and 320 °C 1hr.at N_2
Planarization of polyimide	O_2 plasma, 100 W, 4 Pa, 10 min.
Wiring electrode	Cr-Au, thickness: 0.5 µm

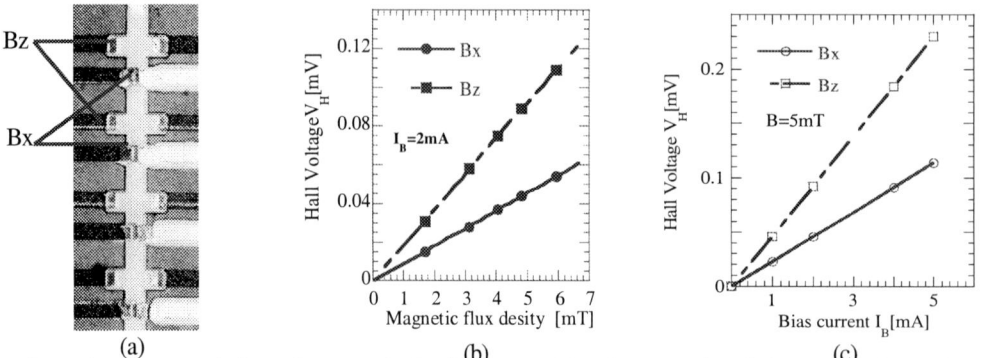

(a)   (b)   (c)

Figure 7 Basic characteristics of two-axis Hall sensor (a) Photograph of the two-axis Hall sensor fabricated on Si substrate (b) Hall voltage vs. magnetic field magnitude (c) Hall voltage vs. bias current.

CONCLUSION

We studied the basic problems with the semiconductor film bonding technology. We used film photoresist as protective material for the semiconductor film and successfully released a large number of thin GaAs film and bonded them on the Si, $LiNbO_3$ substrates or metal surface. We also investigated the release conditions of GaAs film with different thickness. We estimated the crystallinity of the bonded film with an X-ray diffractometer and Raman spectra. The results confirmed the effectiveness of the proposed release and bonding method. As an application of this method, we fabricated a two-axis Hall sensor of GaAs on a Si substrate. The basic sensing characteristics demonstrated the feasibility of this two-axis Hall sensor and we will study its application in estimating of high-temperature superconducting material. We also developed a novel sensing devices using semiconductor film bonding technology.

ACKNOWLEDGMENTS

This work was supported by fund (No. 18560703) from the Ministry of Education, Culture, Sports, Science and Technology. The authors wish to thank Mr. Deguchi and Mr. Ohkage for their contributions to the experiments

REEFERENCES

1. E. Yadlonovitch, T. Gmitter, J. P. Harbison and R. Bhat, *Appl. Phys. Lett..* **50**. 2222-2224 (1987).
2. A. Ersen, I. Schniter, E. Yadlonovitch and T. Gmitter, *Solid-State Electronics.* **36**, 1731-1739 (1993)
3. D. Bhattacharya, P. S. Bal, H. R. Fetterman, D. Streit *IEEE Photon. Techno Lett.* **7**, 1171-1173 (1995).
4. K. Hohkawa, H.Suzuki, K. Koh and S. Noge, *Jpn. J. Appl. Phys.* **39**. 1554-1558 (1997)
5. C. Camperi-Ginestet, M. Hargis, N. Jokerst and M. Alln, IEEE Trans. Photo. Tech. Lett. **3**, 1123—1126 (1991)
6. C. Hong, K. Koh, C. Kanashiro, Y. Aoki and K. Hohkawa, *Jpn. J. Appl. Phys.* **39**. 3666-3670 (2000)
7. R. S. Popovic, *Hall effect devices*, Bristol and Philadelphia ,IoP, ch. 5 (2004)
8. T. Tamegai, Y. Iye, Oguro and K. Kishio, *Physica C*, **213**, 33-42 (1993)
9. K. Shimohata, S. Yokoyama, T. Inaguchi, S. Hakamura and Y. Ozawa, *Cryogenics*, **43,** 111-116 (2003).

Mater. Res. Soc. Symp. Proc. Vol. 1079 © 2008 Materials Research Society 1079-N10-05

Organization of Magnetic/Noble Metal Heterostructures by an Applied External Magnetic Field

Nicolás Pazos-Pérez[1], Dmitry Baranov[1], Michael Hilgendorff[1], Jorge Pérez-Juste[2], Luis. M. Liz-Marzán[2], and Michael Giersig[1]

[1]center of advanced european studies and research (caesar), Bonn, 53175, Germany
[2]Department of Physical Chemistry, University of Vigo, Vigo, 36310, Spain

ABSTRACT

This paper describes the synthesis of binary nanoparticles consisting of a noble metal and a magnetic component. These heterostructures were produced by a seeded-growth approach in aqueous solution. FePt nanoparticles, as the magnetic component, were first synthesized in an organic medium, subsequently transferred into water, and finally used as seeds for the growth of the noble metal Au. This procedure results in FePt-Au heterostructures. Moreover, the synthesized heterodimers were organized into mesoscopic lines under the influence of an externally applied magnetic field. The produced heterostructures were characterized by transmission electron microscopy (TEM), scanning electron microscopy (SEM), and UV-vis spectroscopy.

INTRODUCTION

Over the past few years, metallic nanoparticles have been the subject of a large number of studies due to their extraordinary physical and chemical properties. For example, their electric and optical properties highly depend on the specific particle size, shape, and surrounding environment [1-4]. Gold nanoparticles, in particular, are in the focus of interest because of their potential use in microelectronic [5, 6], optical [7], and biomedical applications [8]. The possibility to synthesize gold nanoparticles linked to a magnetic component is very appealing for the formation of organized gold nanostructures via the application of magnetic fields to nanoparticles with different magnetic properties. The present work is concerned with the optimization of the wet-chemical synthesis of novel metal-magnetic heterostructures in aqueous environment, so that the size and shape of the individual components determine their overall optical and magnetic response. The production of such heterostructures was carried out using a modified seeded-growth method, where the seeds were made of magnetic nanoparticles (FePt), previously produced in high-boiling-point organic solvents [9, 10]. Phase transfer of FePt into water was achieved using the surfactant cetyltrimethylammonium bromide (CTAB) [11, 18]. The growth step on the magnetic seeds requires the reduction of a gold salt with a weak reducing agent (ascorbic acid) in the presence of CTAB, yielding stable composites of magnetic seeds with diverse shapes and sizes attached onto gold, which can also be produced with various sizes and morphologies, so that a wide variety of optical and magnetic properties can be obtained.

EXPERIMENT

Chemicals and apparatus. Diphenyl ether (99%), benzyl ether (99%), $NaBH_4$ (99%), platinum(II)-acetylacetonate (97%, $Pt(acac)_2$), oleic acid (90%), oleylamine (70%), iron pentacarbonyl (99.9%, $Fe(CO)_5$), octadecene (90%), and cetyltrimethylammonium bromide (99%, CTAB) were purchased from Aldrich. Ethanol (99%), hexane (95%), chloroform (99%), and ascorbic acid (99.5%) were purchased from Roth; tetrachloroauric acid ($HAuCl_4 \cdot 3H_2O$) from Fluka. All reactants were used without further purification. Milli-Q water (18 MΩ cm^{-1}) was used in all aqueous solutions, and all glassware was cleaned with aqua regia before the experiments. UV-vis-NIR spectroscopy (Varian, Cary 5000), transmission electron microscopy, TEM (Leo 922A EFTEM, operating at 200 kV), and scanning electron microscopy (SEM, Supra 55) were used to characterize the optical response, composition, structure, and organization degree of the synthesized nanostructures.

Synthesis of magnetic seeds. FePt alloy nanocrystals were used as seeds and prepared by the methods of Sun and co-workers [9, 10].
• Star-shaped FePt nanocrystals were synthesized using a slightly modification from the procedure described in Ref. 9, using a combination of oleic acid and oleyamine to stabilize the monodisperse FePt colloids.
In a typical synthesis (~13 nm star-shaped $Fe_{53}Pt_{47}$ nanocrystals), $Pt(acac)_2$ (0.5 mmol) was mixed with phenyl ether (10 mL) under a nitrogen flow and heated to 100 °C. $Fe(CO)_5$ (1 mmol), oleylamine (20 mmol), and oleic acid (20 mmol) were then added under a blanket of nitrogen. The mixed solution was subsequently heated up to 240°C, at a heating rate of ~0.5 °C/min and kept at this temperature for 1 h to assure complete decomposition of $Fe(CO)_5$, and then heated to reflux (~260°C). The mixture was kept refluxing at this temperature for 2 h. During this time, a slow nitrogen flow was introduced from time to time to remove some low boiling-point by-products and maintain the refluxing temperature between 255~260°C. The reaction mixture was allowed to cool down to room temperature by removing the heating source and the FePt nanoparticles were separated and purified as described in Ref. 12 and stored in hexane.
• FePt nanowires (NWs) and nanorods (NRs) were produced as described in Ref. 10. For the synthesis of 200-nm NWs, oleylamine (60 mmol) was mixed with $[Pt(acac)_2]$ (0.5 mmol) at room temperature. Under a gentle nitrogen flow, the mixture was heated to 60°C to form a light yellow solution. The solution was subsequently heated to 120°C in less than 5 min and kept at this temperature for 30 min. The color of the solution changed to dark yellow. $[Fe(CO)_5]$ (1.14 mmol) was then injected into the hot solution and the temperature rose to 160°C. After 30 min, the solution was cooled down to room temperature by removing the heating mantle from the reaction flask. The NWs were separated by precipitation with hexane (10 mL) and ethanol (50 mL) and subsequent centrifugation (6000 rpm). The obtained product was dispersed in hexane (10 mL). Other NWs/NRs were synthesized under similar reaction conditions but with different oleylamine/octadecene volume ratios. For example, 20-nm FePt NRs were made from 10 mL oleylamine and 10 mL octadecene.

Transfer of FePt nanocrystals into water. The transfer of FePt nanoparticles from an organic into an aqueous medium was accomplished using CTAB as a phase transfer agent [11, 18]. In a typical procedure, a 5 mL aliquot of the prepared magnetic nanoparticles (previously washed as mentioned above and redispersed in 30 mL of a non-polar solvent) was again centrifuged (7500 rpm, 20 min) upon addition of a non-solvent such as ethanol (80%vol). After redispersion in an organic solvent (with a lower boiling point and a higher density than water) such as chloroform (15 mL), an aqueous CTAB solution (30 mL, 0.1 M) was added, forming a two-phase system with the organic phase below and the aqueous phase on top. This mixture was transferred into a typical distillation setup, followed by evaporation of the organic solvent under vigorous mechanical stirring. After several hours at 90 °C, the chloroform was completely removed from the mixture and a clear aqueous nanoparticle solution was obtained. A final centrifugation (4000 rpm, 10 min)-redispersion step was performed to remove possible aggregates.

Growth of noble metals. For growing gold, a seeded-growth approach [13- 18] was used. The general procedure was as follows: 10 mL of a growth solution containing CTAB (0.1 M) and $HAuCl_4$ (5.00×10^{-4} M) was thermostated at 25-30 °C, and ascorbic acid (0.075 mL, 0.1 M) was added, causing a color change from orange (the Au^{3+}-CTAB complex) to colorless (the Au^+-CTAB complex). Finally, selected volumes of the seed solution were added. After 1 h, the color of the solution had changed to violet or red, depending on the amounts of reagents reflecting the reduction of Au^+ to Au^0 and the formation of the heterostructures. A schematic illustration of the seeded-growth process for the heterodimer formation is shown in Figure 1.

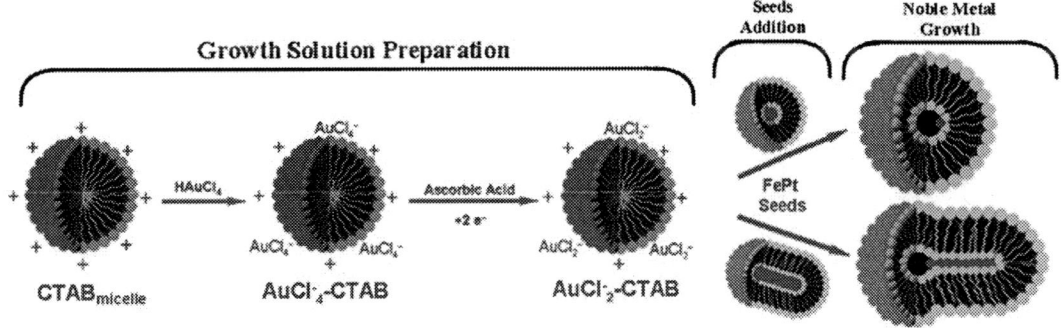

Figure 1. Schematic illustration representing the seeded-growth process, showing, that by the addition of two different morphologies of FePt seeds (star-shaped and rod), heterostructures with different morphologies are obtained.

Heterostructure assembly. To organize the heterostructures, they were washed by centrifuging (9000 rpm, 10 min) and redispersed in the same volume (10 mL) of an aqueous CTAB (0.1 M) solution to remove any remaining unreacted gold salt; this process was repeated twice. $NaBH_4$ (2.5 M, 10 mL) was then added and the mixture was left open under vigorous stirring overnight (~12 h) in an open flask to reduce any oxide that may have formed on the FePt surface. Subsequently, the excess surfactant present in the particle solution was removed by centrifuging 1 mL of the particle solution at 9000 rpm for 10 min. The surfactant-containing supernatant was discarded and the precipitate was redispersed in 1 mL of pure water. It was then centrifuged again and redispersed in 1 mL of water. Next, for TEM and SEM investigations a drop of 0.5 mL of the particle solution was placed on a carbon-coated copper grid or a silicon substrate, respectively, and allowed to dry at room temperature between two permanent magnets separated by ~1 cm and producing a magnetic field of ~ 1 T.

RESULTS AND DISCUSSION

Heterostructure and seed characterization. High resolution transmission electron microscopy (HRTEM, not shown) investigations of the FePt nanocrystals confirmed their crystallinity, showing lattice fringes without stacking faults [18].

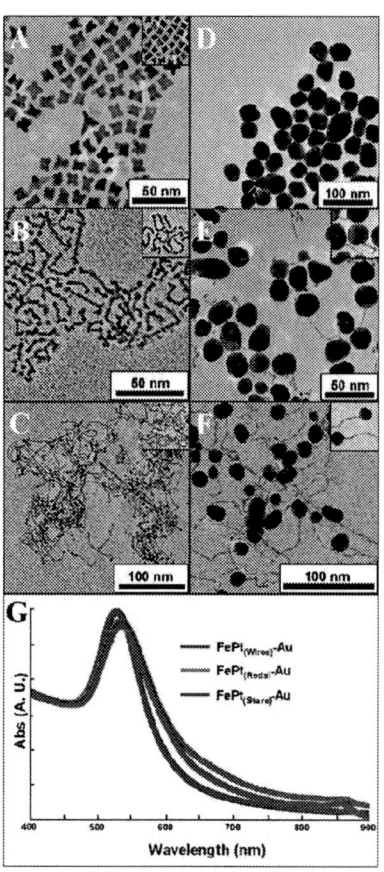

Figure 2. A, B, C: TEM images of FePt nanoparticles with diverse morphologies, stars (~ 13 nm, A), rods (~50 nm, B), and wires (~ 100 nm, C) in aqueous solution after transfer from CHCl₃ into water. Insets of A, B, C are the corresponding TEM images before transferring the heterodimers into water. D, E, F: TEM images of FePt-Au heterostructures produced from FePt seeds with different shapes. Insets of E and F are corresponding colored images showing the different components. G: absorbance spectra corresponding to three different aqueous solutions of FePt-Au heterostructures, where the Au component has a spherical shape in all cases and the FePt has diverse morphologies (wire in red, rod in green, and star in blue).

Figure 2 (A, B, C) presents TEM images of FePt stars (~ 13 nm), rods (~50 nm), and wires (~ 100 nm), respectively, after their transfer from organic into aqueous solution without any appreciable change in their morphology (cf. the insets of the corresponding organic-based nanoparticles). The images also reveal that the FePt nanocrystals are rather monodisperse and well separated on the TEM grid. Figure 2 (D, E, F) shows TEM images of the heterostructures prepared from the seeds shown in A, B, C, as described in the experimental section. The FePt-Au heterostructures comprise (in the cases of Fig 2 E, F) a spherical part corresponding to the gold nanoparticle component and an elongated part corresponding to FePt. However, in the case of the star-shaped FePt (Fig 2D), only the gold spheres are seen which may indicate that the FePt stars are incorporated into the gold. The exact composition of this kind of heteroparticles has to be clarified in the future. Nevertheless, these particles show magnetic behavior as will be shown below. Figure 2 G shows the absorbance spectra corresponding to the three different aqueous based heterostructures shown in Figure 2 (D, E, F). The typical plasmon absorbance of the spherical gold nanocrystals can be clearly seen, with a maximum peak ranging between 530-540 nm due to small deviations from the spherical shape.

Magnetic alignment. The heterostructures were aligned on carbon-coated copper grids and silicon substrates as described in the experimental section. TEM and SEM examinations of the $FePt_{star}$–Au particles (Fig. 3) confirmed their successful assembly into parallel lines over very long distances (up to 0.175 mm length), as can be seen in the SEM image (Fig. 3, lower left). At a higher TEM magnification (Fig. 3, right), in particular, it becomes clear that the observed lines are formed by individual heterodimers that are aligned into quasi-1D (linear) assemblies. These heterostructured arrays were also produced through the alignment of heterodimers containing other FePt morphologies. As an example, Figure 4 shows TEM images of heterodimers produced from FePt nanorods as the magnetic component.

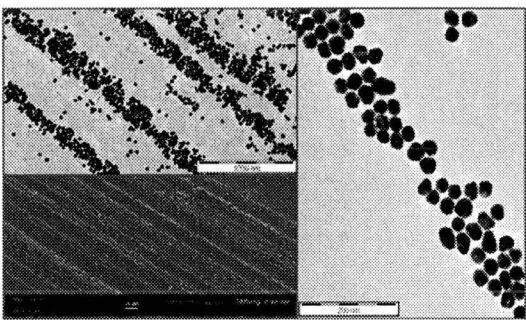

Figure 3. TEM and SEM images of $FePt_{stars}$–$Au_{spheres}$ heterostructures, organized into linear arrays by the influence of an external magnetic field applied during drying.

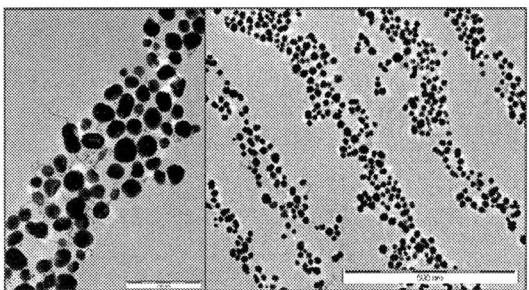

Figure 4. TEM images of $FePt_{rods}$–$Au_{spheres}$ heterostructures organized into lines by means of an external magnetic field applied during drying.

The mean line width of these mesoscopic lines are ~160 nm with variations between ~40 – 300 nm depending on the number of particles forming the lines (1-8 nanoparticles). The lines were formed on two different pitches, copper girds for TEM and silicon oxide substrates for SEM investigations.

CONCLUSIONS

Our attention has been focused on the preparation of FePt-Au heterostructures with diverse FePt morphologies and their controlled organization. It has been shown that different FePt morphologies can be used as seeds for the production of FePt-Au heterostructures. Applying external magnetic fields during the drying process induces the assembly of FePt-Au heterostructures into very long and highly ordered linear particulate Au nanostructures. The alignment mechanism involve in the alignment of the heterostructures, are the magnetic lines forces formed due to the magnetic gradient between the two permanent magnets.

No electrical conductivity investigations have been performed yet on the formed structures, nevertheless the plasmon resonances exhibited by Au nanoparticles can be coupled at such small distances between the particles. Thus, they may be used to transmit optical signals and consequently, information by processing them.[19, 20] Thus, they may be good for the production of advanced interconnects for microelectronics.

ACKNOWLEDGMENTS

The authors thank Izabela Firkowska for the SEM image, the Kurnakov Institute of General and Inorganic Chemistry from the Russian Academy of Sciences in Moscow for the internship allowance of D. Baranov. This work was supported by the Marie Curie Research Training Network SyntOrbMag (Contract MRTN-CT-2004-005567).

REFERENCES

1. Henglein, A. and Meisel, D., Langmuir **14**(26), 7392 (1998).
2. Link, S. and El-Sayed, M. A., Journal of Physical Chemistry B **103**(21), 4212 (1999).
3. Alvarez, M. M., Khoury, J. T., Schaaff, T. G., Shafigullin, M. N., Vezmar, I., and Whetten, R. L., Journal of Physical Chemistry B **101**(19), 3706 (1997).
4. Heath, J. R., Knobler, C. M., and Leff, D. V., Journal of Physical Chemistry B **101**(2), 189 (1997).
5. Schmid, G., Chemical Reviews **92**(8), 1709 (1992).
6. Andres, R. P., Bielefeld, J. D., Henderson, J. I., Janes, D. B., Kolagunta, V. R., Kubiak, C. P., Mahoney, W. J., and Osifchin, R. G., Science **273**(5282), 1690 (1996).
7. M. A. Hayat, *Colloidal gold: principles, methods and applications*, 1. ed.(1991), Vol. 3.
8. Castner, D. G. and Ratner, B. D., Surface Science **500**(1-3), 28 (2002).
9. Chen, M., Liu, J. P., and Sun, S., Journal of the American Chemical Society **126**(27), 8394 (2004).
10. Wang, C., Hou, Y. L., Kim, J. M., and Sun, S. H., Angewandte Chemie-International Edition **46**(33), 6333 (2007).
11. Fan, H. Y., Yang, K., Boye, D. M., Sigmon, T., Malloy, K. J., Xu, H. F., Lopez, G. P., and Brinker, C. J., Science **304**(5670), 567 (2004).

12. Sun, S. H., Murray, C. B., Weller, D., Folks, L., and Moser, A., Science **287**(5460), 1989 (2000).
13. Jana, N. R., Gearheart, L., and Murphy, C. J., Chemical Communications (7), 617 (2001).
14. Nikoobakht, B. and El-Sayed, M. A., Chemistry of Materials **15**(10), 1957 (2003).
15. Perez-Juste, J., Pastoriza-Santos, I., Liz-Marzan, L. M., and Mulvaney, P., Coordination Chemistry Reviews **249**(17-18), 1870 (2005).
16. Liz-Marzan, Luis M., Materials Today **7**(2), 26 (2004).
17. Perez-Juste, J., Correa-Duarte, M. A., and Liz-Marzan, L. M., Applied Surface Science **226**(1-3), 137 (2004).
18. Pazos-Perez, N., Gao, Y., Hilgendorff, M., Irsen, S., Perez-Juste, J., Spasova, M., Farle, M., Liz-Marzan, L. M., and Giersig, M., Chemistry of Materials **19**(18), 4415 (2007).
19. Nelayah, J., Kociak, M., Stéphan, O., García de Abajo, F. J., Tencé, M., Henrard, L., Taverna, D., Pastoriza-Santos, I., Liz-Marzán, L. M., Colliex, C., Nature Phys. **3**, 348-353 (2007).
20. Kim, J. T., Ju, J. J., Park, S., Kim, M., Park, S. K., Lee, M. H., Optics Express **16**(17),13133 (2008).

Mater. Res. Soc. Symp. Proc. Vol. 1079 © 2008 Materials Research Society 1079-N05-02

In-situ early stage electromigration study in Al line using synchrotron polychromatic X-ray microdiffraction

Kai Chen[1,2], N. Tamura[1], and K. N. Tu[2]

[1]Advanced Light Source, Lawrence Berkeley National Laboratory, 1 Cyclotron Road, Berkeley, CA, 94720

[2]Department of Materials Science and Engineering, UCLA, Los Angeles, CA, 90095

ABSTRACT

Electromigration is a phenomenon that has attracted much attention in the semiconductor industry because of its deleterious effects on electronic devices (such as interconnects) as they become smaller and current density passing through them increases. However, the effect of the electric current on the microstructure of interconnect lines during the very early stage of electromigration is not well documented. In the present report, we used synchrotron radiation based polychromatic X-ray microdiffraction for the *in-situ* study of the electromigration induced plasticity effects on individual grains of an Al (Cu) interconnect test structure. Dislocation slips which are activated by the electric current stressing are analyzed by the shape change of the diffraction peaks. The study shows polygonization of the grains due to the rearrangement of geometrically necessary dislocations (GND) in the direction of the current. Consequences of these findings are discussed.

INTRODUCTION

Electromigration (EM) refers to the mass transport phenomenon of materials due to electric current when the current density reaches high values. This phenomenon can have deleterious effects on microelectronic devices such as integrated circuits (ICs) [1]. In very-large-scale integration (VLSI) of circuits on a Si device, Al or Cu thin-film line can experience current density as high as 10^6 A/cm^2, which is enough to cause the atomic motion from the cathode end to the anode end at the device working temperature. Thus voids may form at the cathode end and extrusion may be squeezed out at the anode end, respectively [2]. Although a lot of models have been proposed to understand the mechanism of EM [3-6], the effect of the atomic transport on the dislocation movement and local material microstructure, especially in the early stage of EM is still unclear.

High brilliance X-ray beam with submicron to micron beam size can be produced by focusing optics such as Kirkpatrick-Baez (KB) mirrors at synchrotron facilities [7, 8]. X-ray microdiffraction has been developed for the crystal microstructure study. In this technique, since the crystal grains are larger than the X-ray beam size, a single crystal Laue diffraction pattern is obtained when polychromatic (white beam) X-ray impinges on the sample. By analyzing the Laue diffraction pattern, the orientation and strain information of the material irradiated can be obtained, and the spatial resolution is only limited by the size of the focused X-ray beam.

Previous works have shown the potential of Laue X-ray microdiffraction for EM study on interconnect lines [9-15]. In the present report, we report a more in-depth microstructure study in an Al (0.5 wt % Cu) interconnect during the early stage of electromigration.

EXPERIMENTAL

The EM test sample is a sputtered 30 μm long, 4.1 μm wide and 0.75 μm thick Al (0.5 wt % Cu) two-level structure. A 450 Å thick Ti layer at the bottom of the Al line, as well as another layer with 100 Å thick on the top, is used as shunt. The Al line is constrained by a 0.7 μm thickness SiO_2 passivation layer. Several W vias are used for electrical connection between this Al line and unpassivated Al (Cu) pads at each end. The sample is annealed at 390 °C in a rough vacuum for 30 min to stabilize the crystal structure before EM test.

The *in-situ* X-ray microdiffraction experiment was conducted on beamline 7.3.3 at the Advanced Light Source synchrotron in Berkeley, CA [16]. The X-ray beam size is 0.7 x 0.7 $μm^2$. The wired-bonded sample and an X-ray CCD detector are mounted at 45° and 90° with respect to the beam, respectively. The CCD detector is fixed at 5 cm above the sample when white beam Laue diffractions are recorded.

The EM test is conducted at a 224 °C (heater temperature) and the current density is ramped up to 0.98×10^6 A/cm^2 (I=30 mA) and maintained constant for 45 hours. Before and during the EM test, Laue diffraction patterns are recorded by scanning the sample repeatedly by the submicron-sized white beam. The scanning step size is 0.5 μm, so 15 steps across the width of the line, 65 steps along the length, and a total of 975 CCD frames are needed to cover the whole surface of the sample for each scan. The exposure time is set as 5 seconds so that the intensity of the diffraction peaks is strong but not saturated. The electronic readout time for each frame is about 10 seconds. As a result, for each scan, it takes about 4 to 5 hours. The Laue diffraction data are analyzed using the custom written XMAS software [11, 17].

RESULTS AND DISCUSSIONS

Dislocation active slip systems

Large grains in the Al interconnect line, especially those that span across the width of the line ("bamboo" type), are subject to opposite stress moments on opposite sides across the width. When a certain yield stress was reached, these grains would bend and ultimately polygonize [10].

This aspect of plastic deformation results in broadening of the Laue spots along certain directions compatible with the bending direction due to the arrangement of excess geometrically necessary dislocations (GNDs). Different dislocation slips can result in different shapes of the Laue peaks. As a result, analysis of the Laue spot broadening patterns allows for the identification of the active dislocation glide system that produced the bending.

First the orientation of a given grain is known by indexing the Laue pattern. Secondly, all the possible streaking directions of the Laue peaks are simulated, each of which corresponds to a certain dislocation slip system. Since Al has FCC crystal structure, for each grain, there are 12 possible slip types, all of which are {111}/<110> oriented, which means 12 possibilities for both edge dislocations and screw dislocations. Then all of these 24 simulated diffraction patterns are compared with the experimental result, and the best fitting will tell the real dislocation slip type. In figure 1(a), the white streaking peaks are from one of the typical Laue patterns of a deformed grain after 12 hours EM test, and the yellow patterns are the best matched simulation result. In this example, the slip plane is $(\bar{1}11)$; the Burger's vector is along the [101] direction; and the dislocations are edge typed, so the dislocation line direction is perpendicular to the slip plane normal and Burger's vector, which is along [12$\bar{1}$] for this case. Figure 1(b) is the pole figure of this grain, with all <112> directions marked and the activated direction circled. In this figure, the x direction is along the electric current flow direction and the y direction is along the width direction of the Al line. It is found that the dislocation direction [12$\bar{1}$] is the direction belonging to the <112> family and closest to the current flow direction.

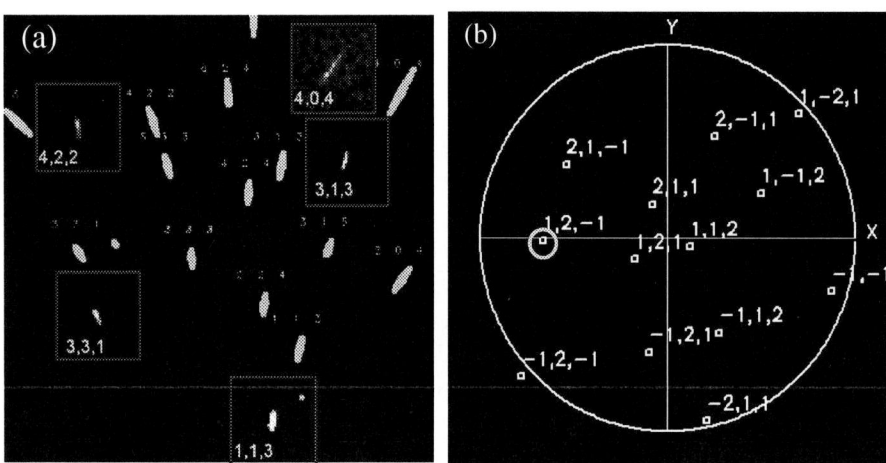

Figure 1. (a) Simulated Laue patterns comparing with the experimental results, and (b) pole figure of this grain with all <112> directions marked

The activated slip systems in 30 grains have been analyzed in this way. All the dislocations are found to be edge typed and about 90 % of the dislocation lines are contained within 5° with respect to the current flow direction. The dislocations which don't obey this rule are located close to the cathode end. We propose that the alignment of the edge type dislocations along the electric current flow direction is resulted from the minimization of the electrical resistance because the scattering between electrons and dislocations is anisotropic. The dislocations at the cathode end are not aligned regularly because electrons enter the Al line here and the electron motion directions are more disordered. The realignment of dislocations will also influence the atomic diffusion inside the line by affecting the diffusion route. The detailed kinetic study is still underway and will be the subject of a forecoming paper.

At higher tensile stress region, in other words, closer to the cathode, a secondary slip system can be activated, and the Laue patterns are broadened in two different directions, as shown in figure 2(a). This grain is located at about 6 µm away from the cathode end. In this example, the two dislocation line directions are [$\bar{1}$12] and [11$\bar{2}$], respectively. The pole figure of this grain, shown as figure 2(b), shows that, as expected, [$\bar{1}$12] is the direction that is the closest to the current flow direction in <112> family, while [11$\bar{2}$] is located on the edge of the circle in the pole figure, which represent the XY plane, rather than the second closest direction. At even higher stress level, multiple slip systems are activated. Figure 2(c) is the enlarged (113) peak of a grain locating at about 2 μm to the cathode end. The Laue diffraction patterns become complicated since dislocations slip in different orientations, as indexed in the figure. Furthermore, the diffraction peak splitting suggests the formation of subgrains. Since all those subgrains have slightly different orientations, each subpeak in the Laue pattern represents a subgrain in the irradiated volume in the sample. From figure 2(d), which shows all the <112> directions in this grain, it is clearly observed that one of the several activated dislocation line directions is along the current direction, while the others are all within the XY plane.

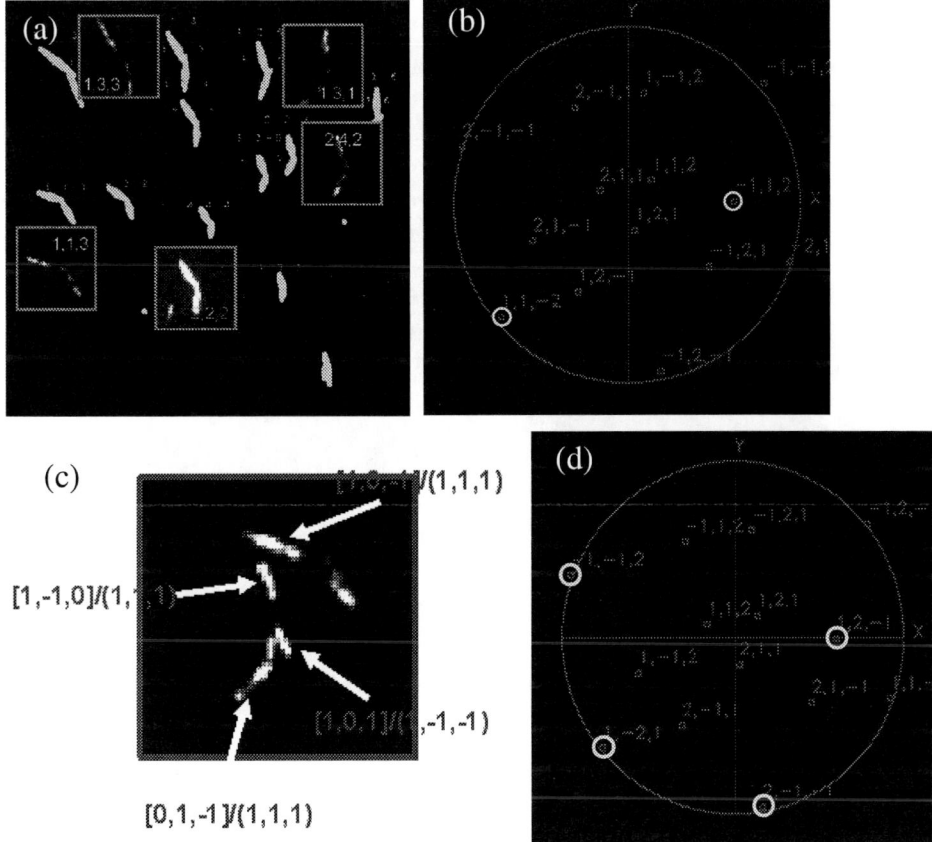

Figure 2. (a) Simulated Laue patterns of a grain which has two slip systems activated and (b) the pole figure of this grain, (c) enlarged (113) diffraction peaks from a grain 2 μm to the cathode end showing the multiple activated slip system and subgrain formation and (b) the pole figure of this grain

The secondary slip system, which is within the XY plane, can be roughly explained by the activation of the slip systems of the subgrains formed by EM. Since each of the subgrains has slightly different orientation and is subjected at different local shear stress status, the slip system activated within each subgrain could be different. In other words, the secondary system is activated at a different location in the grain comparing with the primary system. However, the detailed mechanism is not well understood yet.

Subgrain formation

As mentioned above, polygonization happens on large grains close to the cathode by subgrain formations, as indicated by the observation of the subpeaks in some Laue diffraction patterns. In order to study this process, we have tracked the shape evolution of the Laue peaks of ten grains in which subgrains were formed. Figure 3 shows a series of (022) diffraction peaks produced by the same grain after different period of EM test. Before the EM test the peak is initially sharp, indicating that there are not many dislocations in the grain. As the EM test continues for 15 hours, the peak is broadened more and more in two directions, which indicates that the grain is bended gradually by the electric current in two different directions, but generally there are no obvious subpeaks in this period. In figure 3(d), the diffraction peak becomes broadened after 21.5 hours current stressing, and three subpeaks are observed, though not very clearly separated. Thus at this stage, the dislocations are partially aligned to form the subgrain boundary. After 25 hours EM test, two subpeaks can be very clearly observed and separated. The separation of the two peaks is larger than the maximum broadening in figure 3(d). Both of the two subpeaks are much sharper than the streaking in figure 3(d), indicating that most of the dislocations are concentrated and aligned at the subgrain boundary.

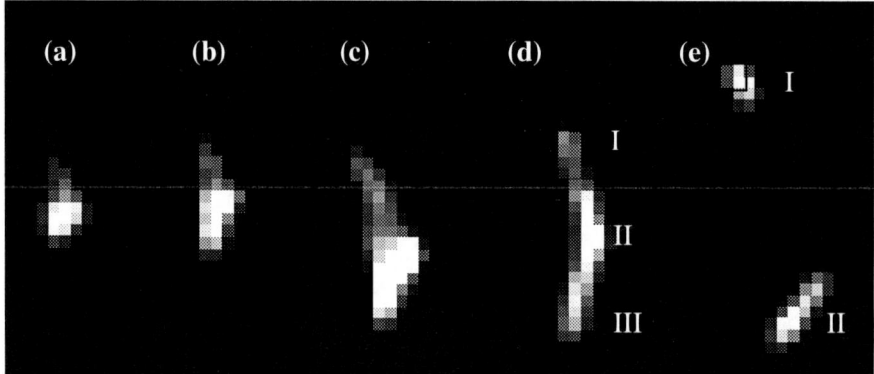

Figure 3. (022) diffraction peaks from the same grain after (a) 0 hr, (b) 6.5hr, (c) 15hr, (d) 21.5 hr, and (e) 25 hr EM test

Generally, subgrain forms by the realignment of the dislocations when they reach certain density. More detailed quantitative study, both experimentally and theoretically, is required to understand the kinetics of this procedure.

CONCLUSION

In this paper, using polychromatic X-ray beam microdiffraction, we have studied the plastic deformation of the Al interconnect line induced by EM. The activated slip systems have been studied from the shape change of the diffraction peaks. It is found that all the dislocations are edge typed. With relative low stress, only the slip system with the dislocation line direction which is the closest to electric current flow direction is activated while under high stress above certain critical value, more slip system can be activated besides the one along the current direction, and the secondary ones prefer to be within the XY plane. Furthermore, the evolution of the diffraction peaks has been carefully studied. The bending of the crystal plane indicated by the elongation of the diffraction peaks and resulted from the rearrangement of the geometrically necessary dislocations results in the peak splitting and subgrain boundary formation by the realignment of the dislocations.

ACKNOWLEDGEMENTS

The Advanced light Source is supported by the Director, Office of Science, Office of Basic Energy Sciences, of the U.S. Department of Energy under Contract No. DE-AC02-05CH11231 at Lawrence Berkeley National Laboratory

REFERENCES

1. I. A. Blech, *J. Appl. Phys.* **47**, 1203 (1976)
2. K. N. Tu, *J. Appl. Phys.* **94**, 5451 (2003)
3. S. Vaidya, and A. K. Sinda, Thin Solid Films **75**, 253 (1981)
4. P. S. Ho, and T. Kwok, *Rep. Prog. Phys.* **52**, 301 (1989)
5. M. A Korhonen, P. Borgesen, K. N. Tu, and C. Li, *J. Appl. Phys.* **73**, 3790 (1993)
6. H. Gan, K. N. Tu, *J. Appl. Phys.* **97**, 063517 (2005)
7. O. Hignette, P. Cloetens, G. Rostaing, P. Bernard, and C. Morawe, Review of Scientific Instruments 76 : Art. No. 063709 (2005)
8. W. J. Liu, G. E. Ice, J. Z. Tischler, A. Khounsary, C. Liu, L. Assoufid, and A. T. Macrander, *Rev. Sci. Instrum.* **76**: Art. No. 113701 (2005)
9. P. C. Wang, G. S. Cargill III, I. C. Noynan, and C. K. Hu, *Appl. Phys. Lett.* **72**, 1296 (1998)
10. B. C. Valek, J. C. Bravman, N. Tamura, A. A. MacDowell, R. S. Celestre, H. A. Padmore, R. Spolenak, W. L. Brown, B. W. Batterman, and J. R. Patel, *Appl. Phys. Lett.* **81**, 4168 (2002)
11. N. Tamura, H. A Padmore, and J. R. Patel, *Mat. Sci. Eng. A*, **399**, 92 (2005)
12. B. C. Valek, N. Tamura, R. Spolenak, W. A Caldwell, A. A. MacDowell, R. S. Celestre, H. A. Padmore, J. C. Bravman, B. W. Batterman, W. D. Nix, and J. R. Patel, *J. Appl. Phys.* **94**, 3757 (2003)
13. A. S. Budiman, N. Tamura, B. C. Valek, K. Gadre, J. Maiz, R. Spolenak, J. R. Patel, and W. D. Nix, *Mater. Res. Soc. Symp. Proc.* **914** (2006)
14. R. I. Barabash, G. E. Ice, N. Tamura, B. C. Valek, J. C. Bravman, R. Spolenak, and J. R. Patel, *J. Appl. Phys.* **93**, 5701 (2003)
15. A. S. Budiman, W. D. Nix, N. Tamura, B. C. Valek, K. Gadre, J. Maiz, R. Spolenak, J. R. Patel, *Appl. Phys. Lett.* **88**, 233515 (2006)
16. N. Tamura, A. A. MacDowell, R. Spolenak, B. C. Valek, J. C. Bravman, W. L. Brown, R. S. Celestre, H. A. Padmore, B. W. Batterman and J. R. Patel, *J. of Synchrotron Radiat.* **10** (2003) 137-143.
17. N. Tamura, R. Spolenak, B. C. Valek, A. Manceau, M. Meier Chang, R. S. Celestre, A. A. MacDowell, H. A. Padmore, and J. R. Patel, *Rev. Sci. Instrum.* **73**, 1369 (2002)

PLASMA MODIFICATION OF Si-O-Si BOND STRUCTURE IN POROUS SiOCH FILMS

Fedor N Dultsev[1], Adam M Urbanowicz[2,3], and Mikhail R Baklanov[2]

[1]ISP SB RAN, Lavrentiev ave., 13, Novosibirsk, 630090, Russian Federation

[2]AMPS, IMEC, Kapeldreef 75, Leuven, 3001, Belgium

[3]Department of Chemistry, Katholieke Universiteit Leuven, Leuven, 3001, Belgium

ABSTRACT

Plasma modification of SiOCH low-k films is analyzed by means of Molecular Mechanics. It is shown that the most probable mechanism of SiOCH modification in He plasma is removal of hydrogen atoms from CH_3 groups. The change of Si–O–Si bond angles depends on the amount of the formed $–CH_2^*$ (CH_x) groups. During the followed exposure in NH_3 plasma, NH_2^* radicals bind CH_x groups with Si forming a $–CH_2–$ Si–O–Si–O–Si–O–Si– chain. The end of this chain gets bound to its beginning through NH_2. This process is the reason of pore sealing.

INTRODUCTION

Porous SiOCH low-k films with carbon containing hydrophobic groups are the most favored class of materials for advanced interconnect technology. Although the matrix of these materials has properties similar to SiO_2, their chemical stability and reactivity strongly depend on porosity, pore size and their interconnectivity. The use of porous materials brings challenges related to the structural properties such as percolation phenomena. In addition to changes in mechanical properties (Young's modulus), porous materials exhibit high chemical activity, which brings complications to their application and technological processing.

The exposure of these films to a plasma leads to the loss of hydrophobic groups and to the densification of their silica backbone. Both effects cause an unwanted increase in the dielectric constant, as well as an increase in the leakage current. The extent of this damage depends on the plasma conditions such as chemistry and power and on the porosity of the film. On the other hand, treatment in certain plasmas [1, 2] may cause pore sealing, which results in a decrease in the chemical activity of these layers. Changing an inert gas and excluding the ion component in treating the surface the authors demonstrated that the film reactivity is affected by the UV radiation of the plasma. For instance, it was shown in [3, 4] that the treatment of SiO_2 films in He plasma caused an increase in the rate of etching these layers in 1% HF solution. Another observation is that sequent exposure of SiOCH low-k film in He and NH_3 plasma results in complete sealing and passivation of the film surface [1] .

The effect of porosity on chemical activity can be considered in two aspects: first, due to the developed surface, with the reaction proceeding not at the boundary but over the whole volume: we will call this effect macrostructural; second, due to changes in the electron structure of surface atoms, which leads to changes in the chemical properties of surface atoms: to be called microstructural effect. It is the investigation into the microstructural effect that will allow us to understand how we can change the properties of the surface atoms and their reactivity. The general idea of the changes in the electron structure under the action of an adsorbed molecule or atom was considered in [5]. The application of such an approach to sensor problem was described in [6]. The use of the idea of the transition adsorption complex proposed in [7] allowed describing adsorption-stimulated surface reconstruction. A theoretical investigation of the dependence of chemical properties of molecules on the molecular structure can be useful because it may result in a substantial decrease in the time necessary for the search for necessary surface groups imparting the required surface properties. Theoretical considerations are also useful in view of the difficulties in the experimental analysis, since the surface groups concentration may be too small to be detected with the help of direct spectroscopic methods. In this situation, modeling may become the basic method to predict the characteristics of thus treated surfaces.

In the present work we consider low-k dielectrics based on silicon dioxide. It is known that the lability of Si–O–Si bonds is the main reason of differences in the adsorption properties of SiO_2 modified with different functional groups. For instance, an interconnection between the reactivity and Si-O-Si bond angle was described in [8]. As the second step of our recent work [1], we are making an attempt to explain with the help of modeling why the SiOCH low-k films change their chemical properties during the exposure in He and NH_3 plasmas.

The experimental data under consideration are those reported in [1] demonstrating how the layer characteristics change under treatment with different kinds of plasma. PECVD carbon-doped SiO_2 low-k films were deposited and UV cured. According to results obtained by ellipsometric porosimetry [9], the final porosity of the films was close to 25% and pore diameter was 2 nm. It was shown that their chemical behavior can change due to differences in Si-O-Si bond mobility, which determines the structure of SiO_2-based films. The IR spectra of the films show that they have different structures, namely, different Si-O-Si angles; an increase in the average angle value causes an increase in the intensity of the band at 1070–1150 cm^{-1}. Film reactivity is determined by the electronic structure and thus by bond geometry.

MODELING

Modeling is being increasingly frequently used for theoretical description of the properties of solids, including low-k dielectrics. For instance, the authors of [10] presented a theoretical investigation of the mechanical and dielectric properties of SiOCH films. In some cases, quantum-chemical and semiempirical methods can well predict or confirm the structure of these layers [11, 12]. The approach proposed in [7] involving calculations of an adsorption complex allows one to obtain additional data on the reactivity of these layers. The main idea of the proposed approach is that a reaction is considered through the formation of a definite intermediate state. The structure of this intermediate state referred to as an adsorption complex (AC) is calculated. This state is searched on the basis of the minimum energy; calculation results allow us to conclude whether such an adsorption complex exists and what are the directions of its further transformations.

Modeling was carried out by means of molecular mechanics MM2. A structure described below was used in modeling. The structure was optimized to the minimum of total energy, and changes in Si–O–Si bond angles were monitored; a distribution over the values of this angle was plotted. In order to do this, the range 135-153 ° was divided into 10 intervals. Total number of Si–O–Si bond angles in the structure under calculation was 58. Thus obtained distribution is directly connected with the frequency of Si–O–Si bond vibrations in the IR spectrum. In other words, we modeled the IR spectrum depending on dπ-pπ bonding. The experimentally determined vibration frequencies of Si–O bond [1] depending on Si–O–Si bond angle are shown in Table 1.

Table 1. A connection between Si-O-Si angle and the vibration frequency in FTIR spectra.

Si–O–Si angle, deg	<140	144	155
FTIR frequency, cm^{-1}	1000	1070	1150

Molecular structure. The structure formed by the rings composed of –Si–O–Si–O atoms was used in modeling. Each ring contained 6 silicon atoms. These rings formed a structure shaped as a cylinder with the inner diameter corresponding to the pore size. The number of rings comprising the structure was chosen so that a pore 1 nm in radius could be obtained. The free bonds of the surface silicon atoms were considered to be filled with hydrogen atoms; a part of H atoms was substituted by methyl groups. Their amount corresponded to the method by which real films were prepared. The fraction of methyl groups was 20 – 25% of the number of silicon atoms. This amount corresponds to the films with the porosity about 25 % and pore radius 1 nm [12]. The structure is shown in Fig. 1.

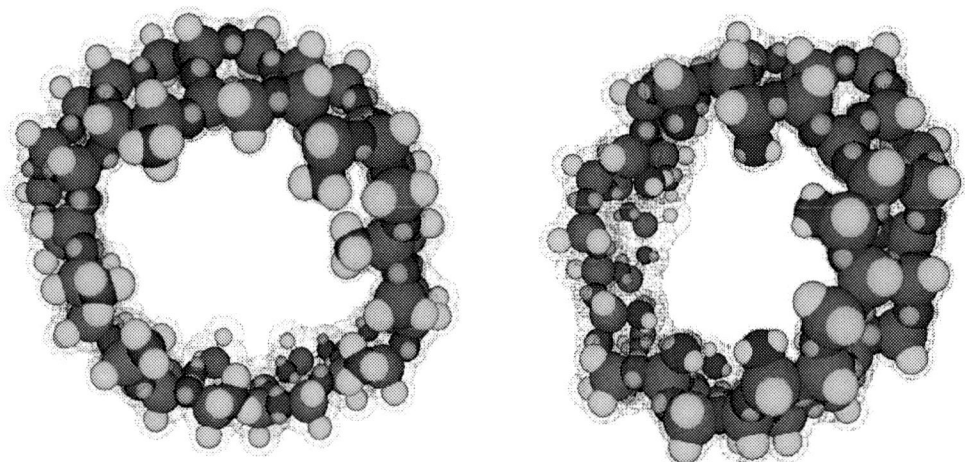

Fig. 1. Initial SiOCH structure (a) used for calculations, and the structure after treatment with ammonia plasma.

He plasma. The effect of He plasma on silicon dioxide was considered mainly as the interaction with CH$_3$ group because this group is the most available one from the viewpoint of energy and stoichiometry. It was assumed in calculations that a rupture of H atom or CH$_3$ group as a whole occurs. Calculation was carried out through the adsorption complex. This approach allowed us to determine the most probable reaction route.

It is shown in Fig. 2a-b how the calculated distribution of Si–O–Si bond angles changes depending on the amount of the introduced CH$_2$* (CH$_x$) groups. One can see in this distribution that the removal of hydrogen atoms results in an increase in the number of bonds characterized by the angle 144-146 $^\circ$. Considering this surface treated in ammonia plasma one may assume that NH$_2$* interacts with a CH$_2$* group forming a Si – CH$_2$ – NH$_2$ chain. After the plasma is switched off, an unshared electron pair at the nitrogen atom forms a bond, due to Van der Waals forces, with a silicon atom inside the pore. The probability of the formation of such a bond in a pore is higher than at an open surface. It is most probable that the following reaction occurs: NH2* binds CHx groups with Si, forming a chain looking like $–CH_2– Si–O–Si–O–Si–O–Si–$; the end of this chain gets bound to its beginning through NH_2. Changes in the distributuion of Si–O–Si angles in this situation are shown in Fig. 2-c. In order to explain the IR spectrum observed after treatment in the ammonia plasma without preliminary treatment in helium, we calculated several reasonable reaction models. The changes observed in experiments are best described with the following model: rupture of CH_3 groups, formation of $Si–NH_2$ and $Si–CH_2–NH_2$ bonds. Changes in Si–O–Si angle distribution during the interaction with ammonia are shown in Fig. 3. One can clearly see here that the most probable angle value is about 144-146°. It should be noted that the number of angles equal to 142 ° and smaller has decreased.

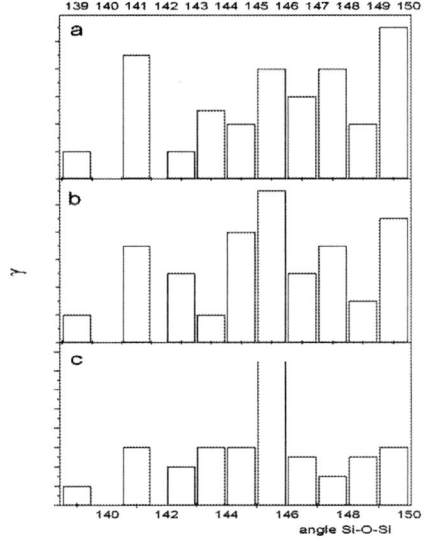

Fig. 2. Calculated distribution of Si–O–Si angles: a) initial structure (see Fig. 1a), b) formation of CH* groups, c) interaction with NH₂* after treatment in He plasma.

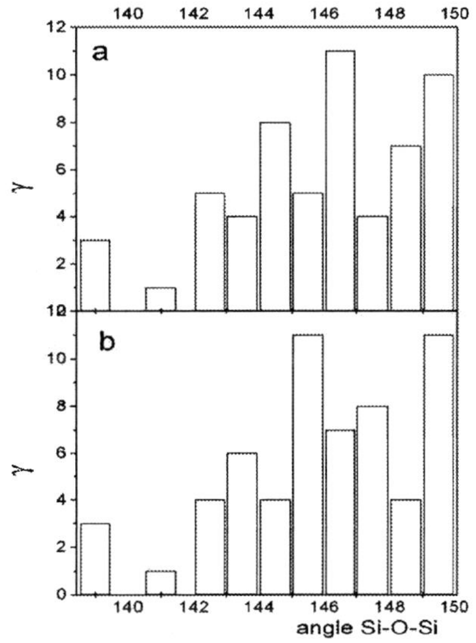

Fig. 3. Distribution of Si–O–Si angles: a) substitution of CH_3 for NH_2 groups, b) formation of $CH_2–NH_2$ bonds.

DISCUSSION

He plasma: since no increase in the intensity of the band at 1000 cm⁻¹ is observed in experiments, one should not be sure that the process occurring in He plasma may be described mainly as the rupture of oxygen atom and its removal from the silicon atom with the formation of a dangling bond. Quite contrary, the observed decrease in the frequency of CH_3 group vibrations is connected either with the rupture of hydrogen or with changes in the dπ-pπ overlapping during the adsorption at the silicon atom, which increases the rigidity of the Si–O–Si bond. In turn, this causes weakening of the bond holding CH_3 group, which correspondingly results in a decrease in the frequency of CH_3 vibrations. A fact evidencing in favour of overlapping is that the intensity of the band at 1070 cm⁻¹ and higher increases, which confirms an increase in the Si-O-Si angle. So, the IR spectra show that the treatment in He causes a decrease in the intensity of the band at 1000 cm^{-1} and an increase in the intensities of the bands within the range 1070-1150 cm^{-1}. These changes are due to the CH$_3$ groups. It is these groups that are able to render (give) the electron density: a removal of a proton causes an increase in the Si-O bond order, that is, an active center is formed at the carbon atom: for example, CH$_2$* or CH$_2$. In both cases, the electron density shifts from the CH$_x$ group ensuring an effective negative charge δ^- at Si atom. This causes changes in the geometry, namely an increase in the Si-O-Si bond angle. This model is in good agreement with our calculations. One can see in Fig. 2 a, b that, indeed, if we assume that a rupture of a proton from the CH$_3$ group occurs in He plasma, this results in a shift of the distribution toward larger angles. Subsequent treatment of the film in NH$_3$ plasma, after He, promotes cross-linking over these CH$_x$* groups, as well as Si atoms; NH$_x$* may serve as a bridge. It is clear that pore size decreases, or the pores get sealed, because the reaction proceeds at the very pore entrance. Because of this, if porosity is measured on the basis of the refractive index, one would not see any essential changes. It should also be noted that the number of built-in NH$_2$ groups may be small, since building-in occurs only on the surface; this may be the reason why these groups do not manifest themselves in the IR spectra. Modeling carried out by us shows that the angle distribution should not strongly differ from the initial state. It should be stressed that the angle distribution became more uniform, with the prevalence of one angle value (Fig. 2c). This also agrees with the IR spectrum in which we actually observe a small narrowing of the bands.

In the case of treatment with NH3 (without preliminary treatment in He), there are no active CH$_x$ centers; the interaction proceeds mainly due to the formation of a bond between NH$_x$* and Si atom. The Si-O-Si angle increases, which causes an increase in the frequencies to 1070 and to 1150 cm^{-1}. The amount of adsorbed molecules can be estimated on the basis of the intensities of these bands. The band related to CH$_3$ group shifts to smaller frequencies, which is also an evidence of a strong increase in π-bonding in the Si-O-Si system. Since an increase in π-bonding causes an increase in Si-O-Si angle, a pore 1 nm in radius can increase by 0.4-0.5 Å (that is, by about 2 %), which is observed experimentally. The experimental data are in good agreement with modeling results. For instance, one can clearly see in the distributions shown in Fig. 3 a,b that the amount of bonds with 146-148 $^\circ$ angle increases; the angles smaller than 142 $^\circ$ disappear almost completely. Such a behavior is characteristic both of the case of the replacement of CH$_x$ groups with NH$_2$ (Fig. 3a) and of the case of the addition of NH$_2$ groups at the silicon atoms (Fig.3b).

Note. Rupture of Si-O bonds in the volume, induced by the VUV radiation of He plasma, cannot result in a noticeable removal of the fragments formed; a most probable process may be recombination followed by redistribution of the excess energy over the lattice. It may be assumed that this process would not have any direct effect on changes in porosity. Because of this, here we do not consider the effect of VUV irradiation on the rupture of stronger bonds like Si-O-Si; however, we do not exclude this process. It should also be noted that an additional appearing dπ-pπ overlapping strengthens bonding between silicon and oxygen, which leads to an increase in the stability of this bond and a decrease in its reactivity.

CONCLUSIONS

Plasma modification of porous SiOCH films is analyzed by means of Molecular Mechanics. It is shown that the most probable mechanism of the modification in He plasma is formation of CH$_x$ groups by removal of hydrogen atoms from CH$_3$ groups. This increases the number of Si–

O–Si groups having bond angles in the range of 144-146 $^\circ$. Subsequent exposure in NH_3 plasma forms $-CH_2-$ Si–O–Si–O–Si–O–Si– chains where the end of the chain gets bound to its beginning through NH_2. This process is the reason of pore sealing.

The results of modeling are in a good agreement with FTIR spectra presented in paper [1]. Additional support was obtained for different types of low-k material (Figure 4). One can see that the absorbance of Si-CH$_3$ groups is decreased after all plasma treatments. Additional NH_3 treatment shifts Si-O-Si frequency to higher wavenumbers. In the case of treatment in NH_3-based plasma, it is possible to incorporate the OSi-N bonds. The energy of those bonds is within the range 874-1042 cm^{-1} [13]. Furthermore, they might be overlapped with Si-O-Si asymmetric stretching vibrations giving rise to Si-O-Si network peak (~144°) as shown in Fig 4. Those observations are in agreement with the results of modeling. A more complex situation is with the pure He and He + NH_3 plasma treatments. We clearly see incorporation of Si-O-Si with bond angle lower than 144 $^\circ$ for both He and He+NH_3 plasma, which demonstrates the possibility of breakage of Si-O bonds.

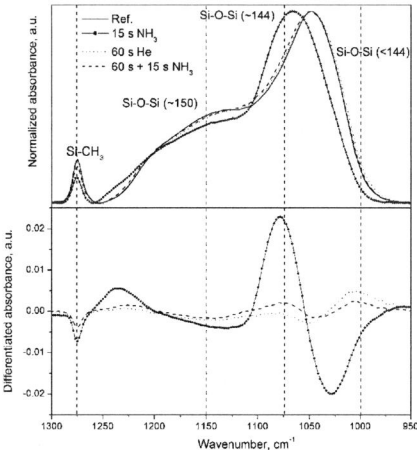

Fig. 4 Conventional (top) and differential (bottom) FTIR spectra of as deposited and plasma treated low-*k*.

References.

1. A. M. Urbanowicz et al. *Electrochem. Sol.-St. Lett.*, 10 /10, G76 (2007) .
2. H.G. Peng et al.. *J. Electrochem. Soc.*, 154(4) G85 (2007).
3. T. Tatsumi, S. Fukuda, S. Kadomura. *Jpn. J. Appl. Phys.* 32, 6114 (1993).
4. T. Tatsumi, S. Fukuda, S. Kadomura. *Jpn. J. Appl. Phys.* 33, 2175 (1994)
5. Wolkenstein T., *Electronic Processes on Semiconductor Surfaces during Chemisorption*, (Consultants, 1991) pp. 35-182.
6. Rothschild A., Komem. *Sensors and Actuators*. B93. 362 (2003).
7. F.N. Dultsev, L.L. Sveshnikova. *Sensors & Actuators*, B120, 434 (2007).
8. F. Dultsev. *J. Struct. Chem.* 48/2, 236 (2007).
9. M. R. Baklanov et al. *J. Vac. Sci. Technol. B 18*, 1385 (2000).
10. N. Tajimaa et al.. *Appl. Phys. Lett.*, 061907 (2006)
11. K. Zagorodniy, H. Hermann, and M. Taut. *Phys. Rev.,* B 75, 245430 (2007)
12. A.S. Zyubin et al.. *J. Chem. Phys.,* 116/1, 281-294. (2002).
13 J. Olivares-Roza, O. Sanchez, and J. M. Albella, J. Vac. Sci. Technol. A**16**, 2757 (1998)

Mater. Res. Soc. Symp. Proc. Vol. 1079 © 2008 Materials Research Society

Cobalt Silicidation on Sub 100nm Hole Patterned Vertical Diode Formed by Silicon Epitaxial Growth and Its Electrical Properties

Min Yong Lee, K. B. Lee, H. S. Lee, S. J. Chae, I. K. Han, H. S. Kang, and S. W. Park

R&D, Hynix Semiconductor Inc., San 136-1 Ami-ri Bubal-eub Icheon-si, Icheon, 467-701, Korea, Republic of

ABSTRACT

Self-aligned Cobalt silicide as ohmic contact layer on sub 100 nm hole patterned Si vertical diode formed by silicon epitaxial growth (SEG) is investigated and silicon epitaxial growth of higher than 4000 Å thickness and good crystalinity for PN diode has been successfully developed. Also, electrical isolation of 100 nm pitch size between diode and diode, and removal of unreacted Co/Ti/TiN layer have been realized by dip-out process without CMP simultaneously. Through the mechanism of void formation due to the variation of Si consumption rate during silicidation at limited hole pattern dimension, critical Co and Capping Ti thickness are investigated as various hole dimensions (80~120 nm), and then with p+ type dopant species ($49BF_2$, $11B$). The ratio of Co thickness to hole dimension demonstrates void free cobalt silicidation on various pattern sizes of silicon epitaxial growth. Silicon epitaxial growing PN diodes including void free $CoSi_2$ show excellent electrical performance, especially lower than 10 pA reverse off leakage current.

INTRODUCTION

Ohmic contact layer forming by Co salicide process is useful for low resistive and high performance device characteristics in very large scale integration (VLSI) circuit. The use of Co silicide has been limited by the degradations of junction leakage and operation current [1-3]. The degradation is caused by a pattern dependent silicidation rate and void formation. The Co silicidation process is very sensitive to the cobalt layer thickness, Ti capping layer quality, P-type dopants and hole dimension [3-6]. Therefore exact understanding about silicidation mechanism is necessary for getting good diode properties.

In this work, we investigate the influence of Co/capping Ti Thickness, dopant species and hole dimension on silicidation rate and void generation at silicon epitaxial growing (SEG) diode hole contact.

EXPERIMENT

We investigated the variation of Co silicidation as a function of thickness of cobalt with capping Ti/TiN which was deposited by PVD. Silicon epitaxial growth (SEG) was prepared as a substrate on the a patterned hole structure with a diameter of 100nm. After interlayer dielectric(ILD) oxide layer 100 nm hole patterning, Vertical diode was formed by silicon epitaxial growth(SEG) higher than 4000 Å, and then CMP was performed for isolation between diode to diode. To compare the dopant effect, BF_2+ and B+ were implanted for P-type formation of vertical diode, respectively. And then RTA was preformed for dopant activation and implantation induced damage reduction. Co/Ti/TiN layers were deposited subsequently in a conventional sputter just after HF wet cleaning. 1st low temperature RTA(500 °C, 60 s) was processed to form silicide and dipped out in $H_2SO_4:H_2O_2$ solution to remove unreacted Co/Ti/TiN layer and electrically isolate between diode and diode simultaneously. 2nd high temperature rapid thermal anneal(RTA) (750 °C, 30 s) was performed to make $CoSi_2$. Samples with various Co, capping Ti thickness values, dopant species and hole dimensions were inspected by using secondary electron microscope(SEM) and transmission electron microscope(TEM).

DISCUSSION

Effect of Co & capping Ti layer thickness on cobalt silicidation at hole

Fig. 1 shows Co silicidation ratio ($CoSi_2$ per Co Thickness) comparison with various Co thicknesses. Generally solid state reaction Ratio for Co to $CoSi_2$ transition with thermal processes is known 3.5 $CoSi_2$ per Co thickness. This ratio was also stable with variable Co thickness in this experiment

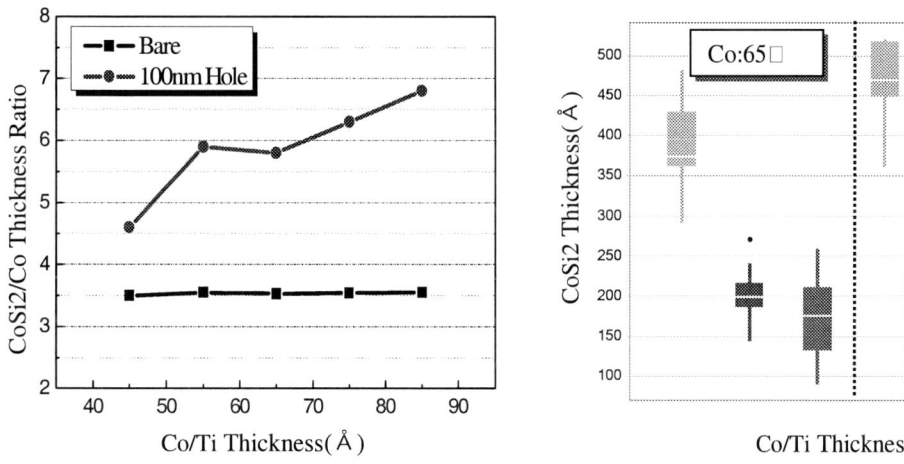

Fig. 1 Co silicidation Ratio($CoSi_2$/Co) between 100nm hole patterned and bare wafer with different Co deposition thickness.
(Capping Ti/TiN: 100/250 Å)

Fig. 2 Thickness effect of Co and capping Ti on $CoSi_2$ Formation at hole contact (Co/Ti/TiN(250Å) multilayer)

The Co silicidation ratio is higher for a patterned SEG as compared to a blanket wafer and increases when the cobalt layer thickness increases. The increased ration is explained by lateral Co diffusion from the ILD top surface into the hole during the 1st RTA step. Therefore, an increased amount of cobalt is available for silicide formation during 2nd RTA step. Ti diffused into Co and Ti-Co phase formation with 1st and 2nd annealing reduced $CoSi_2$ reaction Rate.

Species effect of P-type dopant on cobalt silicidation at hole

Thickness of Co silicide increased as increasing the Co thickness without P-type ion implantation. In addition, we studied the effect of dopants on the silicidation rate where the dopants such as BF_2+ and 11B were implanted as shown in Fig. 3. In case of BF_2+ implanted SEG, the thickness of Co silicide was increased as compared to the un-doped or 11B implanted SEG and there is no significant difference of Co silicide thickness in between un-doped and 11B implanted SEG. On the other hand, it was thought that the void formed on the side wall or the interface between silicide and ILD oxide layer, silicide and SEG plays an important role on the silicide thickness [7, 8]. Void was not observed after 2nd RTA in the BF_2+ implanted SEG below 50 Å thickness of Co and thereby its silicide thickness was similar
with that in the un-doped and 11B+ implanted. It is different rate for a predicted higher trend of silicidation rate including void as shown in dotted line at Fig. 3.

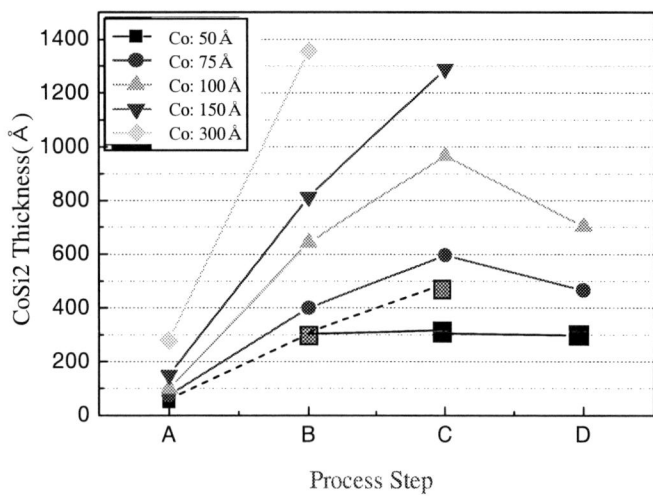

Fig. 3 $CoSi_2$ thickness characterized by A) as Co/Ti/TiN Dep., B) after 2nd RTA without P-type implanted, C) with BF_2+ implanted, D) with 11B+ implanted on 100 nm SEG hole Pattern. (Capping Ti/TiN thickness: 100/250Å)

Void formation during silicidation with Co and capping Ti thicknesses

The dependency of Co and Capping Ti Thickness on the void formation at the 100nm hole size SEG pattern is shown in Fig. 4. As the thickness of Co on silicon epitaxial grown diode hole was decreased, the number of voids during silicide formation was decreased. On the limited SEG hole pattern, the increasing of Co thickness increased Si consumption rate and depletion rate of Si source vs. Co solid state reaction that accelerated the generation of voids. When the Co layer thickness is below 55Å, no voids are detected. At a Co layer thickness of 75 Å, the void generation can be suppressed by increasing the Ti cap layer thickness. This can be explained that Ti diffused into Co and Ti-Co bi-layer formation at 1st RTA reduces the rate of Co silicidation [9, 10]. The silicide capped with increasing Ti thickness had the ohmic layer with suppressed void defects. The void formation was entirely eliminated above capping Ti 300Å.

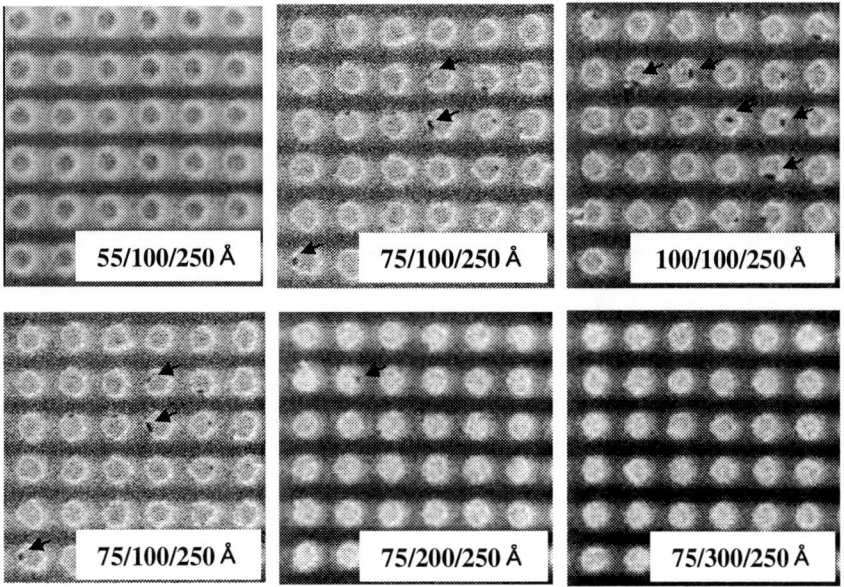

Fig. 4 Void formation during Co silicidation with Co and Capping Ti thickness on SEG hole (Co/Ti/TiN multilayer for Solid state reaction (SSR) at 100 nm hole pattern)

Relation between hole size and Co thickness on void formation

The ratio of Co thickness to hole dimension demonstrated void free cobalt silicidation on various pattern size of silicon epitaxial growth. Table 1 indicates relation between hole size and Co thickness on void formation. For void free silicide formation, the cobalt layer thickness needs to be below 65Å and the recess size above 100nm.

Table. 1 Void observation by TEM analysis with various holes size and Co thicknesses. (Capping Ti/TiN: 100/250 Å)

	Hole size: 85 nm	Hole size: 99 nm	Hole size: 106 nm	Hole size: 116 nm
Co: 45 Å	void free	void free	void free	void free
Co: 65 Å	voids observed	voids observed	voids observed	voids observed
Co: 85 Å	N/A	N/A	voids observed	voids observed

Silicon epitaxial growing PN diodes without void $CoSi_2$ show excellent electrical performance, especially lower than 10 pA reverse off leakage current as shown in Fig. 5.

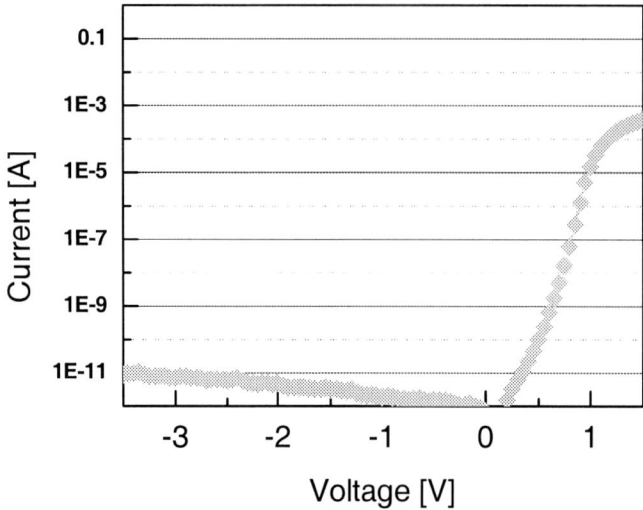

Fig. 5 Off leakage current of void free Co silicidied P (11B+ implanted)-N vertical diode formed by SEG.

CONCLUSIONS

In this work, we investigated the influence of Co/capping Ti Thickness, dopant species and hole dimension on silicidation rate and mechanism of void generation at silicon epitaxial growing (SEG) diode hole contact in Co salicide process. Co silicidation ratio is higher than bare wafer and the increasing with Co thickness increases silicidation ratio that can be explained that lateral Co on ILD oxide diffused into SEG with 1st RTA and more Co is participated in silicide reaction with 2nd RTA. Void is not observed after 2nd RTA in the BF2+ implanted SEG below 50 Å thickness of Co and thereby its silicide thickness is similar with that in the un-doped and 11B+ implanted SEG. More Ti diffused into Co and Ti-Co phase layer formation at 1st RTA can reduce the rate of Co silicidation and void generation is suppressed by the increasing of Capping Ti thickness. Combination with above 100nm hole size and lower than 65 Å Co thickness identifies void free in Co silicide formation. The electrical characteristics of vertical SEG diode including void free CoSi$_2$ as an ohmic contact layer shows an excellent reverse off leakage current.

REFERENCES

1. Y. C.Kim, J. C. Kim, J. H. Choy, J. C. Park, and H. M. Choi, *Appl. Phys. Lett.,* **75**,9 (1999)

2. J. C. Kim, Y. C. Kim, J. H Choy, *J. Korean Ceramic Soc.* **38** (10), 782-786 (2001)

3. J. S. Byun, J. M. Seon, K. S. Youn, H. S. Hwang, J. W. Park, and J. J.Kim J. *Electrochem. Soc.,* **143**, 3 (1996)

4. J. C. Kim, Y. C. Kim, B. K. Kim, *J. Korean Ceramic Soc.* **38** (10) 871-875(2001)

5. K. L. Pey, H. N. Chua, and S. Y. Siah, *Electrochemical and Solid-State Letters*, **3** (9) 442-445 (2000)

6. N. S. Kim, H. S. Cha, N. K. Sung and H.H Ryu J. Vac. Sci. Tech. **20**(4), 2002

7. C. S. Ho,a K. L. Pey,b,z C. H. Tung,c B. C. Zhang,a K. C. Tee,a G. Karunasiri,d,e and S. J. Chuad *Electrochemical and Solid-State Letters*, 7 ~11, H49-H51 (2004)

8. Y. M. Chen, G. C. Tu, *J. Vac. Sci. Tech.* **24**(1) 2006

9. A. Alberti, R. Fronterre, F. La Via, E. Rimini Materials Sceince and Engineering **B 114-115** 232-235 (2004)

10. S. Buschbaum, O. Fursenko, D. Bolze, D. Wolansky, V. Melnik, J. Nieß, W. Lerch, *Microelectronic Engineering* **76** (2004) 311–317

Mater. Res. Soc. Symp. Proc. Vol. 1079 © 2008 Materials Research Society   1079-N02-01

From Process Assumptions to Development to Manufacturing

Theo Standaert[1], Allen Gabor[1], Andrew Simon[1], Anthony Lisi[1], Carsten Peters[2], Craig Child[2], Dimitri Kioussis[2], Edward Engbrecht[1], Fen Chen[1], Frieder Baumann[1], Gerhard Lembach[2], Hermann Wendt[2], Jihong Choi[2], Joseph Linville[2], Kaushik Chanda[1], Kaushik Kumar[1], Kenneth Davis[1], Laertis Economikos[1], Lee Nicholson[1], Moosung Chae[3], Naftali Lustig[1], Oscar Bravo[1], Paul McLaughlin[1], Ravi Prakash Srivastava[4], Ronald Filippi[1], Sujatha Sankaran[1], Tibor Bolom[2], Vinayan Menon[1], Vincent McGahay[1], Wai-Kin Li[1], Wei-Tsu Tseng[1], William Landers[1], Youngjin Choi[1], Glenn Biery[1], and Thom Gow[1]

[1]IBM, Hopewell Junction, NY, 12533

[2]AMD, Hopewell Junction, NY, 12533

[3]Infineon, Hopewell Junction, NY, 12533

[4]Chartered, Hopewell Junction, NY, 12533

ABSTRACT

A tool has been developed that can be used to characterize or validate a BEOL interconnect technology. It connects various process assumptions directly to electrical parameters including resistance. The resistance of narrow copper lines is becoming a challenging parameter, not only in terms of controlling its value but also understanding the underlying mechanisms. The resistance was measured for 45nm-node interconnects and compared to the theory of electron scattering. This work will demonstrate how valuable it is to directly link the electrical models to the physical on-wafer dimensions and in turn to the process assumptions. For example, one can generate a tolerance pareto for physical and or electrical parameters that immediately identifies those process sectors that have the largest contribution to the overall tolerance. It also can be used to easily generate resistance versus capacitance plots which provide a good BEOL performance gauge. Several examples for 45nm BEOL will be given to demonstrate the value of these tools.

INTRODUCTION

When a new technology node gets introduced to its development phase there is typically little or no silicon data. In this concept stage most of the technology definition and elements are therefore based on assumptions. One considers the learning from prior technologies as well as new material and process candidates that are presented by the research community. The result is a set of assumptions that best describes what is expected to happen on wafer for each of the processing steps. This set is commonly referred to as the process assumptions. As real silicon data becomes available it can be used to update and refine the process assumptions.

This work describes how to set up the process assumptions and directly link it to electrical models. The result is a valuable tool that can be used to evaluate various BEOL options and to validate a BEOL technology.

SETTING UP BEOL PROCESS ASSUMPTIONS

Capturing processing details

Each processing step on the wafer can be captured by a schematic and a simple set of equations. Figure 1 shows an example for metal 1 where the contact module is followed by a dielectric deposition, lithography and etch.

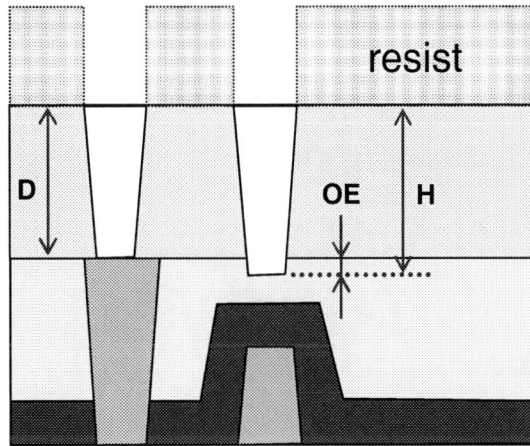

Figure 1: Schematic of how metal 1 could be formed. In this case metal 1 is formed by the deposition of single dielectric film and patterning using lithography and etch.

The metal 1 dielectric deposition thickness is indicated by D. To build in adequate overetch the RIE team targets a thicker film: E. Assuming the etch selectivity to the dielectric in the contact module is s, the following expressions are obtained for the overetch OE and the line height H after metal 1 etch:

$$OE = \frac{1}{s}(E - D) \tag{1}$$

$$H = \left(1 - \frac{1}{s}\right)D + \frac{1}{s}E \tag{2}$$

It is tempting to write the line height as $H=D+OE$ which is of course correct, but D and OE are not independent parameters. Hence, one cannot use this equation to calculate the tolerances. Equation 2 has independent process parameters (D and E) and can be RSSed (Root of the Sum of the Squares) to compute the tolerances in H.

Although the deposition and etch processes are independent in this example, their target values have to be set consistently of course. Not only with respect to each other, but also consistent with the technology targets that need to be achieved. One can continue to capture the next set of processes (and modules) and write the various line and via dimensions as function of the independent process parameters. With these equations it is then possible to solve for some or even all of the process targets by using the technology (final on-wafer) targets and constraints. For example, one constraint could be to set the worst-case metal 1 overetch to 20%. Figure 2 shows the process flow schematically.

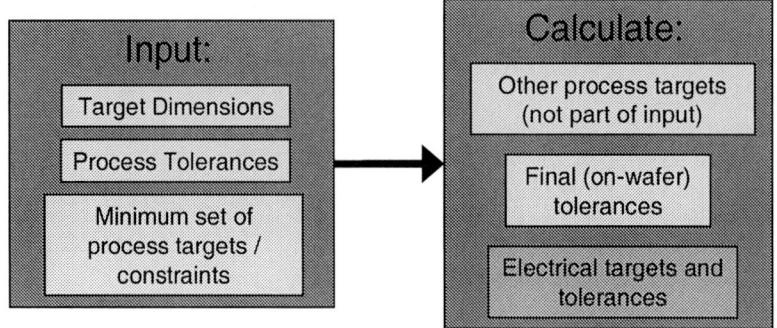

Figure 2: Schematic of how the process assumptions can be setup. The left box indicates the input which includes the target (or final on-wafer) dimensions. The right side indicates the instantaneous feedback from simple calculations

This setup is very powerful and can easily be implemented in a spreadsheet such as Excel. It automatically adjusts the various process targets when a change is made in target dimensions and or/tolerances. The whole set of process assumptions is self-consistent.

Including electrical models

Electrical models can also be incorporated in the flow that is shown in Figure 2. Perhaps the most difficult parameter is the resistance of narrow Cu lines. The Cu resistivity starts to increase as the interconnect dimensions approach the electron mean free path in Cu. This phenomena has been studie since the 1940s [1-3]. The most conventional and elegant approach is to solve the Boltzmann transport equation. Although detailed knowledge of the scattering interactions is not required in this case, the values of certain scattering parameters still have to be determined by experimental data. We follow the work of Steve Rossnagel who has looked in detail at these scattering parameters. The electron mean free path in Cu is assumed to be 39nm. The other two scattering parameters are p=0.1-0.3 and R=0.3 for surface and grain boundary scattering, respectively. We will discuss the choice of scattering parameters and experimental data in a separate publication [4].

Figure 3a shows a cross-sectional TEM of a 45nm BEOL Cu wire of known resistance. The Boltzmann equation can be solved for this particular geometry and results in a Cu conductivity profile which is shown in Figure 3b. The resistance is then obtained as the reciprocal value of the conductivity integral and is in good agreement with the measured resistance. These rather complicated profiles are of course hard to work with in the process assumptions. A profile as drawn in Figure 3c) is more desirable. Fortunately, the modeled resistance of the rectangular profile is very similar to one in Figure 3b). Please note that the profiles in Figure 3 have the same Cu area and width at half-height.

Before resistance can be included in the process assumptions flow (Figure 2), one more simplification is required. Instead of solving the Boltzmann equation which is rather time-consuming, a closed expression is required. For extreme cases (wire dimensions much larger than the electron mean free path) the following expression can be derived [3]

$$\rho \approx \rho_0 \left[1 + \tfrac{3}{8}(1-p)\left(1+\frac{w}{h}\right)\frac{\lambda}{w} + 1.4\frac{R}{1-R}\frac{\lambda}{g} \right] \tag{3}$$

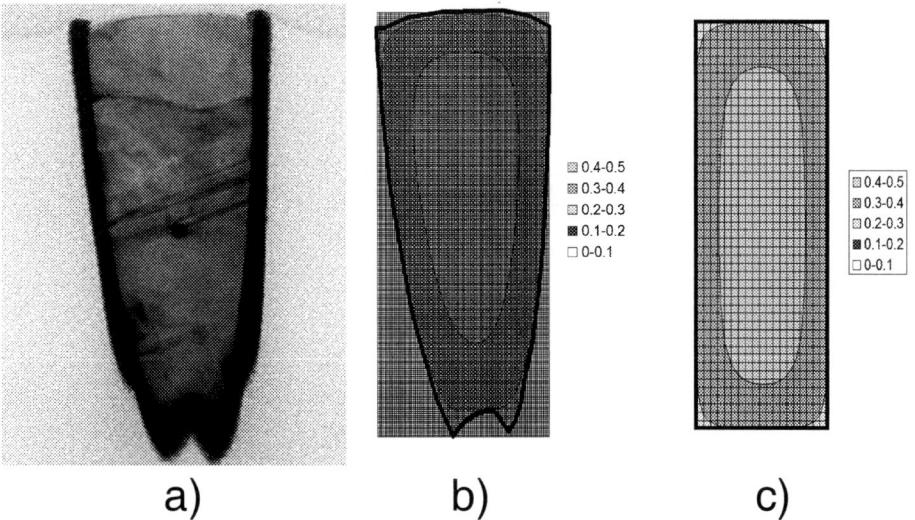

a)   b)   c)

Figure 3: Cross-sectional TEM of a 45nm BEOL wire (a), modeled conductivity profile (b), modeled conductivity profile for rectangular cross section with same Cu area and width at half-height (c).

where ρ is the Cu resistivity, ρ_0 the bulk resistivity, λ the electron mean free path and g the average Cu grain size.

Figure 4 shows Cu resistivity data as function of the Cu width at half-height. The data was obtained by measuring the resistance of various wires (different levels/wafers/lots). The Cu resistivity was then calculated by multiplying it with the Cu cross-sectional area as determined by TEM. The solid green line shows the solution of the Boltzmann transport equation. The grain size g is assumed to be proportional to the linewidth w. The agreement between the two is very good. At the smaller Cu dimensions (~50nm), the deviation is 3-4%. This is still small compared to the spread in the experimental data (appr. ±5%). Note that this data is also consistent with a bulk Cu resistivity of 1.71 $\mu\Omega$-cm.

Capacitance of BEOL wiring levels is typically obtained by using a field solver and is rather time consuming. In order to include capacitance calculations into the process assumptions flow, the parameter space was mapped with a field solver and fit by a simple function. This is very similar to what is for done extraction decks and accuracy within 1-2% is easily obtained.

Tolerance Pareto

The first section explained how the final on-wafer dimensions can be expressed as function of the individual process parameters. The resistance and capacitance can be expressed as function of these process parameters as well. Taking the total capacitance (C_{tot}) as an example, one can calculate the tolerance as

$$\Delta C_{tot}^2 = \sum_i \left(\frac{\partial C_{tot}}{\partial p_i} \Delta p_i \right)^2 \qquad (4)$$

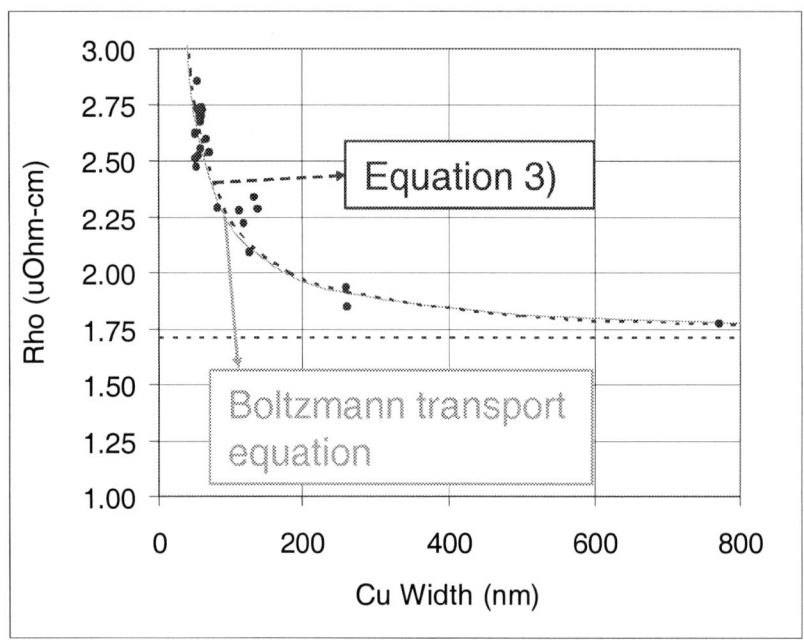

Figure 4: Cu resistivity data of various BEOL interconnects at 45nm-node. The data is compared to solutions of the Boltzmann transport equation as well as Equation 3).

where the set of p_i represents the independent process variables. This equation can be rewritten as

$$\sum_i \alpha_i = 1 \ \text{ with } \ \alpha_j = \left(\frac{1}{\Delta C_{tot}} \frac{\partial C_{tot}}{\partial p_j} \Delta p_j \right)^2 \tag{5}$$

If the α_j terms are ranked by their value they can be interpreted as a tolerance pareto where each of the terms quantifies how much process j contributes to the overall tolerance.

A hypothetical example of a metal 2 wire is shown in Figure 5 where the α_j terms are presented in a bar graph. The assumptions for the significant unit process sectors are listed next to the bars. The etch rate of the metal trench is 15%. Litho control the metal CD within 8.5%. The removal by CMP has a relatively large tolerance of 60%. The sidewall angle of the trench has a tolerance of 0.5° and the dielectric deposition has a tolerance of 10%. In this example, the etch depth is the largest contributor to the variation in capacitance and an effort to reduce this variation should therefore focus on improving the etch depth control. The reason that CMP ranks lower in the pareto despite having the largest (relative) tolerance is that total thickness removed by CMP is smaller than the etched depth in this example.

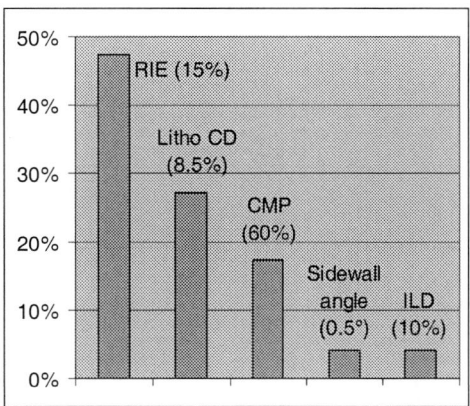

Figure 5: Capacitance tolerance pareto for a metal 2 wire at 45nm node. The numbers in parentheses indicate the unit process tolerances which were made up for this example.

One last note for the resistance tolerance pareto. Resistance is a non-linear parameter as it is inversely proportional to the Cu area and RSS is therefore not appropriate to calculate the resistance tolerances. The Cu area is the the product of the Cu height and width and therefore non-linear as well. However, compared to resistance the area is better behaved and more normally distributed then the resistance as will be shown later. It is therefore a fair assumption to assume that the Cu area and reciprocal resistance is normally distributed. Instead of RSS-ing the resistance and calculating the resistance pareto this way, it is better to calculate the Cu area tolerance pareto (which is identical to the resistance tolerance pareto).

DISCUSSION

Using the process assumptions as a tool to compare integration schemes

The tolerance pareto is a useful tool when comparing different integration schemes. To illustrate this, the examples in Figure 6a) and b) were made up. The stack highlighted in a) represents a conventional metal 2 stack in an ILD that does not have a hardmask when it enters the CMP sector. The stack in b) integrates a lower dielectric constant film that does require a hardmask and the assumption is made that it needs to be polished off to obtain the lowest k_{eff} possible. The additional hardmask in the second stack requires a larger etch depth and a larger CMP removal. The assumption was made that tolerances for litho CD, etch depth and CMP removal are the same for both integration schemes: 7%, 8% and 30%, respectively. For this example, we further assumed that the hardmask required is quite thick: ~40nm. The final target dimensions are assumed to be 150nm tall and 70nm wide. By following the flow outlined in Figure 2, the tolerances are immediately obtained for the two stacks. The metal 2 resistance has a tolerance of ~15% without hardmask, but a ~30% with hardmask. As the paretos indicate, variation in the CMP sector has become the dominant contributor to the resistance tolerances whereas the RIE and litho CD tolerances have fallen to second and third place.

Again, the stacks and tolerances in Figure 6a) and b) were made up to illustrate the use of the tolerance pareto. The increased resistance tolerances with the hardmask can of course be addressed by minimizing the hardmask thickness as well as increasing the polish selectivity between the hardmask and the ILD.

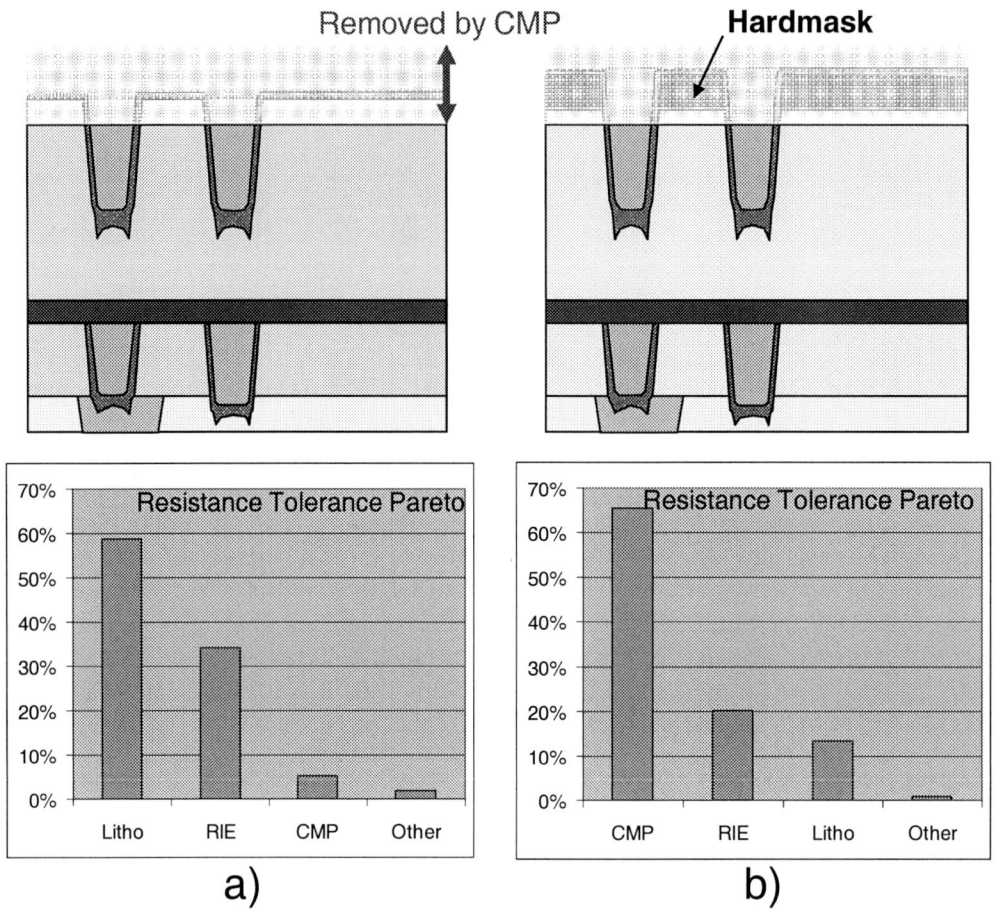

Figure 6: Comparison of two integration schemes: a) ILD without hardmask, b) ILD with hardmask.

Resistance versus capacitance

Resistance is often plotted versus capacitance. In this section it will be shown that the location of the RvsC curve is quite 'universal' and provides a good measure whether or not the dielectric constants and Cu volume of the wires are on target.

When a random generator is added to the process flow in Figure 2 it is very easy to produce an expected RvsC curve as is demonstrated in Figure 7. The shape and location of this curve can now be instantaneously be verified by changing the input assumptions. Changes in line height and width move the datapoints along the curve. Changes in liner thickness do not impact the capacitance and the RvsC cloud is expected to move to lower or higher resistance values. In contrast, changes in via height and dielectric constants do not impact resistance and the RvsC cloud now moves to higher or lower capacitance values. Since only a third of the capacitance consists of vertical coupling, large changes in viaheight are required to significantly move the RvsC cloud to different capacitance values.

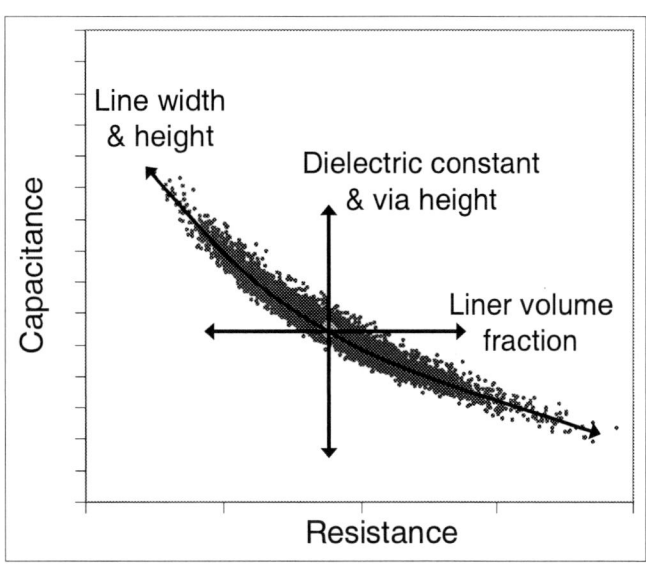

Figure 7: Resistance vs capacitance plots and dependency on various parameters.

Figure 7 was generated by assuming a trapezoidal shape for the metal wires. In reality, however, the profiles can be a lot more complicated as is shown in Figure 3a). To verify the impact of a more complicated shape, wires were considered where the bottom of the wire has an increased taper angle. A few random cases were generated which nominally had the same width at half-height and liner volume fraction as the wires in Figure 7. The capacitance was calculated by using a field solver and the resistance was obtained by solving the Boltzmann transport equation. The results are shown in red and compared to the trapezoidal shapes in Figure 8.

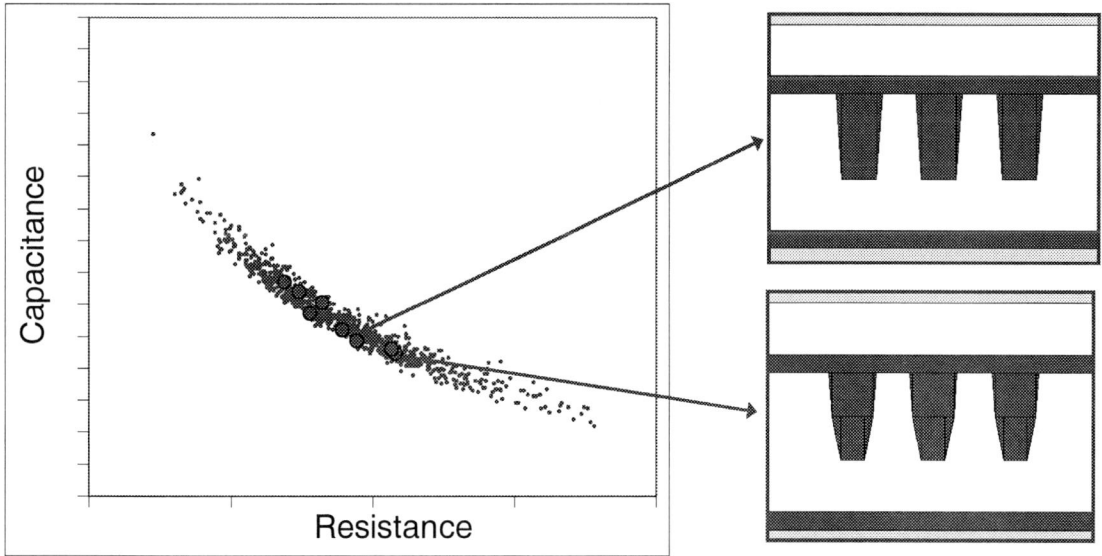

Figure 8: Resistance vs capacitance plots for two different profiles.

This simulated data confirms the earlier statement where the location of the RvsC curve depends mostly on the dielectric constants of the various films in the stack and the Cu volume and resistivity. This property is quite useful. Early silicon data is typically off target for both physical dimensions as well as electrical properties. However, when the resistance is plotted versus resistance, the RC cloud should intersect or approach the technology target, at least within a few percent.

Figure 9 shows a RvsC comparison of experimental data for a 45nm interconnect in an ultra-low k dielectric [5] and simulated data for a dielectric constant of 2.4 and 3.0. The location of the RvsC clouds validate that the assumptions of k=2.4 and Cu volume and resistivity were met. The shape of the RvsC cloud (thickness and width) are determined by the process tolerances. One cannot validate each process tolerance, but the shape of the cloud should be consistent with the whole set of tolerances.

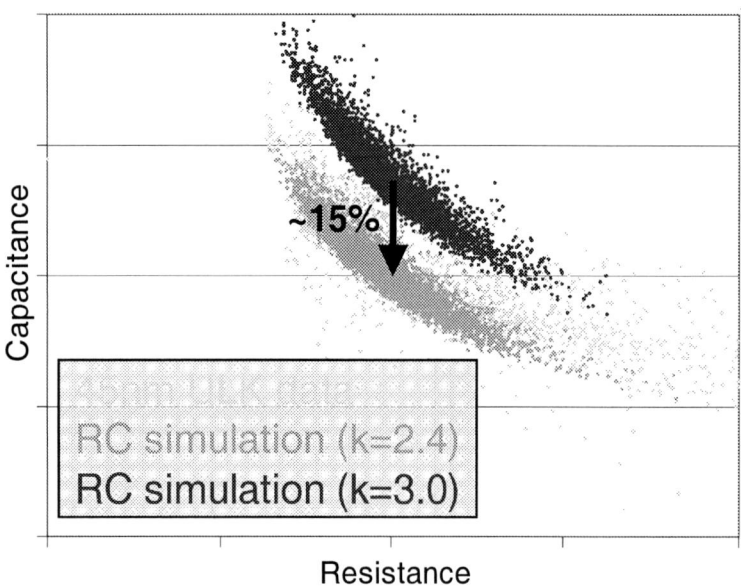

Figure 9: 45nm ULK RvsC data compared to process assumptions for k=2.4 and k=3.0.

Figure 10 shows experimental RvsC data for two 45nm populations which differ in liner thickness as can be seen from the two TEMs. The thinner liner shifts the RvsC cloud to lower resistances as indicated by the solid black arrow. The wire dimensions for the blue population , however, were adjusted to maintain similar resistance values compared to the red population. The adjustment in cross sectional area is indicated by the dashed black arrow. The net change is a drop of 8% in capacitance. This data demonstrates that the liner thickness is an important parameter for the interconnect performance. Care must be taken that new dielectrics are not introduced at the expense of a thicker liner.

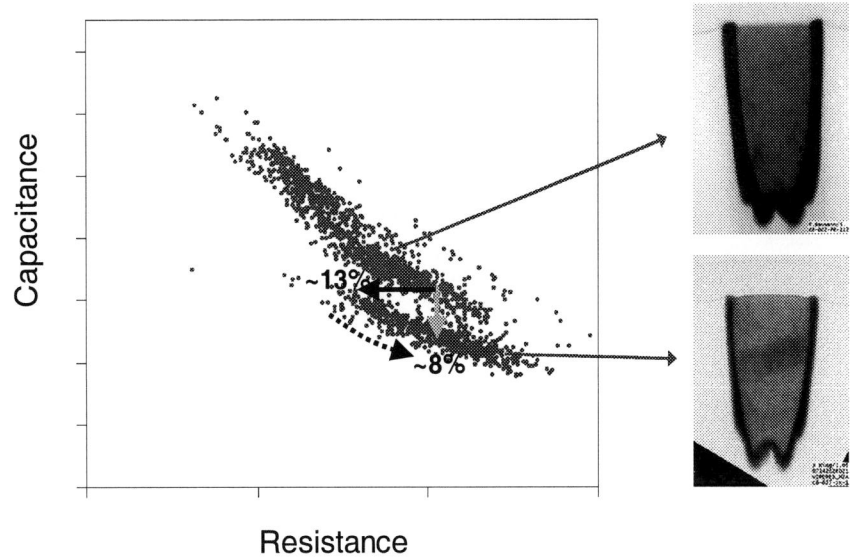

Figure 10: 45nm ULK RvsC data for two different liner thicknesses and line dimensions.

Including reliability considerations

As the dimensions continue to shrink it has become a real challenge to meet reliability requirements for interconnects. Reliability considerations therefore have to be included when setting up process assumptions. This section demonstrates how time-dependent dielectric breakdown (TDDB) and electro-migration (EM) have competing demands that cannot be ignored during the definition phase of a BEOL interconnect technology.

Figure 11 shows EM data for the via-depletion mode for two populations that have 70nm wide wires with different via sizes. Each of the distributions is clearly bimodal, but this is beyond the scope of this discussion. Focusing simply on the t_{50}, a reduction of 37% is observed when the via size is reduced from 80 down to 70nm.

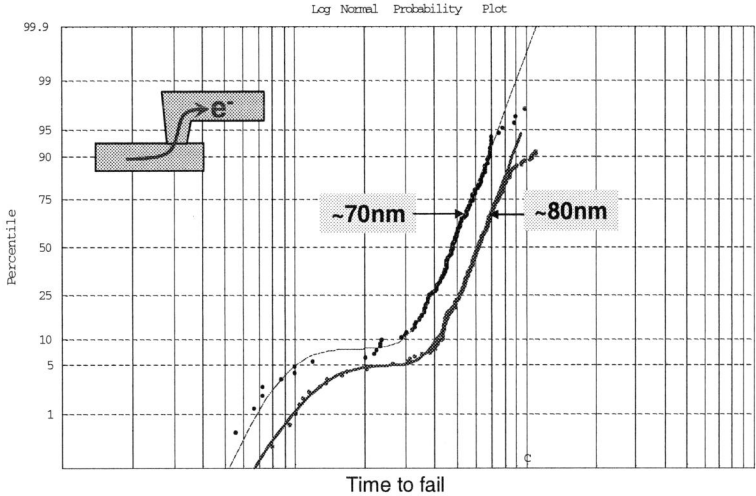

Figure 11: Electro-migration data for 45nm interconnects with 70 and 80nm via sizes.

While EM favors a larger via, TDDB stressing of large dense viachains shows that an increase in via CD results in a significant reduction in the projected lifetimes (~100-1000x !). This effect is illustrated by the diagrams in Figure 12. The diagram in Figure 12a) represents the 70nm vias. As the bottom CD is increased from 70 to 80nm, Figure 12b, the top of the vias increase accordingly and increases the electric field and the chance for catastrophic breakdown. The operating voltages have not been keeping pace with the shrink factors between technology nodes. This increases the burden on process integration to improve the spacing control between different nets and keep the worst-case electric field at an acceptable level.

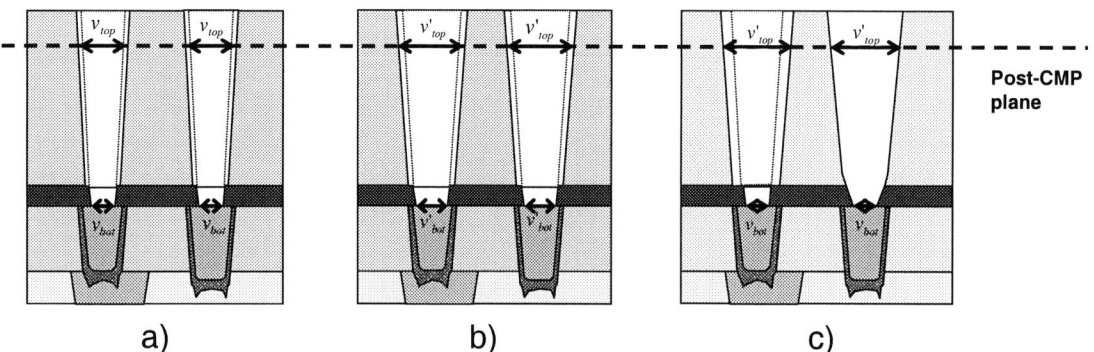

Figure 12: Schematics illustrating how the top CD increases due to larger bottom CDs (b) or new dielectric films that may be more prone to strip damage or RIE challenges at the via bottom (c).

Figure 12c) shows what could happen as a lower-k dielectric material is introduced. These films may be more prone to sidewall damage due various plasma exposures. Higher carbon-content in the film may also result in different etch characteristics that alter the profile near the via bottom. Assuming we want to keep the bottom on target, these two effects would result in larger top CDs and may therefore compromise TDDB performance.

Process Capability

Many chip fabrication plants have been raising their standards when it comes to process capability. Typically Cp/Cpk calculations assume normal distributions. However, non-linear parameters such as resistance should be approached carefully. Figure 13a) shows the resistance histogram and a normal fit for 45nm resistance data. It does not look so bad at first glance, but Figure 13b) shows otherwise. The solid red line is a fit to the blue datapoints assuming the resistance is normally distributed and a poor match is obtained at the tails of the distribution. Since the resistance is inversely proportional to the Cu area, one can get a better fit by assuming the area and, in turn, 1/resistance is normally distributed. As the solid green line shows a much better fit is obtained.

To highlight the importance for process capability, a hypothetical reliability spec limit is placed as indicated by the red dashed line. This spec limit could, for example, correspond to a Cu area which is too small to support the currents used in the designs. The process capability as extracted from a single sided Cpk would be significantly different for the two different fits (R normal and 1/R normal): 1.5 versus 1.1.

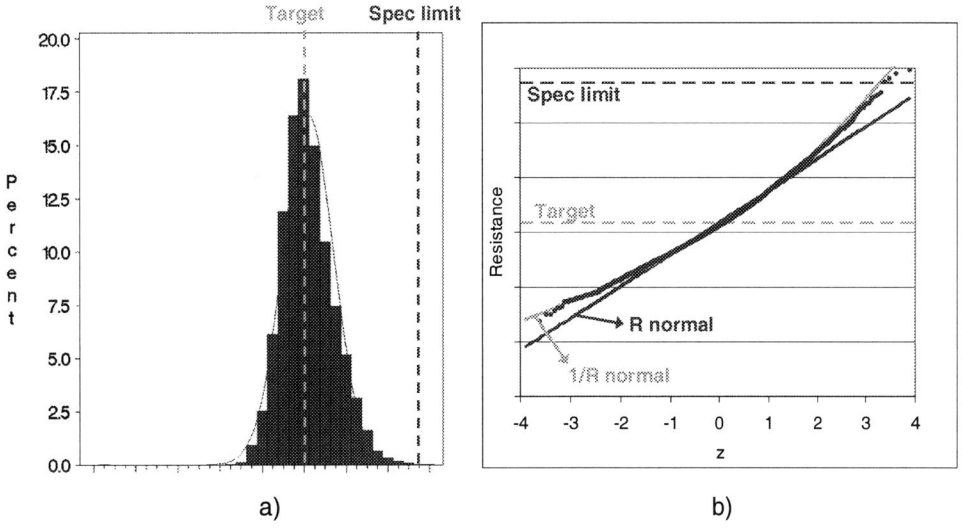

Figure 13: 45nm resistance data: a) histogram, b) distribution with a normal and non-normal fit.

CONCLUSIONS

It has been shown how the process assumptions can directly link the unit processes to the electrical targets and tolerances. The process assumptions can be made self-consistent by implementing proper constraints. The result is a tool where one can change final on-wafer target dimensions and unit process tolerances and get immediate feedback on physical tolerances and electrical parameters. A tolerance pareto calculation was added to identify those processes that have the largest impact on the overall tolerance. The tool can also be used to generate resistance versus capacitance curves. It has been shown that the location of these curves are rather insensitive to the wire dimensions and therefore a good performance gauge for the dielectric constants in the BEOL stack as well as the Cu area and resistivity. The Cu resistivity is a complex parameter and depends on electron scattering at surface and grain boundaries. Experimental data from 45nm and simple models were found to be in good agreement.

ACKNOWLEDGMENTS

This work has been supported by the independent Bulk CMOS and SOI technology development projects at the IBM Microelectronics, Div. Semiconductor Research & Development Center, Hopewell Junction, NY12533

REFERENCES

1. E. H. Sondheimer, Advances in Physics, 2001, 50 (6), 499-537
2. A. F. Mayadas and M Schatzkes, Phys. Rev. B, 1970, 1 (4), 1382-1389
3. S. M. Rossnagel and T. S. Kuan, JVST B, 2004, Vol. 22 (1), 240-247
4. T. Standaert, N. Lustig, and N. Lu, to be published.
5. S. Sankaran, *et al*, Electron Devices Meeting, 2006, IEDM '06 International.

Mater. Res. Soc. Symp. Proc. Vol. 1079 © 2008 Materials Research Society   1079-N11-04

Copper CMP with Composite Polymer Core - Silica Shell Abrasives: A Defectivity Study

Silvia Armini[1,2], Caroline M. Whelan[3], Mansour Moinpour[4], and Karen Maex[5]

[1]ESAT, KULeuven, Leuven, 3001, Belgium

[2]IMEC, Kapeldreef 75, Leuven, 3001, Belgium

[3]IMEC, Leuven, 3001, Belgium

[4]INTEL, Santa Clara, CA, 95052

[5]KULeuven, Leuven, 3001, Belgium

ABSTRACT

The results of copper Chemical Mechanical Planarization (CMP) experiments with a model slurry chemistry based on the combination of Glycine-water-Benzotriazole (Gly-H_2O_2-BTA), and different types of composite A (silane coupling agents between the polymer core and the silica shell) and B (electrostatic attraction between the polymer core and the silica shell) abrasives, are presented. While the presence of BTA allows a ten-fold reduction in the static etch rate from 95 nm/min. to 10 nm/min., combining oxidizer and complexing agent leads to removal rates higher than 400 nm/min. Different surface morphology and RMS roughness are observed after polishing with composite abrasives and different peroxide concentrations. Oxidizer concentrations as low as 0.1 vol.% lead to high non-uniformity and defectivity values. In particular, composite B performs better than pure colloidal/fumed silica during copper CMP using the IC-1000 pad, giving comparable Material Removal Rate (MRR), but a better surface finish due to the contribution of the elasticity of the polymer in gently transferring the applied load to the wafer surface. Cu CMP with pure polymer particles is a promising alternative to the hard inorganic material especially if combined with suitable surfactants that act from both particle stabilization and friction reduction / lubrication improvement perspectives.

The use of the medium/high-hardness pad IC-1000 is compared to the use of a soft Politex pad. In the former case, differences in terms of MRR, MRS roughness, and total defects are observed between the composite abrasives A and B; in the latter case, the behaviour of the two composites is similar. In the case of a soft pad in combination with composite abrasives, there is a remarkable improvement in the defectivity without any loss in MRR.

As revealed by SEM inspection of the composite particles collected in the slurry drain after CMP, for all the composites, the silica shell coverage is not disrupted by the shear forces and chemistry during the 1 min. polishing. Consequently, the stability and agglomeration properties of the particles in the complex Cu CMP chemistry can be helpful in explaining the experimental results in terms of MRR and surface finishing.

INTRODUCTION

In the copper deposition sequence, there are three main areas where defects can be introduced and have an impact on yield: i) during barrier/seed deposition; ii) during electrofill; and iii) during CMP. The most important yield limiting defect types after copper CMP are scratches, voids and pitting since they can create shorts in dense metal lines and long-term reliability failures. Choice of abrasive and chemistry in the slurry is considered as one of the main contributing root causes. A bad combination of abrasive and chemistry and hard foreign matters from the slurry or environment, are the root causes. Furthermore, Cu is known to easily corrode. A Cu film with very large grain size may be easier to corrode along the grain boundaries, where the grain size is dependent mainly on the annealing conditions. Corrosion activity is enhanced under high mechanical stress (stress-induced corrosion). Usually, a corrosion inhibitor is necessary in the slurry. Therefore, the purpose of this study is to evaluate if the composite abrasives already extensively evaluated through oxide CMP experiments [1], can constitute a useful alternative to replace the standard silica and alumina particles, in order to improve defectivity, without loosing their beneficial properties, e.g. MRR and Non-Uniformity (NU). While the effect of the chemical component in the CMP process has been studied extensively [2-5], the role of the mechanical component, that has a strong impact on defect formation, remains unclear. Despite a high number of theoretical models available in literature [6-13], based on different kinds of interaction between pad, particle and wafer, experimental data is scarce. The sheer number of variables associated with the abrasives complicates our understanding of how particles contribute to the MRR, defectivity and surface finishing. Furthermore, despite the volume of patent literature and commercial slurries available, due to the proprietary nature of most CMP compositions, the information that can be gleaned from these sources relevant for the development of next-generation slurries is limited.

Our research focuses on entirely novel composite structures comprising polymer particles coated by smaller silica colloids, achieved by either creating chemical bonds by silane-coupling agents (composite A) or tuning the pH in order to form electrostatic attractive interactions between the core and the shell (composite B). In this way, it is possible to combine several advantageous properties in the same structure. The polymer core shows mechanical properties that are highly tunable by variation of its synthesis parameters [14, 15], while the major advantage of the silica coating is that it can be easily modified, in terms of its surface chemistry and morphology.

In particular, our composite particles are aimed at improving the CMP process of easily damaged materials such as copper, due to the spring-like effect coming from the elastic properties of the core, which allows the composites to easily adapt to the pad asperities.

The main advantage of our bottom-up approach to particle synthesis is the predetermined control of the chemical composition of the system and physical properties of the individual components. Post-synthesis characterization of our particles includes bulk properties, such as the elastic compressive modulus [16], size, morphology, and surface properties, such as surface charge, functional groups. In particular, the ζ potential of the particles *vs.* changes in ionic strength and pH gives essential information on the stability and agglomeration behaviour of the particles in the polishing solution, which is expected to correlate closely with scratch formation during CMP [17].

EXPERIMENT

Materials and set-ups. For all the copper CMP experiments, 200 mm copper wafers were obtained by depositing a 25 nm Ta (15 nm)/TaN (10 nm) diffusion barrier, and a 100 nm Cu seed layer deposited in a physical vapor deposition (PVD) chamber (Electra™, AMAT) on a standard dielectric stack (500 nm SiO_2 on 50 nm Si_3N_4). 1 μm Cu film is deposited by electrochemical deposition (Electra-ECP-Cu, AMAT) using Shipley Nanoplate plating bath chemistry. To allow post-CMP characterization in Cu-shared tools, some of the wafers received a backside 50 nm nitride layer. The polishing tool was a rotary polisher (E 460 Mecapol polisher, Alpsitec). The details of the materials used are summarized in Table 1. The particles used as abrasives in the slurry are presented in Table 2. For the preparation of the slurry, a mechanical shaker was applied for 10 min. before the CMP experiments. During the polishing, the slurry was stirred by a magnetic bar at a speed of 400 rpm.

Table 1. Materials used for the copper CMP experiments.

	Backing film type	DF200, II W/40 CNC Holes, Rohm&Haas Electronic Materials
	Pad type	IC1000 xy-k grooved pad, Rohm&Haas Electronic MaterialsPolitex pad, Rohm&Haas Electronic Materials
	Conditioner	A4-55 diamond pad conditioner, 3M
	DI water	Conductivity 0.055 □S/m
Cu slurry chemical composition	**Oxidizer1vol%**	H_2O_2, Hydrogen peroxide, Air Products, 30 wt%
	Cu complexingAgent1wt%	$C_2H_5NO_2$, Glycine (Gly), ≥ 99%, Sigma Aldrich
	Cu	$C_6H_5N_3$, Benzotriazole (BTA), ≥ 99%, Sigma Aldrich
	corrosionInhibitor0.018wt%	
	Cationic surfactant	Hexadecyltrimethylammonium bisulfate, ≥ 97%, Fluka

213

Table 2. SEM image of the composites abrasives before and after the copper CMP experiments. The average size of the core is also indicated.

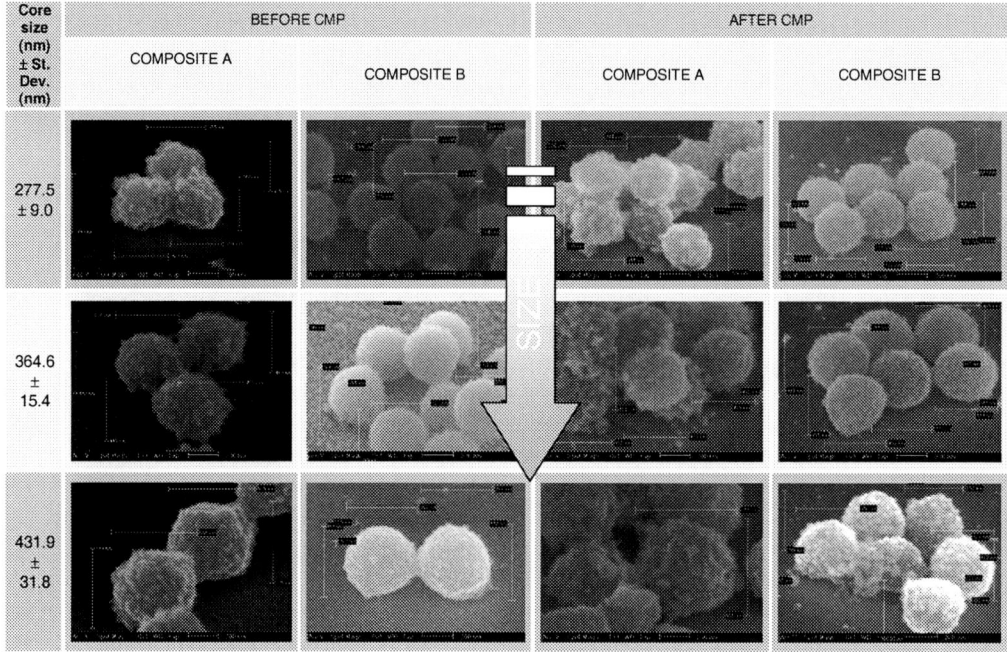

CMP experimental conditions. The general experimental polishing conditions for the copper CMP experiments are shown in Table 3. An overview on all the performed copper CMP experiments is summarized in Table 4.

Table 3. Experimental conditions for the copper CMP experiments.

STEP #1 **Landing** **3 s**	Head rotational speed (HS) (rpm)	50
	Platen rotational speed (HS) (rpm)	48
	Down pressure (DP) (psi)	0.11
	Slurry flow rate (ml/min.)	200
	Back pressure (BP) (psi)	0.35
STEP #2 **Polishing**	Head rotational speed (HS) (rpm)	65
	Platen rotational speed (HS) (rpm)	40
	Down pressure (DP) (psi)	4
	Slurry flow rate (ml/min.)	200
	Back pressure (BP) (psi)	0.35
STEP #3 **Rinsing** **60 s**	Head rotational speed (HS) (rpm)	65
	Platen rotational speed (HS) (rpm)	40
	Down pressure (DP) (psi)	0.11
	Slurry flow rate (ml/min.)	DI water
	Back pressure (BP) (psi)	0.35

Table 4. Overview on the copper CMP experiments.

Experiment set # [a]	1
Studied abrasive system (motivation: investigation of the effect of the silica size)	colloidal silica 15 nm colloidal silica 30 nm colloidal silica 90 nm colloidal silica 600 nm
Total # of wafers	12 (3 for each condition)
pH before-after the addition of the chemicals	8.5-8.2 (colloidal silica)
Polishing time (min.)	1
Abrasive solid content (wt%)	5
Experiment set # [a]	2
Studied abrasive system (motivation: investigate the effect of different silica particle size- composite A: effect of different ratio core size/shell-thickness)	composite A (core 300 nm/shell 15 nm) composite A (core 300 nm/shell 30 nm) composite A (core 300 nm/shell 90 nm) composite A (core 600 nm/shell 90 nm) composite A (core 120 nm/shell 15 nm) [c] composite A (core 120 nm/shell 30 nm) composite B (core 300 nm/shell 15 nm) composite B (core 300 nm/shell 30 nm) [a] composite B (core 300 nm/shell 90 nm) composite B (core 600 nm/shell 15 nm) composite B (core 600 nm/shell 30 nm) composite B (core 600 nm/shell 90 nm) PMMA 300 nm PMMA 300 nm [b]
Total # of wafers	36 (3 for each condition)
pH before-after the addition of the chemicals	2-3.8 (composite A) 4.5-5.0 (composite B) 5.5-5.0 (pmma)
Polishing time (min.)	1
Abrasive solid content (wt%)	5
Experiment set # [a]	3
Studied abrasive system (motivation: measurement of dynamic etch rate)	Only chemical composition – no abrasives
Total # of wafers	3 (3 for each condition)
pH of the slurry chemical composition	5.5-6.0
Polishing time (min.)	1
Abrasive solid content (wt%)	-
Experiment set # [a]	4
Studied abrasive system (motivation: copper CMP in DI water)	colloidal silica 90 nm
Total # of wafers	3 (3 for each condition)
pH	8
Polishing time (min.)	1
Abrasive solid content (wt%)	5
Experiment set # [a]	5
Studied abrasive system (motivation: investigate the effect of different oxidizer concentration)	composite A (core 300 nm/shell 30 nm) composite B (core 300 nm/shell 30 nm)
Total # of wafers	8 (1 for each condition)
Chemical composition [c]	0.018 wt% BTA + 1 wt% Gly + 0.1 vol% H_2O_2 0.018 wt% BTA + 1 wt% Gly + 1 vol% H_2O_2 0.018 wt% BTA + 1 wt% Gly + 5 vol% H_2O_2 0.018 wt% BTA + 1 wt% Gly + 10 vol% H_2O_2
Polishing time (min.)	1
Abrasive solid content (wt%)	5

Experiment set #[a]	6
Studied abrasive system (motivation: investigate the single contribution of the chemicals in the slurry)	colloidal silica 90 nm
Total # of wafers	6 (1 for each condition)
Chemical composition/pH	1 wt% Gly / pH = 7.6-7.8 1 vol% H_2O_2 / pH = 8.6 0.018 wt% BTA / pH = 8.6 0.018 wt% BTA + 1 wt% Gly / pH = 7.6 1 wt% Gly + 1 vol% H_2O_2 / pH = 7.8 0.018 wt% BTA + 1 vol% H_2O_2 / pH = 8.2-8.4
Polishing time (min.)	1
Abrasive solid content (wt%)	5

[a] Additional CMP experiments were done with 3x TEOS concentration in the shell

[b] Hexadecyltrimethylammonium bisulfate 10^{-4} M was added to the slurry in order to compare the CMP performances of the same abrasive system with and without surfactant.

[c] The pH stays constant both for composite A (pH=3.5-3.8) and composite B (pH=5.2).

Post copper-CMP cleaning step. After CMP the copper wafers were kept wet in a DI water tank in order to avoid that the particles and polishing residues irreversibly stick to the wafer surface. After a constant waiting time of longer than 1 h, to avoid serious corrosion of the copper surface, the wafers were processed in the cleaning station of the AMAT Mirra-Mesa polisher through the wet que. The wafers from the input station passed through the Megasonic cleaning station, two brush stations, the spin-rinse dry (SRD) station, and finally they reached the output station. The recipe used was Marathon and only DI water was used for the rinsing.

Defectivity and surface morphology investigation. The defectivity on non-patterned full copper wafers was determined by a light scattering method. For the experiments made in the first part of this work with a Politex pad, silica, polymer and composite particles, the SP1[DLS] tool was used and a recipe was specifically optimized for copper wafers after polishing (IP: s; CP_W: s; CP_N:s; Aperture: 20 degree, calibrated). This recipe allowed a minimum defect size of 254 nm. For the defectivity measurements after copper CMP with the harder pad IC 1000 and a range of chemistry/abrasives we used the Surfscan 6400 but, since the number of defects in this case was much higher leading to the overload of the detector, we had to lower the minimum defect size threshold to 700 nm. This should be constantly considered when comparing the defectivity data.

To investigate the surface morphology we used the AFM technique. WSxM software [18] assisted the extraction of cross-sectional data and facilitated 3D elaboration of the images.

Copper layer Rs measurement. Sheet Resistance (Rs) measurements on copper unpatterned wafers were performed using the Prometrix OmniMap RS75 four probe system (conditions: 49 points measured – 10 mm edge exclusion). The Rs of a layer with resisitivity ρ, and thickness t, is given by their ratio: Rs=ρ/t

RESULTS AND DISCUSSION

Model copper slurry chemical composition. Compared to CMP for silicon dioxide, the metal (particularly copper) CMP process is poorly understood, due mainly to electrochemical interactions between the slurry and the metal film during polishing, and the coupled effect of these on the mechanical properties of the surface. Optimization of the slurry chemistry for Cu is beyond the scope of this work. Nevertheless, a meaningful interpretation of the experimental results in terms of MRR and surface finishing can only arise from the understanding of the slurry system and the awareness of the complex interactions between all the slurry components.

Copper CMP mechanism in the model slurry chemistry. The most commonly accepted planarization mechanism in metal CMP is Kaufman's tungsten CMP model [19]. In this model, the slurry chemistry is such that it induces formation of a protective passivating film on the metal surface. Passivation could be caused by formation of native metal oxides on the surface using oxidizers such as H_2O_2. Chemically adsorbed layers can also be formed using organic inhibitors. For example, benzotriazole (BTA), a common corrosion inhibitor for copper, could be utilized in copper CMP slurries [20]. In an ideal CMP process, the protective film is removed by the mechanical action of the abrasive slurry only at the protruding regions where the wet etch of unprotected metal occurs until metal repassivates. Oxidizers are chemical reagents that change the oxidation state of the metal. They have a dual function in the slurry. They can form oxides which protects the recessed regions from the wet etch as well as induce dissolution of metal at the protruding regions. Continuous cycles of formation, removal and reformation of the passive layer continue until the desired final thickness of planar metal surface is attained. In copper chemical mechanical planarization, in addition to wet etching of the unprotected metal surface, the chemical reagents in the slurry should also be able to completely dissolve all the abraded material [21]. This is necessary to prevent precipitation of abraded copper particles as well as to avoid redeposition of copper ions back to the wafer. In order to preserve the desired passivation and dissolution behaviour, the slurry pH should be always controlled. Copper CMP in a highly acidic pH regime leads to corrosion problems, while Cu CMP in alkaline conditions is faced with an unfavorable polish rate selectivity with respect to SiO_2, leading to interlayer dielectric (ILD) erosion. Thus, an intermediate pH range 4-7 appears to be a better choice for Cu CMP [22-24]. One of the more attractive slurries in this intermediate pH range consists of hydrogen peroxide, glycine (an amino acid), and an abrasive.

We measured the MRR and defectivity after each CMP experiment with colloidal silica 90 nm 5wt.% and only one of the three chemicals (oxidizer, complexing agent, and corrosion inhibitor) or with a combination of two by two of them.

Our results plotted in Figure 1a show that although the MRR of copper in a pure H_2O_2 solution is as small as 7 nm/min., it increases dramatically (by a factor of 60) in the presence of glycine. Interestingly, all the other combinations of chemicals give lower MRR. The combination of the three chemicals without any abrasive leads to a MRR of ca. 150 nm/min.

The effectiveness of glycine is attributed to hydroxyl radicals, which are responsible for the removal of copper from the surface through intermediate oxidation steps, generated copiously through the catalytic action of a complex formed by the glycine with the copper ions in solution. The static etch rate for copper in the presence of the chemical composition of the model slurry (1vol.% H_2O_2, 1wt.% Gly, and 0.018wt.% BTA) in the same pH 5-6 was ca. 10 nm/min., while in the absence of BTA it dramatically increases until 95 nm/min. Benzotriazole (BTAH-$C_5H_4N_3H$) has one replaceable H atom attached to the N_1 atom in the triazole group. The BTA⁻ ion forms a solid cuprous complex. This complex is known to be responsible for the corrosion inhibiting properties of BTAH [25, 26].

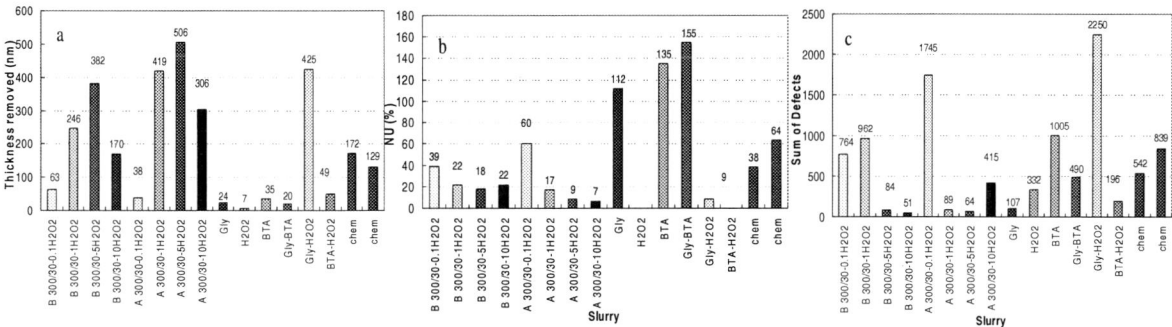

Figure 1. a) Thickness removed, b) Non-Uniformity, and c) Sum of defects vs. slurry abrasive type and chemical composition.

It is now worth discussing the role of the hydrogen peroxide as oxidizer during copper CMP. The following mechanism for copper CMP in H_2O_2-containing acidic slurries has been proposed: 1) When the

H_2O_2 concentration is low, e.g. ca. 1vol.%, the MRR during CMP is controlled by electrochemical dissolution; 2) when the H_2O_2 concentration is high, e.g. ca. 10vol.%, the formation of copper oxide is fast enough, and copper CMP is controlled by mechanical removal of copper oxide and its subsequent dissolution; 3) when the H_2O_2 strength is medium, both mechanisms operate and compete [19].

Copper CMP with the composite abrasives: MRR, NU, defectivity. We experimentally found that very low oxidizer concentrations lead to high NU and defectivity values. Particularly interesting is the AFM analysis of the copper surface after CMP with 5wt.% composite B abrasive in the presence of 0.1 and 10vol.% H_2O_2 shown in Figure 2 a and b, respectively. The different morphology of the two surfaces is immediately evident. The former exhibits higher RMS roughness and copper rip-out kind of defects very similar to the defects appearing on the copper surface after etching for 1 min. in pure chemicals, Figure 2 c. The depth of the voids created is more than double the value of the scratches observed in the presence of 10wt.% oxidizer. The plot of the RMS and Ra roughness versus oxidizer concentration shown in Figure 3 highlights the increase in the surface roughness with decreasing oxidizer concentration. From the shape of the defects shown in Figure 2 a and c we can formulate the hypothesis that there is a preferential attack and penetration of the chemicals along the copper grain boundaries. This effect sums up to the frictional forces acting during CMP (i.e. stick-slip) [25, 26].

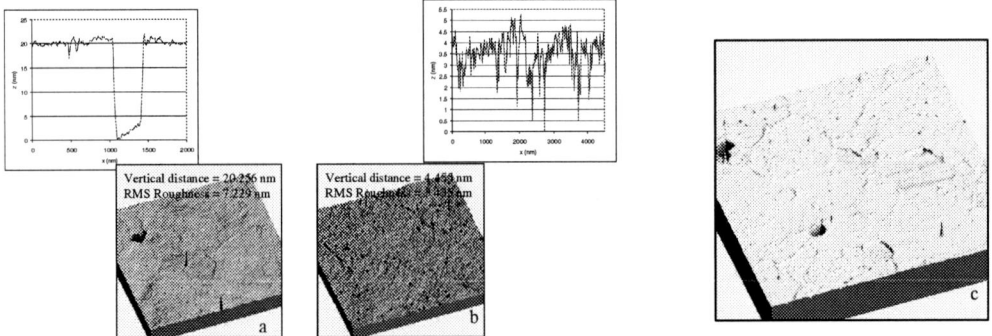

Figure 2. AFM images (5 μm x 5 μm) of the copper surface after CMP with a) 5wt.% composite B abrasives, 0.1wt.% Gly, 0.18wt.% BTA, 0.1vol.% H_2O_2, b) 5wt.% composite B abrasives, 0.1wt.% Gly, 0.18wt.% BTA, 10 vol.% H_2O_2 (5μm x 5μm), and c) pure chemicals (no particles).

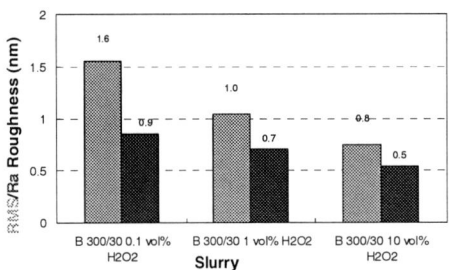

Figure 3. RMS and Ra roughness measured by AFM vs. oxidizer concentration.

Using colloidal silica particles of 15, 30, 90 and 600 nm diameter as abrasives, the MRR increases more than four fold, from 115 nm/min. to 520 nm/min. (Figure 4a), while the RMS roughness increases more than three fold, from 1.7 until 5.8 nm (Figure 4c).

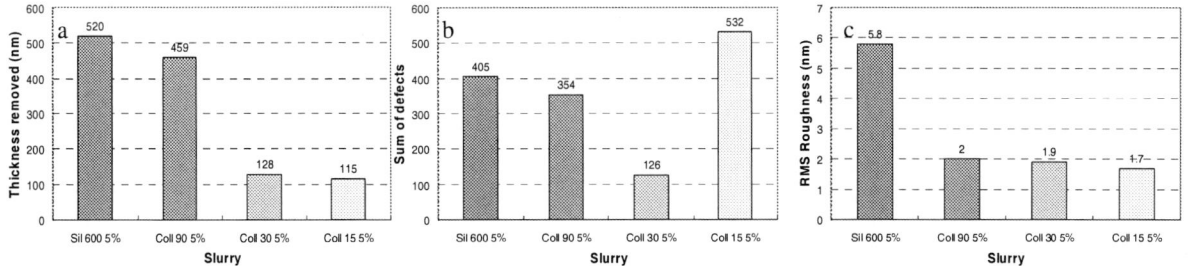

Figure 4. a) Thickness removed, b) Sum of defects, and c) RMS Roughness vs. slurry abrasive type.

After polishing with 5wt.% 600 nm colloidal silica the 3D AFM image shows the clear pattern of severe surface plastic deformation, e.g. roughness and scratches as deep as ca. 20 nm and as wide as ca. 1 μm (Figure 5).

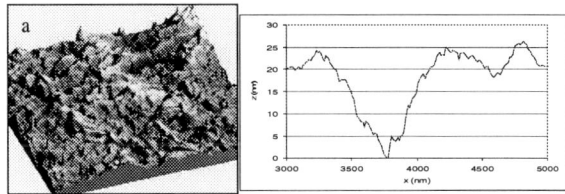

Figure 5. 3D AFM images of the Cu surface after CMP in the model slurry chemistry with 600 nm colloidal silica.

The use of the composite abrasives A during copper CMP, while allowing the achievement of MRR as high as 650 nm, does not lead to a significant improvement in terms of NU and MRS roughness with respect to bare silica particles. Nevertheless, the total number of defects is below 2200 defects.

As we already found for oxide CMP defectivity vs. composite core/shell size ratio [27], the use of the largest size polymer as a core in composite A leads to the lowest amount of defectivity and lowest maximum scratch depth. When used in the composite shell, the small 15 nm silica particles remain on the wafer after the post-CMP cleaning step. They affect both the total defect number and the RMS roughness. Moreover, smallest particles are more prone to agglomeration and larger agglomerates cause an increase in pit and scratch counts.

Improvement in the NU and RMS roughness (Figure 6) is achieved by using electrostatic interactions to create a bond between the polymer core and the silica shell (composite B). A MRR higher than 300 nm/min. is achieved using composite B with 5wt.% 300 nm polymer core - 90 nm silica shell, 5 wt% 300 nm polymer core - 15 nm silica shell, and 5wt% 300 nm polymer particles in the presence of 10^{-4} M CTAHS. The use of surfactants during copper CMP is particularly attractive due to the effect of MRR enhancement that allows polishing with softer abrasives such as polymer particles.

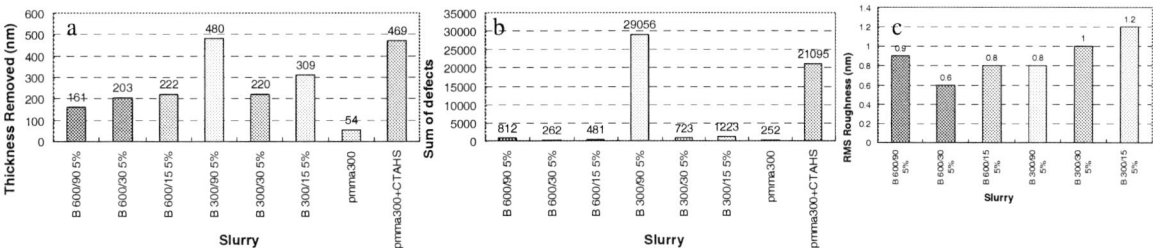

Figure 6. a) Thickness removed, b) Sum of defects, and c) RMS Roughness vs. composite B/pmma abrasive type.

The cationic surfactant CTAHS acts in synergy with the chemicals in the slurry composition improving the material removal mechanism with increased lubrication. This leads to better NU after polishing (< 4%). Unfortunately the presence of particles sticking to the wafer surface after the post-CMP cleaning step increases the defect number to ca. 21000. In addition, the RMS roughness and the AFM images of the copper surface after polishing with the polymers in the presence of CTAHS reveal a surface completely covered by polymer particles.

As already said for composite A, larger polymer cores lead to a low defect count, while the presence of the largest 90 nm colloidal silica at the shell increases the defect number. In Figure 7 a, b , and c the 3D AFM images of the copper surface after polishing with composite B abrasives for decreasing diameters of the colloidal silica in the shells are shown. The surface damage and scratch/pit depths are lower for the smaller size silica particles.

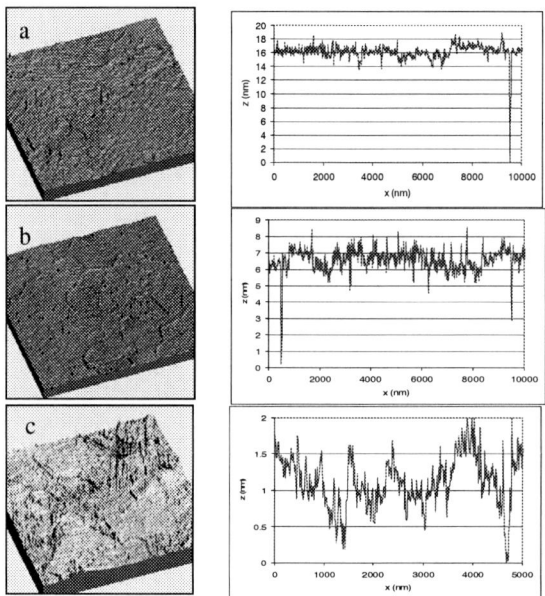

Figure 7. 3D AFM images of the copper surface after CMP with the model slurry chemical composition and a) B 600-90 5wt%, b) B 600-30 5wt%, and c) B 600-15 5wt%

Copper CMP with a soft pad. The effect of the soft Politex pad using the same model slurry chemistry already described for the CMP experiments done with the harder IC-1000 pad and the composite abrasives, has been investigated.

While for the IC-1000 pad a 2 min. diamond conditioning step was foreseen before each polishing run, the Politex pad was rinsed for 1 min. with DI water coming from a high pressure spray gun. In this case, to validate the effectiveness of this procedure, 10 wafers were processed with 5wt.% composite B abrasives in the model slurry chemistry and the MRR monitored. A drop in the MRR would indicate pad surface deterioration as a consequence of a non-optimal conditioning step and the presence of polishing residues in the pad asperities. No drop in the MRR (ca. 450 nm/min.) was observed, the NU was also constant and below 6%, and the defect level maintained below 6000 counts.

Three different polymer core sizes (277, 365, and 432 nm) were investigated and combined with 30 nm colloidal silica to form the composites A and B. 360 nm polymer particles and 30 nm silica colloids were compared to the composites in terms of MRR and defectivity. The MRR values of both composites A and B are comparable (Figure 8 a). In both cases an increase in the MRR is observed for the largest polymer core size. The surprisingly small MRR reported for the composite A with intermediate core size is explained by a non-systematic experimental error due to an inefficient stirring of the slurry composition. As a consequence, after polishing the three wafers, we noted the presence of a dense abrasive mass deposited on the bottom of the slurry container. Therefore, the effective abrasive solid content in the slurry was much smaller than the claimed 5wt.% leading to a MRR more than 4 times smaller with respect to the expected MRR. In addition, using a soft pad, the MRR observed for composites A and B with a larger core size almost doubles the MRR value found for the 30 nm colloidal silica.

Figure 8 b presents the total defect counts after CMP with composites A and B with increasing polymer core size. For composite B the defectivity is halved going from the smallest polymer core size to the maximum latex diameter. For composite A with the largest core size the photodetector of the SP1^{DLS} went into overload due to the high defect number. This unexpected result has been justified in terms of the very broad size distribution of the composite particles, as revealed by SEM investigation on samples collected after CMP (insert in Figure 8.26 c). A map of the defects after polishing with composite B is shown as insert in Figure 8.26 c. The defects are mainly localized at the wafer edge and a pattern can be recognized especially in one half of the wafer area. This pattern can be attributed to the Mirra-Mesa brush geometry.

To test the stability of the coverage of the composites in the slurry chemistry during the CMP process, SEM investigation was performed before and after CMP. The results are shown in Table 5. For all the composites, the coverage is not disrupted by the shear forces and chemistry during the 1 min. polishing and the particles show stability during the CMP process.

In summary, using a soft pad in combination with the composite abrasives allows the achievement of MRR that can reach 600 nm/min. MRR enhancements can be obtained using larger polymer cores for both composites A and B.

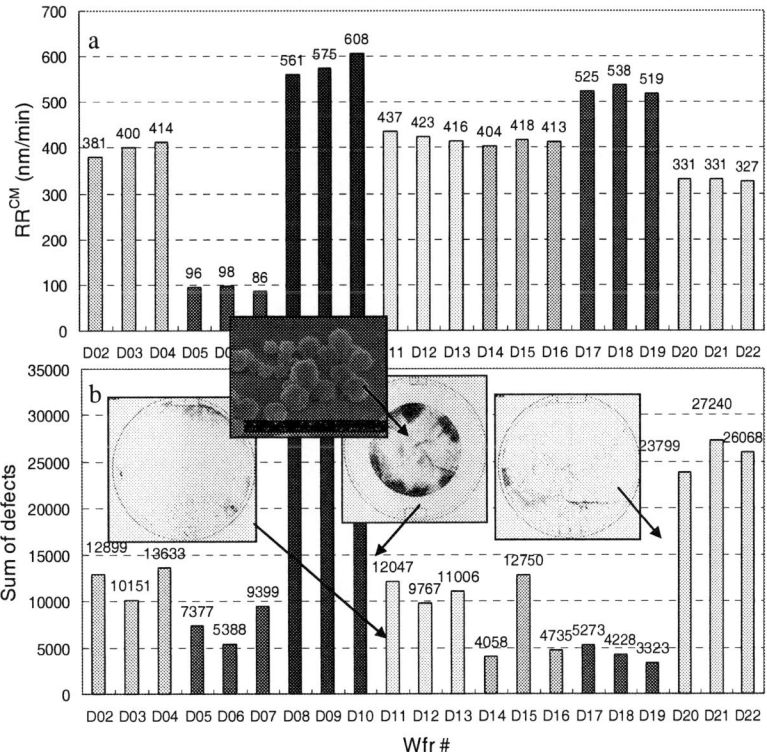

Figure 8. a) MRR, and b) sum of defects for: D02-D04 polished with composite A core 278 nm, D05-D07 polished with composite A core 365 nm, D08-D010 polished with composite A core 432 nm, D11-D13 polished with composite B core 278 nm, D14-D16 polished with composite B core 365 nm, D17-D19 polished with composite B core 432 nm, D20-D22 polished with colloidal silica 30 nm.

221

Table 5. SEM image of the composites abrasives before and after the copper CMP experiments. The average size of the core is also indicated.

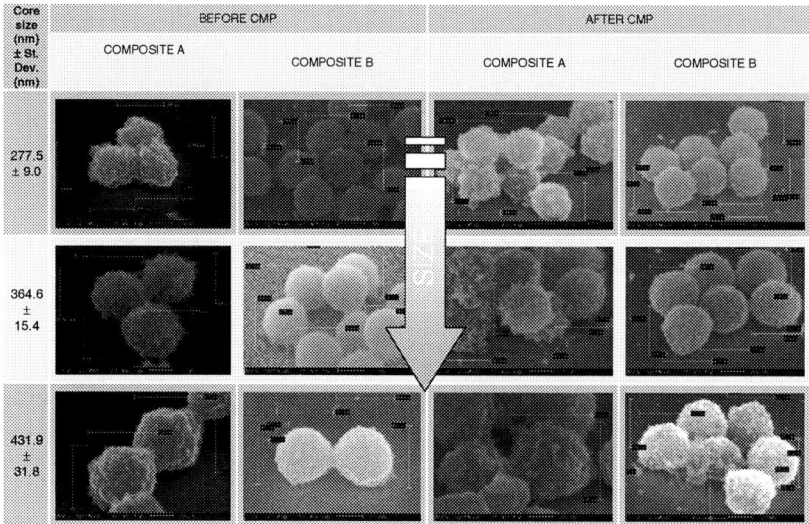

CONCLUSIONS

In this work the role of abrasive particles properties during the copper CMP process was investigated. In particular, the final purpose was to assess if the composite polymer core – inorganic shell system could be a promising direction for copper/low-κ slurry development and design. A model slurry chemistry based on the synergy between a complexing agent (Gly), an oxidizer (H_2O_2), and a corrosion inhibitor (BTA) has been used in combination with composite polymer core – silica/ceria shell abrasives of various sizes. The effect of an increasing oxidizer concentration (from 0.1vol.% to 10vol.%) on the polishing performances of the composite abrasives was also investigated. Copper rip-out types of defects similar to the ones detected on the wafer surface after CMP in pure model slurry chemistry, were observed for low oxidizer concentration, while shallow scratches and cracks develop for higher oxide layer thicknesses.

Experimental data for copper CMP experiments using two kinds of pad of different hardness have been also presented. When a medium-high hardness pad is used, for the colloidal silica particles the MRR increases with abrasive size. Copper polishing with polymer colloids is very promising in improving the surface finish, if an optimized chemistry is combined in the slurry formulation. In particular, the addition of surfactants such as CTAHS is helpful in increasing the MRR. Unfortunately, in this case the presence of a huge amount of polishing debris, polymer particles and surfactant residues on the copper surface after the post-CMP cleaning in DI water made it impossible to draw any conclusions in terms of the impact of the surfactant on defectivity and surface finish.

Remarkably superior surface quality was observed after polishing with composite B with respect to pure silica and composite A. The most frequent defect type associated with composite A is pitting and the measured depth of the holes reaches 80 nm. In addition, a substantial RMS roughness reduction was noted after polishing with composite B and a maximum scratch depth of 20 nm was measured for the largest diameter silica particles in the shell.

A soft pad interacts with the soft composite particles in a completely different mode compared with a medium/high hardness pad. First of all, after polishing with a soft pad the overall number of defects is much lower than the total defects counted in the case of a harder pad. Furthermore, in the former case, the difference between the two composites both in terms of MRR and defectivity is less marked. An increase in the defect count is registered for smaller polymer cores with respect to the larger ones.

ACKNOWLEDGMENTS

The authors wish to thank J. L. Hernandez for his support. Many thanks to J. Urrutia for her help with the synthesis experiments. This work was carried out within an Industrial Affiliation Programme on Advanced Interconnects at IMEC.

REFERENCES

[1] S. Armini, C. M. Whelan, K. Maex, J. L. Hernandez, M. Moinpour, *J. Electrochem. Soc.*, **154**, 8, H667, (2007).

[2] S. Aksu, F. M. Doyle, *J. Electrochem. Soc.*, **149**, 6, G352, (2002).

[3] S. Deshpande, S. C. Kuiry, M. Klimov, Y. Obeng, S. Seal, *J. Electrochem. Soc.*, **151**, 11, G788, (2004).

[4] M. Hariharaputhiran, J. Zhang, S. Ramarajan, J. J. Keleher, Y. Li, S. V. Babu, *J. Electrochem. Soc.*, **147**, 10, 3820, (2000).

[5] Z. Li, K. Ina, I. Koshiyama, A. Philipossian, *J. Electrochem. Soc.*, **152**, 4, G299, (2005).

[6] P. Wrschka, J. Hernandez, Y. Hsu, T. S. Kuan, G. S. Oehrlein, H. J. Sun, D. A. Hansen, J. King, M. A. Fury, *J. Electrochem. Soc.*, **146**, 7, 2689, (1999)..

[7] N. J. Brown, P. C. Baker, R. T. Maney, *Proceedings of SPIE*, **306**, 42, (1981).

[8] F. Zhang, A. Busnaina, *Electrochem. Solid-State Lett.*, **1**, 4, 184, (1998).

[9] F. Zhang, A. Busnaina, G. Ahmadi, *J. Electrochem. Soc.*, **146**, 7, 2665, (1999).

[10] G. Ahmadi, X. Xia, *J. Electrochem. Soc.*, **148**, 3, G99, (2001).

[11] R. Mazaheri, G. Ahmadi, *J. Electrochem. Soc.*, **149**, 7, G370 (2002).

[12] R. Mazaheri, G. Ahmadi, *J. Electrochem. Soc.*, **150**, 4, G233 (2003).

[13] J. Luo, D. A. Dornfeld, *IEEE Transaction: Semiconductor Manufacturing*, **16**, 1, 45, (2003).

[14] S. Armini, C. M. Whelan, M. Smet, S. Eslava, K. Maex, *Polymer Journal*, **38**, 8, 786, (2006).

[15] S. Armini, S. Eslava, C. M. Whelan, V. Terzieva, K. Maex, *Proceedings of the American Chemical Society Colloid and Surface Science Symposium*, June 12-15, 2005, Potsdam, New York, USA.

[16] S. Armini, I. U. Vakarelski, C. M. Whelan, K. Maex, K. Higashitani, *Langmuir*, **23**, 4, 2007, (2007).

[17] S. Armini, R. Burtovyy, I. Luzinov, C. M. Whelan, K. Maex, M. Moinpour, *Electrochemical and Solid-State Letters*, **10**, H74, (2007).

[18] I. Horcas, R. Fernandez, J. M. Gomez-Rodriguez, J. Colchero, J. Gomez-Herrero, A. M. Baro, *Rev. Sci. Instrum.*, **78**, 013705, (2007).

[19] M. Harihaputhiran, S. Ramarajan, S. V. Babu, *MRS Symp. Proceedings Series*, **566**, 129, (1999).

[20] J. M. Steigerwald, S. P. Murarka, R. J. Gutmann, *"Chemical Mechanical Planarization of Microelectronic Materials"*, John Wiley and Sons Ed. New York, 1997.

[21] P. W. Schindler, *Adv. Chem. Series*, **67**, 196, (1967).

[22] C. W. Kaanta, S. G. Bombardier, W. J. Cote, W. R. Hill, G. Kerszykowski, H. S. Landis, D. J. Poindexer, C. W. Pollard, G. H. Ross, J. G. Ryan, S. Wolff, J. E. Cronnin, in *Proceedings of the 8th International VLSI Multilevel Interconnection Conference*, VMIC, 1992, 226.

[23] S. V. Babu, Y. Li, M. Hariharaputhiran, S. Ramarajan, J. Zhang, Y-S. Her, J. E. Prendergast, in *Proceedings of the 15th International VLSI Multilevel Interconnection Conference*, VMIC, 1998, 443.

[24] D. Zeidler, Z. Stavreva, M. Plotner, K. Dresker, *Microelectron. Eng.*, **33**, 259, (1997).

[25] T. Notoya, G. W. Poling, *Corrosion,* **35**, 193, (1979).

[26] M. Hariharaputhiran, J. Zhang, S. Ramarajan, J. J. Keleher, Y. Li, S. V. Babu, *J. Electrochem. Soc.*, **147**, 10, 3820, (2000).

[27] S. Armini, C. M. Whelan, M. Moinpour, K. Maex, *Electrochem. Solid-State Lett.*, submitted.

AUTHOR INDEX

Aki, S. 30
Akiyama, S. 119
Andideh, E. 53
Armini, S. 212
Assous, M. 112
Baker, B. 35
Baklanov, M.R. 156, 188
Bally, L. 112
Bao, J. 150
Baranov, D. 175
Barbe, J. 144
Baumann, F. 200
Beyer, G. 106
Bhatnagar, A. 101
Biery, G. 200
Bolom, T. 35, 200
Bosco, N.S. 59
Bravo, O. 200
Budrevich, A. 53
Chae, M. 200
Chae, S.J. 194
Chanda, K. 200
Chapple-Sokol, J. 162
Charlet, B. 112
Chen, F. 200
Chen, K. 182
Child, C. 200
Chiodarelli, N. 133
Chiteboun, A. 112
Choi, J. 200
Choi, Y. 200
Chung, S. 24
Cioti, I. 106
Cohen, Y. 156
Cott, D.J. 133
Davis, K. 200

De Gendt, S. 133
Delibac, D.A. 162
Dultsev, F.N. 188
Economikos, L. 200
Ee, Y.C. 79
Engbrecht, E. 200
Esconjauregui, S. 133
Escorcia, O. 11
Filippi, R. 200
Flaitz, P.L. 35
Forster, J. 101
Fujii, Y. 47
Gabor, A. 200
Gaillard, F. 144
Gan, C.L. 79
Ghassemieh, E. 125
Giersig, M. 175
Gopalraja, P. 101
Gow, T. 200
Grant, P. 70
Gras, R. 144
Groeseneken, G. 133
Grunow, S. 35
Gstrein, F. 53
Han, I.K. 194
Hayashi, Y. 1
He, Z.X. 162
Heyns, M. 133
Hilgendorff, M. 175
Ho, P.S. 150
Hohkawa, K. 169
Huang, H. 150
Iljima, J. 30, 47
Itani, T. 119
Ito, K. 41
Janczak-Rusch, J. 59

AUTHOR INDEX

Johnston, C. 70

Kang, H.S. 194

Karmarkar, A.P. 95

Kase, M. 119

Kawamura, K. 119

Kennel, H. 53

Kim, D. 85

Kim, Y. 85

Kioussis, D. 200

Kloster, G.M. 150

Koh, K. 169

Kohama, K. 41

Koike, J. 24, 30, 47

Kotaka, Y. 119

Kumar, K. 200

Kumar, N. 101

Kuo, Y. 89

Landers, W. 200

Leduc, P. 112

Lee, H.S. 194

Lee, K.B. 194

Lee, M.Y. 194

Lee, T.C. 162

Lembach, G. 200

Li, W. 200

Lim, M.K. 79

Lin, X. 95, 95

Linville, J. 200

Lisi, A. 200

Liu, G. 89

Liu, J. 150

Liz-Marzan, L.M. 175

Luce, S.E. 162

Lustig, N. 200

Maekawa, K. 41

Maex, K. 212

Mangraviti, G. 106

Manhat, B. 59

Marques, V.F. 70

Marsik, P. 156

Matushita, T. 169

McGahay, V. 200

McKerrow, A.J. 18

McLaughlin, P. 200

McSwiney, M.L. 150

Menon, V. 200

Miner, B. 53

Moinpour, M. 150, 212

Mongeon, S.A. 162

Moore, D. 11

Moreau, S. 144

Mori, K. 41

Moroz, V. 95

Mueller, D.W. 18

Murakami, M. 41

Murphy, W.J. 162

Nakaishi, M. 119

Negreira, A.R. 133

Neishi, K. 30, 47

Nelson, K. 101

Ng, C.M. 79

Nicholson, L. 200

Okubo, K. 119

Park, S.W. 194

Passemard, G. 144

Pazos-Perez, N. 175

Perez-Juste, J. 175

Peters, C. 35, 200

Plombon, J. 53

Reidy, R. 18

Rhoads, B. 35

Rollins, G. 95

AUTHOR INDEX

Sakthivel, P. 11

Sankaran, S. 35, 200

Seo, H. 85

Shi, H. 150

Siddiq, A.M. 125

Simon, A. 35, 200

Smith, C.E. 18

Smith, R.S. 150

Srivastava, R.P. 200

Standaert, T. 200

Sullivan, T.D. 162

Sun, Y. 150

Sunddarrajan, A. 101

Tamura, N. 182

Tan, J.B. 79

Tang, T.J. 35

Thomas, D.C. 162

Tokei, Z. 24, 106

Torres, J. 144

Tseng, J.M. 101

Tseng, W. 200

Tsukimoto, S. 41

Tu, K.N. 182

Urbanowicz, A. 156, 188

Vanslette, D.S. 162

Vereecken, P.M. 133

Verhulst, A.S. 133

Vinogradova, E. 18

Vinokur, K. 156

Waldfried, C. 11

Watatani, H. 119

Weis, J.E. 133

Wendt, H. 200

Whelan, C.M. 133, 212

Xu, X. 95

Yanai, K. 119

Yang, H. 101

Zhang, B.C. 79

Zhang, F. 101

Zussy, M. 112